核能科学与工程系列译丛

# 超临界水冷反应堆材料与水化学

Materials and Water Chemistry for Supercritical Water-cooled Reactors

[加]大卫·古佐纳斯(David Guzonas)
[捷克]拉德克·诺沃特尼(Radek Novotny)
[芬]萨米·彭蒂莱(Sami Penttilä)　著
[芬]阿基·托伊沃宁(Aki Toivonen)
[加]郑文跃(Wenyue Zheng)

周艳民　谷海峰　孟兆明　译

国防工業出版社
·北京·

**著作权合同登记　图字：军-2021-024 号**

**图书在版编目(CIP)数据**

超临界水冷反应堆材料与水化学 / (加) 大卫・古佐纳斯等著 ; 周艳民, 谷海峰, 孟兆明译. -- 北京 : 国防工业出版社, 2024. 6. --(核能科学与工程系列译丛).

ISBN 978-7-118-13253-3

Ⅰ. TM623.91

中国国家版本馆 CIP 数据核字第 2024G39Y24 号

Materials and Water Chemistry for Supercritical Water-cooled Reactors, 1 Edition

David Guzonas, Radek Novotny, Sami Penttilä, Aki Toivonen, Wenyue Zheng

ISBN:978-0-08-102049-4



Authorized Chinese translation published by National Defense Industry Press

超临界水冷反应堆材料与水化学(周艳民,谷海峰,孟兆明　译)

ISBN:978-7-118-13253-3



**注意**

本书涉及领域的知识和实践标准在不断变化。新的研究和经验拓展我们的理解,因此须对研究方法、专业实践或医疗方法作出调整。从业者和研究人员必须始终依靠自身经验和知识来评估和使用本书中提到的所有信息、方法、化合物或本书中描述的实验。在使用这些信息或方法时,他们应注意自身和他人的安全,包括注意他们负有专业责任的当事人的安全。在法律允许的最大范围内,爱思唯尔、译文的原文作者、原文编辑及原文内容提供者均不对因产品责任、疏忽或其他人身或财产伤害及/或损失承担责任,亦不对由于使用或操作文中提到的方法、产品、说明或思想而导致的人身或财产伤害及/或损失承担责任。

※

国防工业出版社 出版发行

(北京市海淀区紫竹院南路 23 号　邮政编码 100048)

北京虎彩文化传播有限公司印刷

新华书店经售

*

**开本** 710×1000　1/16　插页 2　**印张** 16¼　**字数** 280 千字

2024 年 6 月第 1 版第 1 次印刷　**印数** 1—1200 册　**定价** 168.00 元

**(本书如有印装错误,我社负责调换)**

国防书店:(010)88540777　　书店传真:(010)88540776

发行业务:(010)88540717　　发行传真:(010)88540762

# 译者序

超临界水冷反应堆(SCWR)是目前世界上6种第四代反应堆中唯一以轻水做冷却剂的反应堆,同时也是与目前主流第三代水冷反应堆最为接近的第四代反应堆,具有系统简单、装置尺寸小、热效率高和安全性高等优点,并且在设计经验和运行管理上较第三代反应堆以及超临界火电厂有更多的可借鉴之处。因此,超临界水冷反应堆是替代第三代核动力系统的理想堆型,在军事和民用核动力领域占有重要地位。

超临界水冷反应堆在高温、高压同时叠加强辐照条件下的材料稳定性,是现阶段困扰研究者和设计者的主要技术瓶颈。本书专门针对超临界水冷反应堆的材料辐照损伤机理和常规水化学进行了较为系统的阐述,内容涵盖了堆内材料在辐照作用下的结构和力学特性、辐照作用下的水化学特性、超临界条件下冷却剂与材料的相容性机制,以及常规水化学问题。书中大部分内容来源于国际上专业研究团队获得的实验结果,这些研究成果对于我国超临界水冷反应堆结构材料的选择乃至运行策略的制定都具有重要的指导价值。

目前,英国、美国、日本、法国等国家均在积极开展SCWR的基础研究和工程应用研究工作,研究工作的重点集中在耐高温、耐腐蚀和耐辐照的新材料研发设计上。我国在相关领域的研究尚处于起步阶段,无论是知识体系还是关键技术均与国外有一定差距。本书恰好能够作为这方面的参考资料,它的翻译出版对于促进和推动我国在超临界水冷反应堆方面的研究具有重要现实意义,将成为国内科研人员、工程技术人员等研究超临界水冷反应堆的经典译著。

周艳民、谷海峰、孟兆明负责本书的翻译工作,周艳民负责全书的统稿工作。丁铭教授对本书内容进行了审校,马钎朝、刘新星、刁寒、杨静豪等多位博(硕)士参与了书稿的核对工作,在此表示感谢。同时,感谢装备科技译著出版基金资助本书出版。

本书学术思想新颖,内容多专业、多领域交叉融合。鉴于译者水平有限,书中翻译难免存在不妥之处,恳请读者给予批评、指正。

译者

2023年12月

# 原著致谢

本书内容虽然由作者所著,但是很多研究成果都来自于过去10年中我们的团队以及各国联合开展的国际合作项目。在此对我们的同事致以最真挚的感谢,感谢他们长期以来致力于第四代反应堆国际论坛所做的工作,以及对于超临界水冷反应堆进行的有价值且富有启发性的讨论、信息共享、相互关怀和为了超临界水冷反应堆发展而做出的奉献,他们是 Sylvie Aniel、Patrick Arnoux、Yoon-Yeong Bae、Carlos-Kjell Guerard、Liisa Heikinheimo、Akos Horvath、Jinsung Jang、Junya Kaneda、Alberto Saez-Maderuelo、Hideki Matsui、Henri Paillère、Thomas Schulenberg、Gary Was、Katsumi Yamada 和 Marketa Zychova。这里要特别感谢2015年去世的 Jan Kysela,他为本书的创作构思做出了非常重要的贡献。Jan 是一位非常敬业的核化学与水化学学者,也是 SCWR 的忠实倡导者,同时还是一位热情、慷慨的人,他对艺术和音乐的热情不亚于对科学的热情。还要感谢国际原子能机构主办的标准物质信息共享会议以及所有参加者所做的宝贵贡献,他们是 A. Bakai、M. Fulger、B. Gong、Y. Huang、Q. J. Peng、V. Yurmanov、A. Zeeman 和 Q. Zhang。对于已故者 Roger Staehle 所做的诸多贡献在这里一并表示感谢,特别是他对腐蚀科学的热情、对环境辅助开裂机理的深刻理解以及对核材料发展历史的深刻认识,让人敬佩。

Dave Guzonas 要感谢他在加拿大乔克里弗原子能实验有限公司(现在的加拿大核实验室)工作期间的许多前同事:在临界转变物理现象中提供指导性帮助的 V. Anghell;2001年全身心投入 SCWR 研究的 K. Burrill;早期支持 SCWR 水化学研究的 S. Bushby 和 H. Khartabil;鼓励探索、追求挑战且严谨求实的 R. Duffey;具有工程视野且对 SCWR 具有辩证思维的 M. Edwards;对化学专家传授热工水力学知识的 L. Leung;深度讨论 SCWR 中热工水力学和溶解度的 L. Qiu;提供了实验室、设备和技术人员的 J. Semmler;破解水辐照分解奥秘的 C. R. Stuart;给予广泛而富有前瞻性的讨论和鼓励的 R. Tapping;在午后头脑风暴中提出关于 SCWR 表面结构设想的 C. Turner; L. Deschenes、M. Godin 和

S. Rousseau 负责实验室的日常工作，监督学生、修复漏洞和指出错误；J. Wills 多年来监管实验室，感谢他在试验中尤其是在涂层研究方面的密切合作；同时感谢 J. Ball、T. Chiasson、P. Chow、J. Festarini、W. Kupferschmidt、J. Licht、S. Livingstone、C. Macdonald、D. McCracken、D. Mancey、G. McRae、I. Muir、J. Pencer、D. Rodgers、R. Rogge、R. Sadhankar、S. Sell、J. Smith、R. Speranzini、Z. Tun、L. Walters、O. -T. Woo、D. Wren、M. Wright 和 M. Yetisir 提供的意见和大力支持。还要感谢在夏令营活动中参与项目的学生，他们在样片抛光、分析数据、电力输出和防止高压反应釜泄漏等工作中付出了大量劳动，他们分别是 K. Chu、M. Chutumstid、H. Dole、S. Fessahaye、A. Fraser、M. Haycock、S. Jang、J. Michel、S. Sullivan、S. Wren 和 M. Yetisir。

特别感谢同 Peter Tremaine（圭尔夫大学）和 Jean-Paul Jay-Gerin（舍布鲁克大学）在 SCWR 水化学方面开展的早期合作。合作期间，本书第 4 章提出的大多数想法都初具雏形。F. Brosseau 和 K. Bissonette（圭尔夫大学）在 David 实验室完成了合作教育理学硕士学位的各项工作，并做出了重要贡献。W. Cook（新不伦瑞克大学）在 Chalk River 实验室时，实实在在地把 SCWR 循环化学以及超临界水化学与腐蚀过程之间的关系作为自己学术研究的一部分。这些研究基础促成了与 I. Svishchev（特伦特大学）的卓有成效的合作，他们利用分子动力学研究了 SCWR 的表面结构，使理论猜想变成了长期的科学研究方法。

David 感谢“水和蒸汽性质国际协会”的同事就超临界化石燃料发电厂中超临界水的性质和化学控制方面的许多讨论，特别是对 Barry Dooley、Derek Lister、Digby Macdonald 和 Ian Woolsey 提供的见解表示感谢。David 还要感谢：同 T. Allen、D. Bartel 和 G. Was 的讨论以及实验室参观之旅；同 J. Kysela（捷克共和国核能研究所）在水化学方面许多愉快且富有启发性的讨论，还有美味的比尔森啤酒、苹果馅奶酪卷和美妙的爵士乐；R. Staehle（Staehle 咨询公司）在他造访荷兰佩滕并共进晚餐时，介绍了腐蚀、破裂以及核蒸汽过热反应堆早期工作中的知识；Y. Katsumura 和 M. Lin 向他展示了日本最先进的脉冲辐射分解设施，并参加了加拿大的 SCW 辐射分解研讨会；V. Yurmanov 就俄罗斯水化学方面的研究进行了热情介绍，并分享了见解和数据；接受 L. Zhang 的邀请，参观了他的实验室并为其学生做即兴报告，同 CanmetMaterials 的许多同事进行了讨论和数据分享，并讲解了一些材料科学方面的知识；P. Ampornrat、J. Bischoff、S. Teysserre 和 E. West 在他还是初学者时就在会议上讨论了 SCWR 的相关材料和化学问题。最后，David 感谢 Margaret 和 John 的鼓励，尤其感谢 Jaleh 在本书编写过程中提供的鼓励、支持、包容和热情。

如果没有加拿大大学同事的工作帮助，David 和 Wenyue 在本书创作过程中得到的资料将大大减少。衷心感谢由 NSERC/NRCan/AECL Gen-Ⅳ 能源技术计划资助的加拿大材料和化学项目的参与者,他们提供了数据,参加了研讨会和会议,发送了更新后的会议材料,保证了会议质量,有时会对研究进度较慢的领域予以提醒,并进行激烈讨论,相关的参与者如下:A. Alfantazi(不列颠哥伦比亚大学)、A. Anderson(圣弗朗西斯·泽维尔大学)、D. Artymowizc(多伦多大学)、G. Botton(麦克马斯特大学)、W. Chen(阿尔伯塔大学)、W. Cook(新不伦瑞克大学)、M. Daymond(皇后大学)、K. Ghandi(阿里森山大学)、R. Holt(皇后大学)、X. Huang(卡尔顿大学)、R. Hui(NRC-IFCI)、J. -P. Jay-Gerin(舍布鲁克大学)、J. Kish(麦克马斯特大学)、R. Klassen(韦仕敦大学)、J. Kozinski(约克大学)、J. Luo(阿尔伯塔大学)、D. Mitlin(阿尔伯塔大学)、R. Newman(多伦多大学)、J. Noel(韦仕敦大学)、P. Percival(西蒙弗雷泽大学)、C. Pye(圣玛丽大学)、I. Svishchev(特伦特大学)、P. Tremaine(圭尔夫大学)、C. Wren(韦仕敦大学)。许多学生和博士后研究员参加了加拿大材料和化学项目,并做出了宝贵的贡献;感谢以下人员分享数据、讨论实验以及有时容忍对报告和论文的大幅度修改:C. Alcorn、H. Arcis、L. Applegarth、Y. Behnamian、R. Carvajal-Ortiz、K. Choudhry、Y. Jiao、D. Kallikragas、S. Mahboubi、J. Meesungnoen、J. Miles、R. Olive、R. Sanchez、S. Sanguanmith。在 T. Andersen(加拿大自然科学和工程研究理事会)和 D. Brady、K. Huynh、C. MacKinlay、J. Poupore、S. Quinn(加拿大自然资源部)的努力下,这项在加拿大范围内为期 8 年的合作一直进行顺利。我们还认可 C. Rangacharyulu 和 J. Szpunar(萨斯卡彻温大学)及其学生对加拿大 SCWR 研究的贡献。

Wenyue Zheng 感谢在他 CanmetMaterials 工作期间的同事们所做的贡献,他们从 2007 年开始大力支持与材料有关的第四代反应堆工作,当时他在 NRCan 开始了核材料研究计划,直到 2017 年才回到北京科技大学。他特别感谢 Kevin Boyle、Jen Collier、Gordon Gu、Mike Greenwood、Selcuk Kuyucak、Jian Li、Jason Lo、Maggie Matchim、Xu Su、Yimin Zeng 的不懈努力。期间,许多学生和博士后为 Canmet 试验和计算结果的产生做出了贡献。

除上述人员外,Aki Toivonen 和 Sami Penttilä 还想感谢芬兰国家技术研究中心(VTT)的所有同事,他们在腐蚀和 SCC 实验中促成了实验工作的完成并参与了国际合作。特别感谢担任知识中心经理的 Liisa Heikinheimo,他在大约 10 年前向我们委托了与 SCWR 材料相关的项目;感谢实验室工程师和技术人员 P. Väisänen、T. Lehtikuusi、M. Mattila、J. Lukin、T. Ikäläinen、S. Peltonen、J. Ranta,他们致力于解决水化学问题和高压釜泄漏问题;感谢 T. Saario、P. Kinnunen 和 P. Auerkari(VTT)提供的有关近 10 年与实验工作相关的一般腐蚀过程的见解;感谢 M. Bojinov(保加利亚索菲亚化工冶金大学)和 I. Betova(保加利亚索菲亚技术大学)就高温水条件下材料建模方面提供的宝贵意见和建议。

Radek Novotny 向他在联合研究中心(JRC)的几位同事 L. Debarberis、P. Hähner、V. Ranguelova 和 O. Martin 表示感谢,他们支持并帮助了试验工作,并且参与了欧洲和国际项目,特别是关于腐蚀和 SCC 的项目。Radek 在此感谢实验室工程师、技术人员和学生 T. Timke、T. Heftrich、A. Van de Sande、S. Ripplinger、A. Ruiz、Z. Szaraz、P. Janík、D. Prchal 和 K. Mathis,他们为实验设备的开发和高质量数据的生成做出了贡献。Radek 特别感谢 P. Moilanen,他在可用 SCW 进行机械测试的设备设计上参与了长达数小时的讨论,特别感谢 J. Macak 对利用电化学理解亚临界水和在 SCW 腐蚀上展现的兴趣和热情。最后同样重要的是,Radek 要感谢他的妻子 Anna Kotoučová 的支持和给予的无限耐心。

技术上的贡献仅仅是本书完成过程中的一部分。还要特别感谢 Matthew 和 Danielle Edwards 在撰写第 4 章内容时提供的帮助。如果没有 Elsevier 出版社员工的大力支持,本书就无法完成,感谢他们所做的以下工作:Alex White、Sarah Hughes 和 Cari Owen 促成了本书的出版,Natasha Welford 指导了这本书的制作完成,Sheela Bernardine Josy 协助了版权申报,Swapna Srinivasan 及其制作团队则将这一切工作完美地结合在一起。

# 缩略语

| | | |
|---|---|---|
| AC | alternating current | 交流电 |
| AEC | Atomic Energy Commission | 原子能委员会 |
| AECL | Atomic Energy of Canada Limited | 加拿大原子能有限公司 |
| AOA | axial offset anomaly | 轴向偏移异常 |
| ASME | American Society of Mechanical Engineers | 美国机械工程师协会 |
| ASTM | American Society for Testing and Materials | 美国材料试验协会 |
| ATEM | analytical transmission electron microscopy | 分析透射显微镜 |
| BONUS | boiling nuclear superheater | 沸水堆过热器 |
| BOP | balance-of-plant | 电厂配套设施 |
| BORAX-V | boiling reactor experiment V | 沸水实验堆-V |
| BWR | boiling water reactor | 沸水反应堆 |
| CANDU | Canada deuterium uranium | 加拿大重水反应堆 |
| CERT | constant extension rate tensile | 恒定拉伸速率 |
| CF | corrosion fatigue | 腐蚀疲劳 |
| CGR | crack growth rate | 裂纹增长速率 |
| CNL | Canadian Nuclear Laboratories | 加拿大原子能研究所 |
| CO | chemical oxidation | 化学氧化 |
| CUEITF | Canadae-Ukraine electron irradiation test facility | 加拿大-乌克兰电子辐照测试设施 |
| CVR | Centrum Výzkumu Řež | 捷克研究院 |
| CW | cold work | 冷加工 |
| DB | double bellows | 双波纹管 |
| DCT | disk compact tension | 圆盘紧凑拉伸 |
| DHC | delayed hydride cracking | 氢致延迟断裂 |

| | | |
|---|---|---|
| DLEPR | double-loop electrochemical potentiokinetic reactivation | 双环电化学电动势再活化 |
| DO | dissolved oxygen | 溶解氧 |
| dpa | displacements per atom | 单位原子位移 |
| EAC | environmentally assisted corrosion | 环境辅助开裂 |
| EBSD | electron backscatter diffraction | 电子背向散射衍射 |
| EBR-Ⅱ | experimental breeder reactor | 试验增殖反应堆 |
| ECAP | equal channel angular processing | 等径弯曲通道变形 |
| ECP | electrochemical corrosion potential | 电化学腐蚀电位 |
| EDX 或 EDS | energy dispersive X-ray spectroscopy | X 射线能量色散谱 |
| EGBE | effective grain boundary energy | 有效晶界能 |
| EO | electrochemical oxidation | 电化学氧化 |
| EPRI | Electric Power Research Institute | 电力研究院 |
| ESADA | Empire State Atomic Development Associates | 帝国原子能发展协会 |
| EVESR | ESADA Vallecitos Experimental Superheat Reactor | 帝国原子能协会-瓦莱西托斯过热蒸汽试验反应堆 |
| F/M | ferritic-martensitic | 铁素体-马氏体 |
| FAC | flow accelerated corrosion | 流体加速腐蚀 |
| FIB | focussed ion beam | 聚焦离子束 |
| FP | fission products | 裂变产物 |
| FPP | fossil power plant | 火电厂 |
| FWHM | full width at half maximum | 半高宽度 |
| GBE | grain boundary engineering | 晶界工程 |
| GDOES | glow discharge optical emission spectroscopy | 射频辉光放电谱仪 |
| GETR | general electric test reactor | 通用电气试验反应堆 |
| GIF | Generation IV International Forum | 第四代反应堆国际论坛 |
| HAADF | high-angle annular dark field | 高角度环形暗场 |
| HDAC | hydrothermal diamond anvil cell | 热液金刚石砧芯 |
| HDR | heissdampfreaktor | 德国热气反应堆 |
| HKF | Helgesone-Kirkhame-Flowers | HKF 核型 |
| HPLWR | high-performance light water reactor | 高性能轻水反应堆 |
| HPOC | high-pressure optical cell | 高压可视腔体 |
| HWC | hydrogen water chemistry | 氢水化学 |

| | | |
|---|---|---|
| IAEA | International Atomic Energy Agency | 国际原子能机构 |
| IAPWS | International Association for the Properties of Water and Steam | 国际水和蒸汽性质协会 |
| IASCC | irradiation-assisted stress corrosion cracking | 辐照促进应力腐蚀开裂 |
| IC | ion chromatography | 离子色谱仪 |
| ICP | inductively coupled plasma | 电感耦合等离子体 |
| IGSCC | intergranular stress corrosion cracking | 晶间应力腐蚀破裂 |
| KIPT | Kharkiv Institute of Physics and Technology | 哈尔科夫物理与技术研究所 |
| LET | linear energy transfer | 线性能量传输 |
| LINAC | linear electron accelerator | 直线电子加速器 |
| LLRM | liquidlike radiolysis model | 类液体辐照分解模型 |
| LWR | light water reactor | 轻水反应堆 |
| M&C | materials and chemistry | 材料及化学 |
| MA | mill-annealed | 轧制退火 |
| MC | Monte Carlo | 蒙特卡罗 |
| MCM | mixed conduction model | 混合传导模型 |
| MD | molecular dynamics | 分子动力学 |
| MDL | method detection limit | 方法检测极限 |
| MS | mass spectrometry | 质谱 |
| μSR | muon spin spectroscopy | μ介子自旋光谱 |
| NEA | Nuclear Energy Agency | 核能机构 |
| NPP | nuclear power plant | 核电站 |
| NRT | Norgett, Robinson and Torrens | 诺吉特、罗宾逊和托伦斯 |
| NRX | National Research Experimental | 国家研究实验核反应堆 |
| NWC | normal water chemistry | 常规水化学 |
| ODS | oxide dispersion strengthening | 氧化物弥散强化 |
| OECD | Organization for Economic Cooperation and Development | 经济合作与发展组织 |
| OES | optical emission spectrometry | 光学发射光谱 |
| OT | oxygenated treatment | 氧化处理 |

| | | |
|---|---|---|
| PCI | pellete-cladding interaction | 燃料芯块与包壳之间的作用 |
| PCT | peak cladding temperature | 包壳峰值温度 |
| PDM | point defect model | 点缺陷模型 |
| PHWR | pressurized heavy water reactor | 加压重水反应堆 |
| PKA | primary knock-on atom | 主碰撞原子 |
| PMB | Project Management Board | 项目管理委员会 |
| PWR | pressurized water reactor | 压水反应堆 |
| R&D | research and development | 研究与开发 |
| RBMK | Reactor Bolshoy Moshchnosty Kanalny (high-power channel-type reactor) | Bolshoy Moshchnosty Kanalny 反应堆(大功率通道型反应堆) |
| RIS | radiation-induced segregation | 辐照诱导的偏析 |
| SADE | superheat advanced demonstration experiment | 过热先进演示实验 |
| SAM | scanning auger microscopy | 扫描螺旋显微镜 |
| SC | supercritical | 超临界的 |
| SCC | stress corrosion cracking | 应力腐蚀开裂 |
| SCFPP | supercritical fossil power plant | 超临界化石燃料厂 |
| SCW | supercritical water | 超临界水 |
| SCWO | supercritical water oxidation | 超临界水氧化 |
| SCWR | supercritical water-cooled reactors | 超临界水冷反应堆 |
| SEM | scanning electron microscopy | 扫描电子显微镜 |
| SHS | superheated steam | 过热蒸汽 |
| SICC | strain-induced corrosion cracking | 应变腐蚀开裂 |
| SIMFUEL | simulated fuel | 模拟燃料 |
| SIMS | secondary ion mass spectrometry | 二次离子质谱法 |
| SSRT | slow strain rate testing | 慢应变速率试验 |
| TEM | transmission electron microscopy | 透射电子显微镜 |
| TGSCC | transgranular stress corrosion cracking | 穿晶应力腐蚀开裂 |
| TRIGA | training, research, isotopes, general atomic | 训练、研究、同位素、通用原子 |
| TT | thermally treated | 热处理 |
| UKAEA | United Kingdom Atomic Energy Agency | 英国原子能机构 |

| | | |
|---|---|---|
| USC | ultrasupercritical | 超超临界 |
| VBWR | Vallecitos boiling water reactor | 瓦莱西托斯(Vallecitos)沸水反应堆 |
| VGB | Vereinijung der Groβkesselbesitzen | 欧洲热电运营商协会 |
| VLRM | vapourlike radiolysis model | 类蒸汽辐射分解模型 |
| WCR | water-cooled reactor | 水冷反应堆 |
| XAS | X-ray absorption spectroscopy | X 射线吸收光谱 |
| XPS | X-ray photoelectron spectroscopy | X 射线光电子能谱 |
| XRF | X-ray fluorescence | X 射线荧光光谱仪 |
| YSZ | yttrium-stabilized zirconium oxide | 钇稳定氧化锆 |

# 目 录

# 第1章
# 引　言

水是人们熟知并且相对安全的一种传热介质，而且当今世界上商用发电反应堆绝大部分都是水冷反应堆（WCR）①，使用超临界水（SCW）作为反应堆的冷却剂是在现有反应堆技术上的一种合理改进。超临界冷却剂的概念是在20世纪50—60年代提出的，学者们普遍认为冷却剂在临界温度以上工作时能为反应堆带来更高的热效率。目前，很多电力企业同时经营着核电厂（NPP）和超临界（SC）化石燃料电厂（FPP）。第一座超临界化石燃料电厂Philo 6号于20世纪40年代末至50年代中在美国建成，并于1957年投入运行（621℃、31.0MPa）。随后1959年又有两座超临界化石燃料电厂Eddystone 1号（649℃、34.5MPa）和Breed 1号（566℃、24.1MPa）相继投入运行。通用电气公司的Cohen和Zebroski（1959）为了研发实用的商用沸水反应堆（BWR），分别比较了自然循环锅炉、强迫循环锅炉以及核能过热蒸汽锅炉的优、缺点。Dollezhal等（1958、1964）提出了一种以铀为核燃料，石墨为慢化剂的过热蒸汽（SHS）反应堆②。尽管到目前为止还未建成在超临界温度和压力下运行的水冷反应堆，但在20世纪60年代，为了探索采用SHS作为反应堆冷却剂，研究人员付出了巨大的努力。图1.1展示了现役水冷反应堆、超临界化石燃料电厂、60年代概念性设计的过热蒸汽和再热蒸汽反应堆，以及目前设计的超临界水冷反应堆（SCWR）的温度压力工作范围。反应堆在超临界压力下运行还有一个好处就是冷却剂的密度增加，进而使传热系数增大，这样在相同的泵功率条件下冷却剂可以达到更大的质量流速。Oka等在2010年发表的文章中对超临界反应堆堆芯的热工水力特性做了详细的论述。

---

① 水冷式反应堆包括使用轻水或重水冷却剂运行的所有反应堆类型：沸水反应堆（BWR）、压水反应堆（PWR）、水冷式反应堆、水慢化反应堆（VVER）和加压重水反应堆（PHWR）。

② 第4章将探讨超临界水与过热蒸汽的关系。

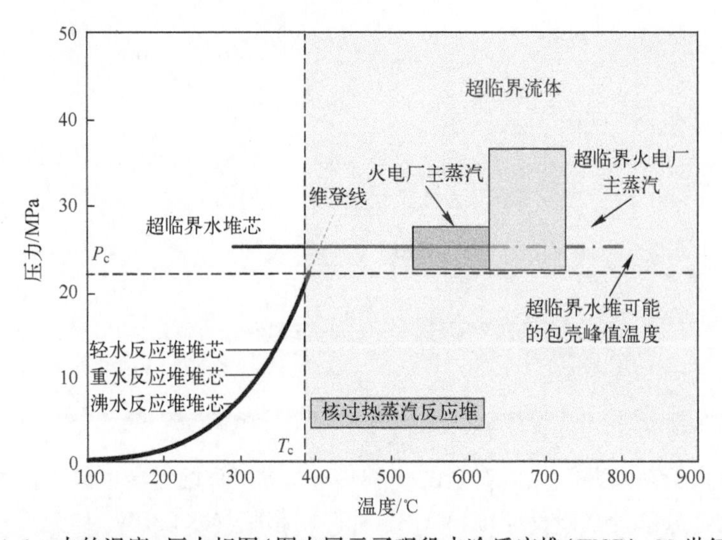


图 1.1 水的温度-压力相图(图中展示了现役水冷反应堆(WCR)、20 世纪 60 年代概念性过热蒸汽(SHS)反应堆、化石燃料电厂(SCFPP)以及几种第四代超临界水冷反应堆概念性堆芯的工作范围。对于超临界水冷反应堆,冷却剂从堆芯入口到出口之间的推荐工作温度范围由水平实线表示,燃料包壳可能达到的峰值温度范围由点画线表示,临界温度 $T_c$ 和临界压力 $P_c$ 由虚线表示,汽液共存曲线由实线表示,维登(Widom)线(见第 4 章)由共存曲线的延长线表示)(引自文献 Heikinheimo, L., Guzonas, D., Fazio, C., 2009. GenIV materials and chemistry research—common issues with the SCWR concept. In: 4th Int. Symp. on Supercritical Water-Cooled Reactors, Heidelberg, Germany, March 8e11, 2009, Paper 81)



## 1.1 早期的研究工作

1959 年,Vallecitos 原子实验室在“过热先进演示实验(SADE)”项目中首次直接通过核能供热的方式产生出过热蒸汽(Barnard,1961)。帝国原子能发展协会(ESADA)开发了 ESADA Vallecitos 过热蒸汽试验反应堆(EVESR)(Barnard,1961),采用了分体式过热器设计,即反应堆产生的 6.9MPa 饱和蒸汽二次通入反应堆的燃料组件部分。其他的设计大都考虑采用整体式方案,即反应堆和过热器同处在一个单独的容器里。例如,由 Allis Chalmers 设计和建造的 Pathfinder 反应堆即为整体式过热反应堆,该反应堆于 1966 年投入运行。美国原子能委员会也出资建造了沸水实验堆-V(BORAX-V),该项目于 1962 年开始实施,沸水堆过热器(BONUS)项目于 1964 年开始实施(Saltanov 和 Pioro,2011)。Ru 和 Staehle(2013a~c)基于美国试验项目数据,整理了一份关于材料性能的详细报告。

德国 100MW 过热蒸汽沸水堆原型堆 HDR(Heissdampfreaktor)设计运行参数为 457℃、9MPa(Kornbichler,1964)。该电站于 1969 年 10 月 14 日达到临界状态,并于 1970 年 8 月 2 日开始商业运行,发电能力达到 2300MW。此后不久,由于堆芯受到了损坏,反应堆被迫于 1971 年 4 月关闭,原因至今尚未公布(Schulenberg 等,2014)。瑞典的 Marviken 沸水冷却重水慢化型反应堆增设了一个独立的蒸汽过热器,可以选择性地与饱和蒸汽隔离使用。英国的研究人员对核反应堆中超临界蒸汽的实用性进行了评价(Moore 等,1964),广泛测试了直接和间接循环压力容器和压力管道的特性。

俄罗斯在压力管式沸水过热蒸汽反应堆方面积累了大量经验,这些经验来源于别洛亚尔斯克的 Kuchatov 四号反应堆(本书中更多称其为 Beloyarsk NPP),该反应堆在压力为 8MPa、冷却剂温度为 500℃的条件下运行了大约 30 年(Dollezhal 等,1958、1964;Emel'yanov 等,1972)。其中的两座反应堆(分别为 $100MW_e$[①] 和 $200MW_e$)采用了相同的汽轮机入口蒸汽参数($P_{in}=8.8MPa$,$T_{in}=500\sim510℃$)。1 号堆于 1964 年投入运行,紧接着 2 号堆于 1967 年投入运行。两个机组的尺寸和设计较为相似,均采用沸腾(蒸发器)通道和过热器通道,但 2 号机组的流量和堆芯布置比 1 号机组简化了很多。值得注意的是,1 号堆的沸腾通道包含在一个闭合环路中,即利用蒸汽发生器的热量交换作用产生过热蒸汽通道所需的饱和蒸汽,而 2 号堆则采用直接循环。关于上述电厂更多详细的阐述可以参照 2011 年 Saltanov 和 Pioro 发表的文章。

表 1.1 列出了一部分用于早期反应堆堆芯构件的合金材料。合金的名称随着时间的推移和划分方法的变化有所不同,如"铟铬镍合金"曾被命名为"600""铟铬镍 600"及"600 合金"。本书采用"600 合金"这一惯例叫法。由于一些合金材料有多个常用的名称,如 HCM12A 和 T122 虽然名字不同但实际组分一致。为了不致混淆,本书采用的合金名称选择其最原始的命名。

表 1.1 20 世纪 60 年代核能过热蒸汽项目中使用的材料汇总

| 堆　型 | 燃料包壳合金 | 参考文献 |
|---|---|---|
| Beloyarsk NPP | Kh18Ni10T 合金:C(最大值)0.1;Si(最大值)0.8;Mn(1~2);Ni(10~11);S(最大值)0.02;P(最大值)0.035;Cr(17~19);5(C−0.02)<Ti<0.6,相当于 321 SS。<br>EI−847:C(0.04~0.06);Mn(0.4~0.8);Si≤0.4;S≤0.010;P≤0.15;Cr(15.0~16.0);Ni(15.0~16.0);Mo(2.7~3.2);Nb≤0.9;N≤0.025;B≤0.001;Co≤0.02;Cu≤0.05;Bi≤0.01;Pb≤0.001;Ti≤0.05 | Emel'yanov 等(1972) |

① 单位 $MW_e$ 表示电功率。

续表

| 堆　　型 | 燃料包壳合金 | 参考文献 |
|---|---|---|
| BONUS | 348 SS | Saltanov 和 Pioro(2011) |
| BORAX-V | 304 SS | Saltanov 和 Pioro(2011) |
| VSER | 800 合金,347 SS,348 SS、301 SS、304 SS、600 合金 | Hazel 等(1965) |
| Pathfinders | 316 SS | Saltanov 和 Pioro(2011) |
| HDR | 环形燃料:625 合金(内部包壳)、合金 1.4981S(X8CrNiMoNb16-16)或 1.4550(X6CrNiNb18-10)、347 SS | Schulenberg 和 Starflinger(2012) |

注:表中只列举出过热蒸汽反应堆使用的材料;合金组分单位为%(质量分数)。

在早期发展阶段,材料性能被认为是最重要的问题。Novick 等(1964)总结了过热蒸汽反应堆研发工作中需要攻克的多项关键问题,其中包括可能是最重要的问题“燃料元件包壳在堆内高温环境下的性能及抗腐蚀能力”。Marchaterre 和 Petrick(1960)提出过“超临界水技术应用于反应堆的主要困难,在于缺乏放射性物质在外部系统中的沉积和辐照条件下内部杂质累积的相关信息”。另外,在早期开展的许多试验中发现,沸腾区内给水中氯化物的沉积是一个严重问题,尽管在试验中已经尽可能地对氯化物进行去除,但这个问题最终仍会导致应力腐蚀开裂(SCC)(Ru 和 Staehle,2013;Bevilacqua 和 Brown,1963)。



## 1.2 近期研究进展

20 世纪 90 年代,学术界重新唤起了对超临界水冷反应堆的研究兴趣。东京大学的研究人员公布了采用热中子谱和快中子谱的超临界水冷反应堆概念性设计方案(Oka 和 Koshizuka,1993;Oka,2000;Oka 等,2010);Dobashi 等(1998)提出了一种采用热中子谱的超临界轻水反应堆(LWR),其平均堆芯出口温度为 455℃。俄罗斯的研究人员提出了一种自然循环一体式超临界压水堆概念(B500SKDI)(Slin 等,1992),堆芯出口压力为 23.5MPa,冷却剂堆芯进出口温度分别为 365℃和 381.1℃。加拿大原子能有限公司(AECL)①也提出了一系列采用直接循环和间接循环的超临界反应堆,设计运行压力为 25MPa,堆芯出口温度为 400~625℃(Bushby 等,2000)。

超临界水冷反应堆在第四代反应堆国际论坛(GIF)中被确定为 6 种最具发

① 现在是加拿大核实验室(加拿大原子能研究所),Canada Deuterium Uranium(CANDU)是加拿大原子能有限公司(AECL)的注册商标。

展前景的创新型反应堆之一。GIF对第四代反应堆提出了4个高水平目标性要求,分别要求在安全性、经济性、可持续发展以及防止放射性扩散方面获得大幅改善。Locatelli等针对6种四代概念堆的近期研究现状提供了一份较为实用的综述报告。Pioro在2011年发表的文章中探讨了超临界水作为冷却剂在核反应堆中的应用,并包含了一套完整的术语定义。Schulenberg等(2012、2014)也在发表的文章中简要概述了SCWR系统概念以及由不同项目管理委员会(PMB)开展的研究工作。GIF超临界水冷反应堆材料及化学(M&C)项目管理委员会(Guzonas,2009)协调了正在开展的各类国家级项目(Buorngiorno等,2003;Anderson等,2009;Bae等,2007;Schulenberg和Starflinger,2009)。2010年加拿大、欧洲原子能联盟和日本签署了GIF M&C项目框架协议,2017年进行了修订并将中国纳入该项目。

关于超临界水冷反应堆的堆芯设计提出了两种概念:①与压水堆和沸水堆设计类似,堆芯布置在反应堆压力容器内部(图1.2(a));②与传统的压力型重水反应堆和RBMK[①]类似,燃料棒位于压力管或者压力通道内,压力管或压力通道采用分散布置(图1.2(b))。在沸水堆中,大部分的冷却剂会在堆芯内发生二次循环;与沸水堆不同,所有超临界水冷反应堆的设计概念都不提供冷却剂的堆芯再循环,即全部都流向汽轮机,大大简化了超临界水冷反应堆的堆芯设计。

尽管Cook和Fatoux(2009)以及Alekseev等(2015)提出过间接循环的概念,但是大多数第四代超临界水冷反应堆概念仍然采用直接循环,即受热的冷却剂直接通过汽轮机。这种方式的部分电厂配套设施与现役的超临界化石燃料电厂非常相似。

大多数超临界水冷反应堆概念中冷却剂系统主要包括:①主泵;②堆芯;③主蒸汽管线、汽轮机和发电机组。只有部分堆芯管路、主蒸汽管线以及高压汽轮机所处环境温度大于临界温度,即$T>T_c$,并且只有前半部分会受到放射性辐照。系统其余部分的运行条件与火电机组相似,运行经验和知识体系都较为成熟。国际水和蒸汽性质协会(IAPWS)、电力研究院(EPRI)以及欧洲热电运营商协会(Vereinijung der Groβkesselbesitzen(VGB)Powertech)等组织将会继续协调、报告并解读这一领域的研究成果。因此,在进行具体设计时,SCWR主泵材料的选择可能主要依赖于沸水堆和超临界化石燃料电厂的使用经验以及超超临

① RBMK反应堆(大功率通道型反应堆)(Reactor Bolshoy Moshchnosty Kanalny(high-power channel-type reactor),RBMK)。

界电厂先进材料方面的进展①。

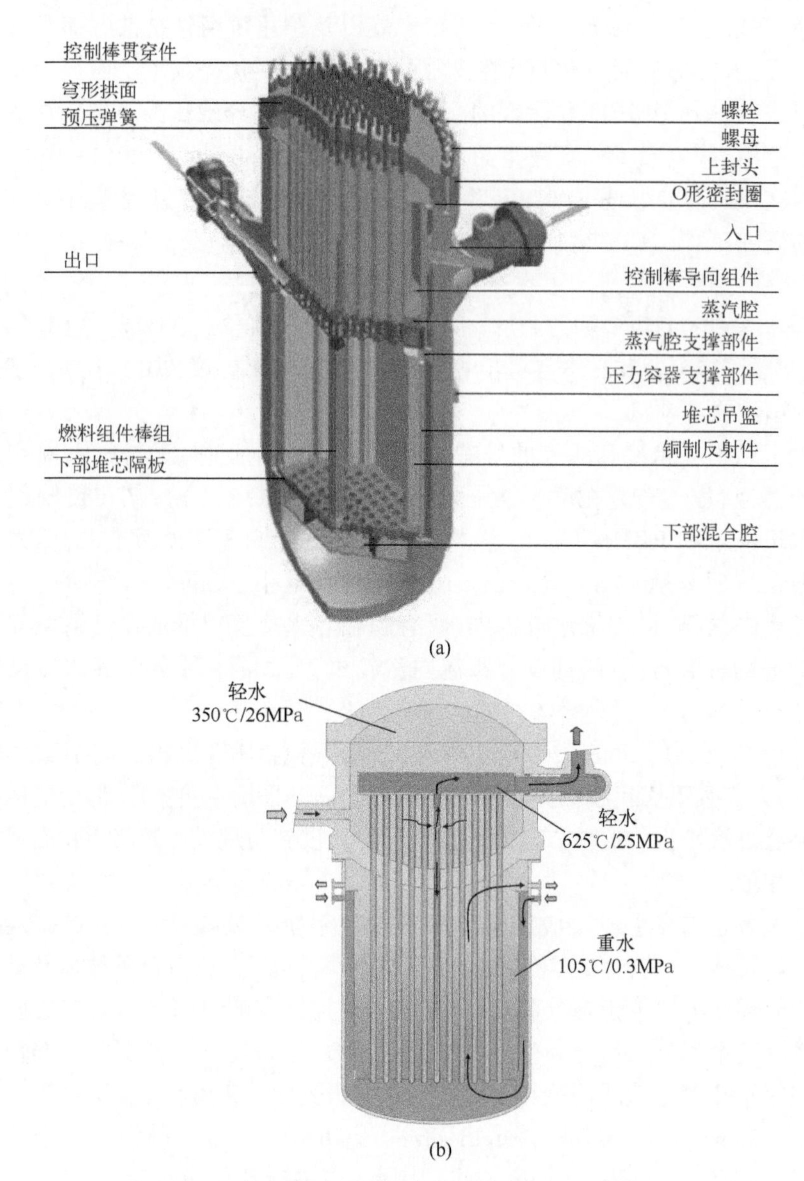


图 1.2 关于超临界水冷反应堆的堆芯设计的两种概念

(a)高性能轻水反应堆堆芯设计概念示意图(Schulenberg 和 Starflinger,2012);

(b) 加拿大超临界水冷反应堆概念设计图(Yetisir 等,2016)。



① “超超临界”这一术语在工业上用来表示超临界化石燃料电厂(SCFPP)运行在蒸汽温度超过566℃(1050℉)的情况。这个温度没有物理意义。

超临界水冷反应堆堆芯在材料和水化学两方面提出了挑战。第一个挑战是中子以及 γ 等射线的存在导致的强辐照效应。辐照效应将会带来两方面的影响(图 1.3)。首先,辐照破坏了受辐照材料的原子结构,进而影响材料的性能,这部分内容详见第 3 章。此外,对冷却剂的辐照会导致水分子的分解(第 4 章),由此产生的辐照溶解产物会影响材料表面过程,如腐蚀等(总体效应,第 5 章;局部效应,第 6 章)。因此,对非辐照系统的研究只能为超临界水冷反应堆堆芯材料的老化现象提供部分支撑,如第 2 章所述。第二个挑战是整个堆芯运行温度的巨大变化,温度梯度之高和工作压力之大都远远超出了目前水冷反应堆的运行经验,因此目前的水冷反应堆数据的参考价值有限,只能用于识别重要的老化现象。

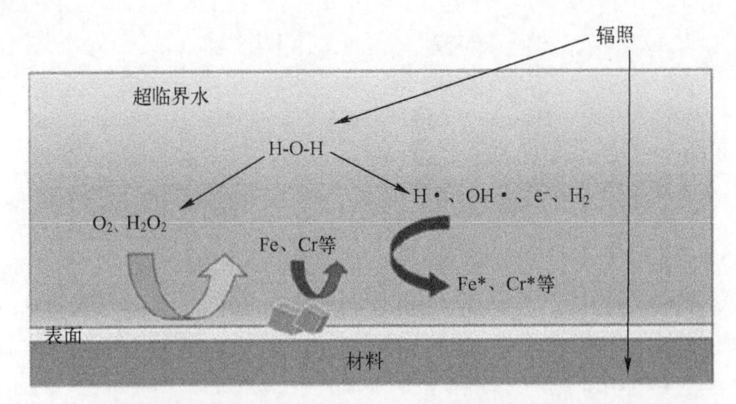


图 1.3　放射性物质与堆芯材料以及充当冷却剂的超临界水之间的相互作用。书中放射性物质与材料之间的相互作用在第 3 章中涉及,与冷却剂的相互作用在第 4 章中涉及。与冷却剂作用产生的中间产物和稳定产物还会与材料表面以及其他类型的溶剂发生反应在第 5 章和第 6 章中涉及。(引自文献 Katsumura, Y., Kiuchi, K., Domae, M., Wada, Y., Yotsuyanagi, T., 2003. Fundamental R&D program on water chemistry of supercritical pressure water under radiation field. In: Genes4/ANP3003, Kyoto, Japan, September 15e19, 2003, Paper 1178)

## 1.2.1 协同发展情况

从 20 世纪 60 年代过热蒸汽反应堆概念的提出到 90 年代超临界水冷反应堆研究的重启和发展,在材料方面尤其是超临界和超超临界火电机组的材料方面取得了较大进展。化石燃料电厂在压力 27~28MPa、蒸汽温度 600~620℃ 的条件下已经运行多年,未来电厂的蒸汽目标温度为 700~760℃。超临界化石燃料电厂的材料研发仍然是热门研究领域,并且将会与超临界水冷反应堆的材料协同发展,这一点对于非直接暴露在高辐照剂量下的堆芯组件尤其重要。而对于强辐照作用下的材料,如燃料包壳,化石燃料电厂中无辐照影响的运行数据,如腐蚀速率等仍然具有价值。但是,化石燃料电厂锅炉中的管道明显比 SCWR 中的燃料包层更大、更厚(图 1.4),这对于允许腐蚀速率和环境敏感开裂效应都会产生显著的影响。此外,在堆芯材料中特定合金元素(如 Co)的允许浓度可能会受到限制,因为它们可以被中子活化为有害同位素(如$^{59}$Co(n,γ)$^{60}$Co)①。在超临界化石燃料电厂中,水在传热管内流动,热源(燃气)位于外部,而在 SCWR 中正好相反,固体燃料位于燃料包层内部,SCW 冷却剂在包壳外部流动。尽管超临界水的化学氧化条件比 SCWR 堆芯中预期的还要糟糕,但在这些条件下获得的腐蚀和 SCC 数据,特别是在非正常化学条件下得到的数据为研究 SCWR 堆芯中的腐蚀现象提供了有价值的参考。

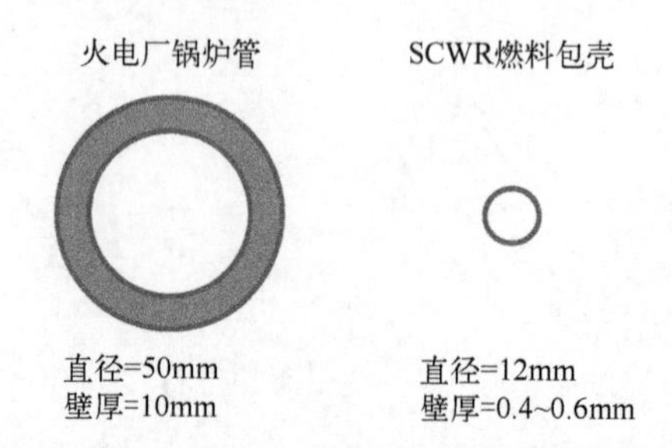


图 1.4 传统化石燃料电厂传热管尺寸与超临界水冷反应堆中设计燃料棒尺寸的对比
(引自文献 Allen, T. R., Chen, Y., Ren, X., Sridhara, K., Tan, L., Was, G. S., West, . E., Guzonas, D. A., 2012. Material performance in supercritical water. In: Konings, R. J. M. (Ed.), Comprehensive Nuclear Materials, vol. 5. Elsevier, Amsterdam, pp. 279-326)



## 1.3 超临界水冷反应堆材料要求

Guzonas 和 Novotny(2014)总结了 SCWR 材料发展面临的挑战。堆芯材料的选择标准可以参照材料在寿命期内不失效的原则。燃料包壳失效可能发生在以下几种情况(Guzonas 等,2016)。

(1) 一般性腐蚀或局部腐蚀穿透包壳壁面。

(2) 氧化物累积形成热阻。

(3) 环境敏感开裂。

(4) 由以下情况导致的应力超过材料屈服极限:

---

① 这种核反应堆的表示方法(从左到右)意为:母体$^{59}$Co 与入射中子(n)反应,随以 γ 射线的形式释放多余的能量,形成产物$^{60}$Co。

① 随着材料老化和辐照剂量累积，材料的屈服极限降低，包括蠕变、膨胀和微观结构变化引起的强度变化进而导致疲劳和脆化；

② 材料受到辐照后应力增加，应力主要来源于裂变气体释放导致的包壳内部压力增加和有限空间内的膨胀效应。

(5) 燃料芯块与包壳之间的作用(PCI)。

Buongiorno 等(2003)对压力容器式 SCWR 的材料要求进行了详细的总结，并提出了备选材料。他们指出，在美国提出的 SCWR 概念中，所有压力容器内部组件(堆芯内部构件)将采用周期性更换的方案，这样就可以不考虑高注量率(大于 20dpa)[①]造成的影响。Zhang 等(2012)从最小屈服强度、最小蠕变强度、允许在役壁面损失、相稳定性和辐照诱导效应敏感性等方面概述了 SCWR 燃料包壳的一般技术要求。Schulenberg(2013)概述了高性能轻水反应堆(HPLWR)燃料包壳和各种堆芯组件的关键材料要求。Guzonas 等(2016)和 Zheng 等(2016)针对加拿大 SCWR 概念中的 8 项性能指标，介绍了 5 种备选燃料包壳合金的评估结果。

截至本书出版，很少有人关注燃料芯块与包壳间的相互作用(PCI)，因为采用类似性质的燃料芯块和包壳材料就可以满足材料的初步选择要求。Oka 等指出，在轻水反应堆中锆合金包壳的热膨胀系数小于 $UO_2$ 燃料芯块的热膨胀系数，而不锈钢或者镍基合金的热膨胀系数在 SCWR 中可能与 $UO_2$ 非常接近。

需要注意的是，本书中提到的包壳峰值温度(PCT)并非是冷却剂的最大温度，这一点对于很多材料的特性是非常重要的。正如 Oka 等(2010)所指出的，PCT 才是堆芯设计的标准，而并非是临界热流密度，因为在超临界压力下，冷却剂为单相流体，不会出现干烧现象。因此，必须限制该温度，以保证在正常运行工况和异常瞬态条件下保持燃料包壳的完整性。

SCWR 燃料原件棒的概念性设计有两种方案。一种是采用独立燃料包壳内部充压的方案，即新燃料元件内部最初充填 8MPa 左右的氦气，一个燃料周期结束时，由于裂变气体的释放，内部压力将增加至 25MPa 左右，与外部环境压力达到平衡。内部加压对材料的屈服强度和蠕变强度提出了更高要求，同时也限制了包壳的最小壁厚，HPLWR 燃料包壳在运行 20000h 后，最大允许腐蚀穿透深度为 140μm(Schulenberg，2013)。另一种是加拿大 SCWR 燃料包壳的设计方案，燃料包壳贴附在燃料芯块上，以此来承受外部 25MPa 的压力(Yetisir 等，2016)，尽管辐照会引起燃料膨胀进而导致燃料包壳产生周向拉应力，合金的高温机械强度和蠕变性都不再是材料选择过程中的首要因素。加拿大的 SCWR

① dpa：材料辐照损伤强度单位，单位原子位移。

规定了在使用寿命内最大允许腐蚀穿透深度为 200μm(包括沿晶界的氧化物穿透),这就意味着在运行 20000h 后的穿透深度大约为 130μm,与 HPLWR 的值基本一致。

在反应堆运行过程中,来自主泵的腐蚀产物将会腐蚀燃料包壳表面并在表面沉降形成氧化物(第 4 章)。少量的氧化物会增加表面的粗糙度,可能强化冷却剂的对流换热效应。但是,当大量的氧化物在包壳表面堆积时,则会降低包壳的热导率,相同冷却剂温度条件下会造成包壳和燃料温度升高。另外,氧化物的脱落(散裂)也需要考虑,因为当氧化物碎片的尺寸大到能够被线圈或垫片拦截时,就有可能会堵塞燃料组件或损坏高压汽轮机,因此需要限制总氧化层厚度(腐蚀膜加上沉积的氧化层)。

对于壁厚 0.4~0.6mm 的燃料原件包壳,为了满足堆芯使用寿命,在换料周期内裂纹深度应小于壁厚的 30%,因为在辐照下壁厚的减小意味着裂纹产生后的断裂韧性大大降低。因此,最大裂纹增长率大概每年 0.05mm(约 $1.4\times10^{-9}$ mm/s)。目前大量 WCR 运行经验表明,避免高残余应力和冷加工,再加上良好的水化学控制将会有效降低应力腐蚀开裂 SCC 的可能性。

### 1.3.1 备选材料



Allen 等(2012 年)综合性地概述了 SCWR 材料的多方面性能。国际上针对 SCW 腐蚀特性开展的研究已经测试了多种材料,如铁素体-马氏体(F/M)钢、奥氏体钢、镍基合金、锆基合金和钛基合金等,所得数据涵盖了 90 多种合金材料(Gu 等,2010)。锆基合金由于具有良好的中子经济性,已考虑作为 SCWR 堆芯的备选材料。但是,目前商用合金在 SCWR 温度条件下的承重强度还不明确,并且腐蚀速率已经达到难以接受的程度。加拿大 SCWR 概念中保留了锆合金压力管的使用(Yetisir 等,2013),即压力管构成了燃料芯块的压力边界,冷却剂与慢化剂通过压力管实现物理隔离,压力管外的慢化剂工作温度约为 80℃,此时包壳的机械强度较高,腐蚀速率也较低。表 1.2 针对 SCWR 中的关键老化机制概述了各类合金的性能评价,这些老化机制将在后面的各章中加以说明。

表 1.2　基于主要的堆芯内降解模式划分的 SCWR 备选材料性能汇总

| 合金类型 | 耐腐蚀性 | | 耐辐照性 | | 高温下的机械完整性 |
|---|---|---|---|---|---|
| | 均匀腐蚀 | 应力腐蚀破裂 | 膨胀 | 脆化 | |
| 奥氏体钢 | 高 | 中 | 低 | 高 | 中 |
| 铁素体-马氏体钢 | 中 | 高 | 高 | 高 | 低 |

续表

| 合金类型 | 耐腐蚀性 | | 耐辐照性 | | 高温下的机械完整性 |
|---|---|---|---|---|---|
| | 均匀腐蚀 | 应力腐蚀破裂 | 膨胀 | 脆化 | |
| 镍基 | 高 | 中 | 低 | 中 | 高 |
| 钛基 | 中-高 | 无数据 | | | 中-高 |

引自文献 Heikinheimo, L., Guzonas, D., Fazio, C., 2009, GenIV materials and chemistry research-common issues with the SCWR concept. In: 4th Int. Symp. on Supercritical WaterCooled Reactors, Heidelberg, Germany, March 8-11, 2009, Paper 81。

在已评估的商用合金中,目前只有奥氏体不锈钢和一些镍基合金能够满足所有的性能要求。表 1.3 列出了 SCWR 最新设计中考虑的燃料包壳备选材料。表中所列的内容还不够详尽,不包括已经过测试的钛基和锆基合金,也不包括 F/M 钢,这些虽然已经过充分研究但目前还不考虑用于堆芯。Allen 等(2012)还给出了 SCWR 中使用的其他合金,Ru 和 Staehle(2013)在表 1.1 中列出了在早期化石燃料电厂和过热蒸汽反应堆中经过检验的合金,包括部分表 1.3 中未列出的合金。

表 1.3 SCWR 包壳备选材料性能测试对比

单位:%(质量分数)

| 合金 | Fe | Cr | Ni | Mo | Mn | Si | C | Ti | 其他 |
|---|---|---|---|---|---|---|---|---|---|
| ODS PM2000 | 其余成分 | 19.0 | | | | | | 0.5 | $Y_2O_3$:0.5, Al:5.5 |
| ODS MA956 | 其余成分 | 20.0 | — | — | ≤0.30 | | <0.1 | 0.2~0.6 | $Y_2O_3$:0.34 |
| 奥氏体不锈钢 | | | | | | | | | |
| 1.4970 | 其余成分 | 15 | 15.3 | 1.18 | 1.68 | 0.53 | 0.095 | 0.45 | |
| 316L | 其余成分 | 16.6 | 10 | 2.0-3.0 | 1.9 | 0.65 | 0.022 | — | |
| 316L(N) | 其余成分 | 18 | 12 | 2.0-3.0 | 2.0 | 1.0 | 0.030 | — | N:0.16 |
| 316Ti | 其余成分 | 16.6 | 12.1 | 2.03 | 1.15 | 0.45 | 0.032 | 0.38 | |
| 316L+Zr(HI) | 其余成分 | 16.54 | 10.71 | 2.22 | | 0.46 | 0.006 | | P:0.016, Zr:0.56 |
| 347 | 其余成分 | 18.0 | 11.0 | — | 2.0 | 1.0 | 0.08 | — | Nb:10×C① |
| 321 | 其余成分 | 18 | 8 | | | | 0.08 | | Ti:0.6 |
| AL6XN | 其余成分 | 20.4 | 23.8 | 6.23 | 0.42 | 0.34 | 0.02 | — | Co:0.24, P:0.024, N:0.21 |
| SAVE25 | 其余成分 | 22.5 | 18.5 | 0.1 | 0.5 | 0.2 | 0.07 | | W:1.5, Cu:3.25, Nb:0.45, N:0.2, V:0.04 |

续表

| 合金 | Fe | Cr | Ni | Mo | Mn | Si | C | Ti | 其他 |
|---|---|---|---|---|---|---|---|---|---|
| Sanicro 25 | 其余成分 | 22.5 | 25 |  | 0.5 | 0.2 | 0.1 |  | Co:1.5,Cu:3.0,W: 3.6, Nb: 0.5,N:0.23 |
| 800H | 其余成分 | 22.5 | 34.8 | — | 1.59 | 0.95 | 0.08 | — | Al:0.45 |
| 310 | 其余成分 | 24.5 | 20.2 | 0.29 | 1.17 | 0.33 | 0.048 | — |  |
| 310S FG(T3F) | 其余成分 | 24.74 | 21.92 | — | — | 0.25 | 0.099 | 0.81 | N:0.0006,P:<0.005 |
| 310S FG(T6F) | 其余成分 | 25.03 | 22.81 | 2.38 | — | — | — | 0.41 | Nb:0.26,N:0.002 |
| 310S+Zr(H2) | 其余成分 | 25.04 | 20.82 | 0.51 | — | 0.51 | 0.034 | — | P:0.016,Zr:0.59 |
| HR3C | 其余成分 | 25 | 21 | 0.1 | 1.1 | 0.4 | 0.07 | — | N:0.25,Nb:0.45,V:0.07 |
| 镍基 |  |  |  |  |  |  |  |  |  |
| C276 | 5.35 | 15.88 | 其余成分 | 15.64 | 0.52 | 0.03 |  | 0.01 |  |
| 214 | 3 | 16 | 75 | — | 0.5 | 0.2 | 0.05 | — | Al:4.5,Y:0.01,B:0.01,Zr:0.1 |
| 625 | 4.9 | 22.6 | 其余成分 | 9.8 | 0.43 | 0.47 | 0.09 | 0.45 | Al:0.47,Nb:3.7,Cu:0.68 |
| Hastelloy C690 | 9.2 | 29 | 72 |  | 1(最大值) | 0.35 |  |  | Al:0.02 |

注:这个清单并不详尽,F/M 钢被省略了。Allen 等(2012)提供了这类合金的详细信息。
① Nb+Ta。

## 1.4 总　结

GIF SCWR 材料和化学管理委员会 PMB 明确了为实现 SCWR 的安全可靠性运行(Guzonas,2009)而需要克服的两个主要知识缺口。

(1) 任何一种单一合金,尤其是用于堆芯组件的合金,还没有足够的数据能够明确地保证其在 SCWR 中的性能。

(2) 由于水的物理和化学性质通过临界点时发生了巨大的变化,加之对水的辐照分解效应了解尚不充分,以目前对 SCWR 化学的了解还难以确定一种化学控制策略。

近年来,得益于各类国际研究和发展项目,在理论知识方面获得了重大进展。日本、欧盟和加拿大分别于 2010 年、2012 年和 2015 年提出了 SCWR 的设计理念,并在研究和开发上取得了不同程度的阶段性进展(日本的研究工作在福岛核事故后被搁置)。

本书后面章节着重介绍了 SCWR 化学和材料方面的研究现状。第 2 章介绍了 SCWR 材料和化学研究中用到的试验设备以及试验中遇到的一些挑战。第 3 章讨论了材料的力学性能(如拉伸和蠕变),介绍了辐照对这些性能的影响,并总结了相关数据。第 4 章以“什么是 SCW”开端,讨论了堆芯环境特性如何影响腐蚀和 EAC 性能。最后两章着重介绍了一般性腐蚀和 EAC。

SCWR 的研究和开发是一项持久性的工作,本书是对书稿撰写期间大约 18 个月内研究现状的总结。后续研究成果会不断在各种会议上被展示出来,如 SCWR 国际研讨会(每两年举行一次)、国际原子能机构(IAEA)主办的技术研讨会以及同行评议的学术论文。由于作者认知范围有限,不可避免地会忽略一些重要的数据,另外,书中各章节内容仅代表作者个人观点。

本书旨在为从事这一领域以及拟开展相关研究的工作人员提供参考,同时对于 SCFPP 行业的研究人员也应该是有价值的。本书的目标读者为研究生及以上学历的人群,读者应该具备良好的本科知识水平,具备包括材料科学、热力学、腐蚀科学、化学动力学以及核能发电的基础等,相关专业感兴趣的大四学生对书中大部分内容也能读懂。

## 参考文献


Alekseev, P., Semchenkov, Y., Sedov, A., Sidorenko, V., Silin, V., Mokhov, V., Nikitenko, M., Mahkin, V., Churkin, A., 2015. Conceptual proposals on reactor VVER-SCW developed on the basis of technologies of VVER and steam - turbine installations at supercritical parameters. In: 7th International Symposium on Supercritical Water-Cooled Reactors (ISSCWR-7), March 15-18, 2015, Helsinki, Finland, Paper ISSCWR7-2055.

Allen, T. R., Chen, Y., Ren, X., Sridharan, K., Tan, L., Was, G. S., West, E., Guzonas, D. A., 2012. Material performance in supercritical water. In: Konings, R. J. M. (Ed.), Comprehensive Nuclear Materials, vol. 5. Elsevier, Amsterdam, pp. 279-326.

Anderson, T., Brady, D., Guzonas, D., Khartabil, H., Leung, L., Quinn, S., Zheng, W., 2009. Canada's NSERC/NRCan/AECL Generation IV energy technologies program. In: 4th Int. Symp. on SCWRs, March 8-11. Heidelberg, Germany.

Bae, Y. Y., Jang, J., Kim, H. Y., Yoon, H. Y., Kang, H. O., Bae, K. M., 2007. Research activities on a supercritical pressure water reactor in Korea. Nucl. Eng. Technol 39(4), 273-286.

Barnard, J., 1961. The Reactor and Plant Design of the ESADA and EVESR. ASME paper 61-WA-223.

Bevilacqua, F., Brown, G. M., 1963. Chloride Deposition from Steam onto Superheater Fuel Clad Materials. General Nuclear Engineering Corporation Report GNEC 295.

Buongiorno, J., Corwin, W., MacDonald, P., Mansur, L., Nanstad, R., Swindemann, R., Rowcliffe, A., Was, G., Wilson, D., Wright, I., 2003. Supercritical Water Reactor (SCWR): Survey of Materials Experience and R&D Needs to Assess Viability. INEEL/EXT-03-00693Rev1, 2003.

Bushby, S. J. , Dimmick, G. R. , Duffey, R. B. , Burrill, K. A. , Chan, P. S. W. , 2000. Conceptual designs for advanced, high temperature CANDU reactors. In: Proc. 8th Int. Conference on Nuclear Engineering, Baltimore, MD, April 2-6, 2000. ICONE- 8470.

Cohen, K. , Zebroski, E. , 1959. Operation Sunrise. Nucleonics 17(3), 63-70. Cook, W. G. , Fatoux, W. , 2009. A CANDU-SCWR with a steam generator: thermodynamic assessment and estimation of fouling rates. In: 4th Int. Symp. on SCWRs, March 8-11, Heidelberg, Germany.

Dobashi, K. , Kimura, A. , Oka, Y. , Koshizuka, S. , 1998. Conceptual design of a high temperature power reactor cooled and moderated by supercritical light water. Ann. Nucl. Energy 25, 487-505.

Dollezhal, N. A. , Krasin, A. K. , Aleshchenkov, P. I. , Galanin, A. N. , Grigoryants, A. N. , Emelyanov, I. Ya. , Kugushev, N. M. , Minashin, M. E. , Mityaev, U. I. , Florinsky, B. V. , Sharpov, B. N. , 1958. Uranium-graphite reactor with superheated high pressure steam. In: Proc. 2nd United Nations Int. Conf. on the Peaceful Uses of Atomic Energy, Geneva, September 1-13, 1958, vol. 8, pp. 398-414.

Dollezhal, N. A. , Emel'yanov, I. Ya. , Aleshchenkov, P. I. , 1964. Development of superheating power reactors of Beloyarsk nuclear power station type. In: Proceedings of the Third International Conference on the Peaceful Uses of Atomic Energy, Geneva, August 31-September 9, 1964, vol. 6, paper 309, pp. 256-265.

Emel'yanov, I. , Shatskaya, O. A. , Rivkin, E. Yu. , Nikolenko, N. Y. , 1972. Strength of construction elements in the fuel channels of the Beloyarsk power station reactors. Atomnaya Energiya 33(3), 729-733(in Russian) Translated in Soviet. Atomic Energy 33(3), 842-847.

GIF, 2002. A Technology Roadmap for Generation IV Nuclear Energy Systems. GIF-002-00.

GIF, 2014. Technology Roadmap Update for Generation IV Nuclear Energy Systems. https://www.gen-4.org/gif/jcms/c-9352/technology-roadmap.



Gu, G. P. , Zheng, W. , Guzonas, D. , 2010. Corrosion database for SCWR development. In: 2nd Canada-China Joint Workshop on Supercritical Water-Cooled Reactors(CCSC-2010) Toronto, Ontario, Canada, April 25-28, 2010.

Guzonas, D. A. , 2009. SCWR materials and chemistry-status of ongoing research. In: Proceedings of the GIF Symposium, Paris, France, September 9-10, 2009.

Guzonas, D. , Novotny, R. , 2014. Supercritical water-cooled reactor materials-Summary of research and open issues. Prog. Nucl. Energy 77, 361-372.

Guzonas, D. , Edwards, M. , Zheng, W. , 2016. Assessment of candidate fuel cladding alloys for the Canadian supercritical water-cooled reactor concept. J. Nucl. Eng. Radiat. Sci. 2, 011016.

Hazel, V. E. , Boyle, R. F. , Busboom, H. J. , Murdock, T. B. , Skarpelos, J. M. , Spalaris, C. N. , 1965. Fuel Irradiations in the ESADE-VBWR Nuclear Superheat Loop. General Electric Atomic Power report GEAP-4775.

Heikinheimo, L. , Guzonas, D. , Fazio, C. , 2009. GenIV materials and chemistry research-common issues with the SCWR concept. In: 4th Int. Symp. on Supercritical Water-Cooled Reactors, Heidelberg, Germany, March 8-11, 2009, Paper 81.

Katsumura, Y. , Kiuchi, K. , Domae, M. , Wada, Y. , Yotsuyanagi, T. , 2003. Fundamental R&D program on water chemistry of supercritical pressure water under radiation field. In: GENES4/ANP3003, Kyoto, Japan, September 15-19, 2003, Paper 1178.

Kornbichler, H. , 1964. Superheat reactor development in the Federal Republic of Germany. In: Proc. 3rd Int. Conf. on the Peaceful Uses of Atomic Energy, Geneva, August 31-September 9, 1964, vol. 6, pp. 266-276.

Lo, K. H., Shek, C. H., Lai, J. K. L., 2009. Recent developments in stainless steels. Mater. Sci. Eng. R 65, 39-104.

Locatelli, G., Mancini, M., Todeschini, N., 2013. Generation IV nuclear reactors: current status and future prospects. Energy Policy 61, 1503-1520.

Marchaterre, J. F., Petrick, M., 1960. Review of the Status of Supercritical Water Reactor Technology. Argonne National Laboratory. Atomic Energy Commission Research and Development Report, ANL 6202.

Margen, P. H., Leine, L., Nilson, R., 1964. The design of the Marviken boiling heavy-water reactor with nuclear superheat. In: Proc. 3rd Int. Conf. on the Peaceful Uses of Atomic Energy, Geneva, August 31-September 9, 1964, vol. 6, pp. 277-288.

Moore, R. V., Barker, A., Bishop, J. F. W., Bradley, N., Iliffe, C. E., Nichols, R. W., Thorn, J. D., Tyzack, C., Walker, V., 1964. The Utilisation of Supercritical Steam in Nuclear Power Reactors, Final Report of the Supercritical Steam Panel. United Kingdom Atomic Energy Authority, TRG Report 776(R).

Novick, M., Rice, R. E., Graham, C. B., Imhoff, D. H., West, J. M., 1964. Development of nuclear superheat. In: Proc. 3rd Int. Conf. on the Peaceful Uses of Atomic Energy, vol. 6, p. 225. Geneva, August 31-September 09, 1964.

Oka, Y., 2000. Review of high temperature water and steam cooled reactor concepts. In: Proc. 1st Int. Symp. on SCWR. Tokyo, Japan, November 6-8, 2000, Paper 104(2000).

Oka, Y., Koshizuka, S., 1993. Concept and design of a supercritical-pressure, direct-cycle light water reactor. Nucl. Technol. 103(3), 295-302.

Oka, Y., Koshizuka, S., Ishiwatari, Y., Yamaji, A., 2010. Super Light Water Reactors and Super Fast Reactors: Supercritical-Pressure Light Water Cooled Reactors. Springer, New York.

Pioro, I., 2011. The potential use of supercritical water-cooling in nuclear reactors. In: Krivit, S. B., Lehr, J. H., Kingery, T. B. (Eds.), Nuclear Energy Encyclopedia: Science, Technology, and Applications. John Wiley and Sons, Inc.

Ru, X., Staehle, R. W., 2013a. Historical experience providing bases for predicting corrosion and stress corrosion in emerging supercritical water nuclear technology: Part 1-Review. Corrosion 69(3), 211-229.

Ru, X., Staehle, R. W., 2013b. Historical experience providing bases for predicting corrosion and stress corrosion in emerging supercritical water nuclear technology: Part 2-Review. Corrosion 69(4), 319-334.

Ru, X., Staehle, R. W., 2013c. Historical experience providing bases for predicting corrosion and stress corrosion in emerging supercritical water nuclear technology: Part 3- Review. Corrosion 69(5), 423-447.

Saltanov, E., Pioro, I., 2011. World experience in nuclear steam reheat. In: Tsvetkov, P. (Ed.), Nuclear Power -Operation, Safety and Environment. In Tech, ISBN 978-953-307-507-5. Available from. http://www.intechopen.com/books/nuclear-power-operation-safetyandenvironment/world-experience-in-nuclear-steam-reheat.

Schulenberg, T., 2013. Material requirements of the high performance light water reactor. J. Supercrit. Fluids 77, 127-133.

Schulenberg, T., Starflinger, J., 2009. European research project on the high performance light water reactor. In: 4th International Symposium on Supercritical Water-cooled Reactors, Heidelberg, Germany, March 8-11, 2009, Paper 54.

Schulenberg, T., Starflinger, J (Eds.), 2012. High Performance Light Water Reactor: Design and Analyses.

KIT Scientific Publishing, Karlsruhe.

Schulenberg, T., Matsui, H., Leung, L., Sedov, A., 2012. Supercritical water cooled reactors. In: GIF Symposium Proceedings/2012 Annual Report. San Diego California, November 14-15, 2012, NEA No. 7141.

Schulenberg, T., Leung, L. K. H., Oka, Y., 2014. Review of R&D for supercritical water cooled reactors. Prog. Nucl. Energy 77, 282-299.

Slin, V. A., Voznessensky, V. A., Afrov, A. M., 1992. The light water integral reactor with natural circulation of the coolant at supercritical pressure B 500 SKDI. In: Proc. ANP'92, Tokyo, Japan, October 25-29, 1992, vol. 1. Session 4.6, 1 7.

Viswanathan, R., Henry, J. F., Tanzosh, J., Stanko, G., Shingledecker, J., Vitalis, B., Purgent, R., 2013. US program on materials technology for ultra-supercriticalcoal power plants. J. Mater. Eng. Perform. 22(10), 2904-2915.

Yetisir, M., Gaudet, M., Rhodes, D., 2013. Development and integration of Canadian SCWR concept with counter-flow fuel assembly. In: 6th International Symposium on Supercritical Water-cooled Reactors (ISSCWR-6), March 3-7, 2013, Shenzhen, Guangdong, China, Paper ISSCWR6-13059.

Yetisir, M., Xu, R., Gaudet, M., Movassat, M., Hamilton, H., Nimrouzi, M., Goldak, J. A., 2016. Various design aspects of the Canadian supercritical water-cooled reactor core. J. Nucl. Eng. Radiat. Sci. 2, 011007.

Zhang, L., Bao, Y., Tang, R., 2012. Selection and corrosion evaluation tests of candidate SCWR fuel cladding materials. Nucl. Eng. Design 249, 180-187.

Zheng, W., Guzonas, D., Boyle, K. P., Li, J., Xu, S., 2016. Materials assessment for the Canadian SCWR core concept. JOM 68, 456-462.

# 第2章
# 试 验 方 法

轻水反应堆(LWR)的一回路工作在高温高压条件下,对一回路冷却剂开展试验具有很大的挑战性,而在超临界水冷反应堆(SCWR)堆芯条件下进行试验更是如此。第四代反应堆国际论坛材料和化学项目管理委员会(GIF M&C PMB)提出了一项支持SCWR概念的四级材料研发方案(图2.1),研发复杂程度和成本随着层级的增加而增长,相应的控制范围和可用的知识库也在减少。使用辐照材料或模拟辐照条件(第3级)的试验挑战性也大幅增加,并且在反应堆内测试(第4级)的技术难度和成本都相应增加。

本章主要讨论SCWR开发中涉及材料和水化学测试所用的试验设备和方法,并重点关注SCW(超临界水)方面的开创性工作。

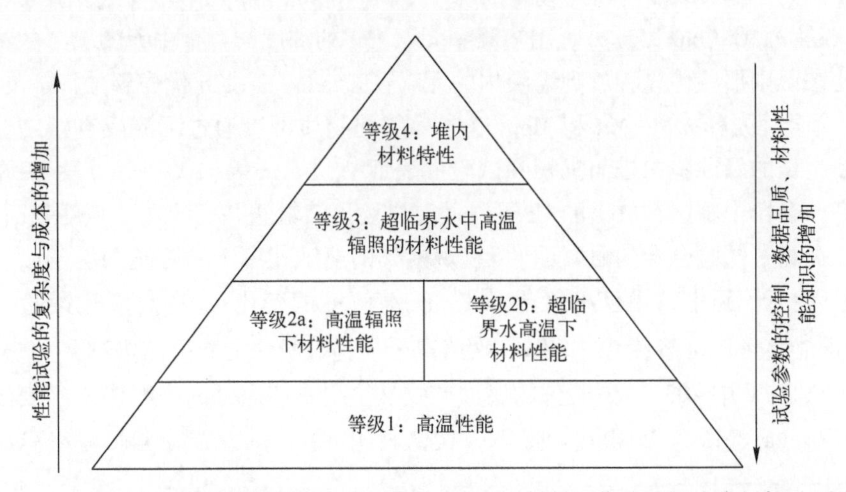


图2.1 第四代国际论坛超临界水冷反应堆(SCWR)材料工作的四级试验方案原理图
(引自文献Guzona, D. A., 2009. SCWR materials and chemistry—status of on-going research. In: Proceedings of the GIF Symposium, Paris, France, September 9-10, 2009)

## 2.1 腐蚀和环境辅助开裂研究的试验装置

对于给定的合金，影响 SCW 中一般性腐蚀和环境辅助开裂(EAC)的关键试验变量是温度、水密度(压力)、溶解氧(DO)浓度、水纯度、流速和表面粗糙度。试验装置首先需要能够在长达几千个小时内，周期性地控制和监控这些参数中的前 5 项。SCWR 试验中采用的水通常是低电导率、无添加剂的水，其氧浓度范围(小于 10μg/kg)从设计工况到 25℃下的溶解饱和状态(约 8mg/kg)；在一些试验中，添加过氧化氢($H_2O_2$)以获得较高的氧浓度(如 20mg/kg)。

SCW 的腐蚀性试验和 EAC 试验可以在静态反应釜中进行，也可以在小容器、冲洗高压釜或者流动系统中进行。后者的尺寸可以从小型实验室规模的容器扩展到大型工程规模的系统。在小容器中测试时，小容器本体就是测试部分。SCWR 测试所需的高温(概念上可高达 800℃)下，用于回路和高温高压容器结构的材料腐蚀速率通常与所测试的材料相似，由于试验件的表面积通常比回路或高温高压管道的表面积小得多，这使数据分析变得异常复杂。因此，除了试验件腐蚀释放出的可溶解腐蚀产物外，试验液还将包含回路和高温高压容器管道表面腐蚀所产生的溶解腐蚀产物。通过对材料为 Hastelloy C-276 和 625 合金的高温高压容器(作为试验样品的高温高压本体)取样测量可知，在 SCW 条件下高温高压本体暴露前后，大量的 Ni、Mo 和 W 被释放到溶液中，如图 2.2 所示(Guzona 和 Cook，2012)。用溶液中元素"X"的浓度与合金中元素"X"的浓度之比来识别那些优先释放到溶液中的元素；Hastelloy C-276 优先释放 Mo 和 W，625 合金优先释放 Al、Mn 和 Mo。这些溶解的物质可以通过沉淀或直接结合到不断生长的腐蚀膜中进而沉积到试验件表面①。Daigo 等(2007 a、b)发现，在高温高压容器中腐蚀释放出的 Cr 可以通过试验溶液输送并沉积在试验件表面，并由此提高了试验合金的耐腐蚀性②。观察到的腐蚀速率下降可能归咎于以下两点原因：①回路中的温度梯度导致 $Cr_2O_3$ 溶解度的变化；②在高氧化试验条件下，可能会形成可溶性 $Cr^{6+}$。基于次级离子质谱分析法的断层扫描结果，充分证明了在 SCW 中 403 SS 试验件的腐蚀产物从金属本体以外的来源渗入到表面膜中(Guzona 和 Cook，2012)。$^{52}Cr/^{56}Fe$ 比值表明，由于 Cr 从高压釜体等来源加入



---

① 杂质或腐蚀产物的迁移和沉积，现象是当前水冷反应堆(WCR)活性迁移的基础，将在第 4 章中进一步讨论。

② 为了开发超临界水氧化工艺，控制温度为 400℃、压力为 30MPa，在 0.01mol/L $H_2SO_4$ 和 0.025mol/L $O_2$(约 800mg/kg)环境中进行了试验。

到最外层氧化物中，导致富铬内层形成于富铁外层之下，外层氧化物逐渐向氧化物-溶液界面富铬。$^{90}Zr^{16}O$ 的微弱信号(来自用于保存测试样品的氧化锆)也在氧化物的外部被观察到，这是材料在 SCW 中从一个表面迁移到另一个表面的有力证据。因此，虽然可以在同一试验容器中对若干不同的合金进行范围界定研究，但稳妥的做法是每次试验(应)使用一种合金或少量成分相似的合金进行限制性测试。

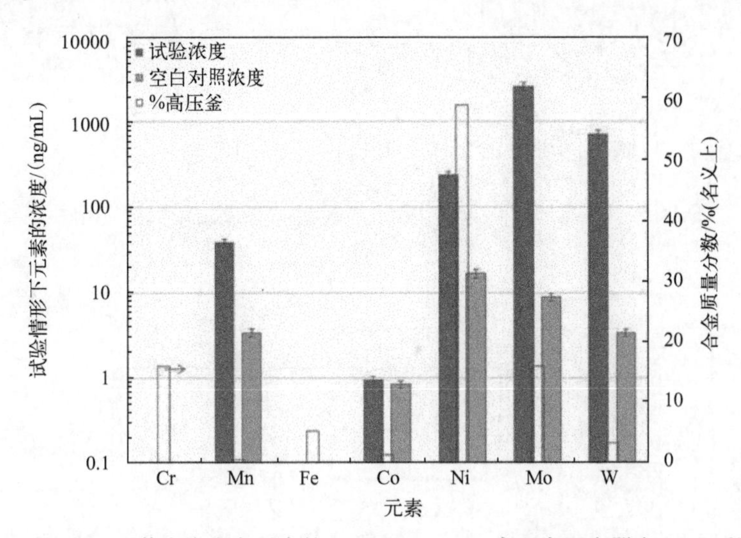


图 2.2 在 450℃下装有含去离子水的 Hastelloy C-276 高温高压容器在 280h 后溶液中各种金属的浓度(图中还显示了合金中各种元素的浓度和测试前去离子水中的浓度)
(引自文献 Guzonas, D. A., Cook, W. G., 2012. Cycle chemistry and its effect on materials in a supercritical water-cooled reactor: a synthesis of current understanding. Corrosion Sci. 65, 48-66)

在静态高压釜或容器中开展试验相对比较简单，但水化学控制(如氧浓度)是困难的，因为容器是密封的。此外，由于试验样品和高压釜的腐蚀，大量可能不具有代表性的溶解腐蚀产物浓度将会随着时间的推移而积累。

使用具有离子交换树脂或者过滤功能的可冲洗高压反应釜或流动系统，能够有效去除离子态杂质和固体颗粒物，进而使腐蚀产物和其他杂质的堆积最小化。在"直通式"循环中，水要么从补充罐中加入(图 2.3(a))，要么在重新引入测试段之前进行冷却和净化以去除杂质(图 2.3(b))。而在流动系统中，要确保测试样品浸没在组分已知且严格受控的水中，试验人员必须对使用的水化学条件予以慎重考虑；反应堆运行中的故障通常发生在非常规条件或瞬态化学条件，而不是正常运行期间。运行中的水冷反应堆(WCR)冷却剂含有杂质(溶解的腐蚀产物以及阴离子杂质 $Cl^-$、$SO_4^{2-}$)，这些杂质可能会影响对腐蚀的响应；除非有意引入，否则在典型浓度的直流回路上游测试段不可能出现类似杂质。而

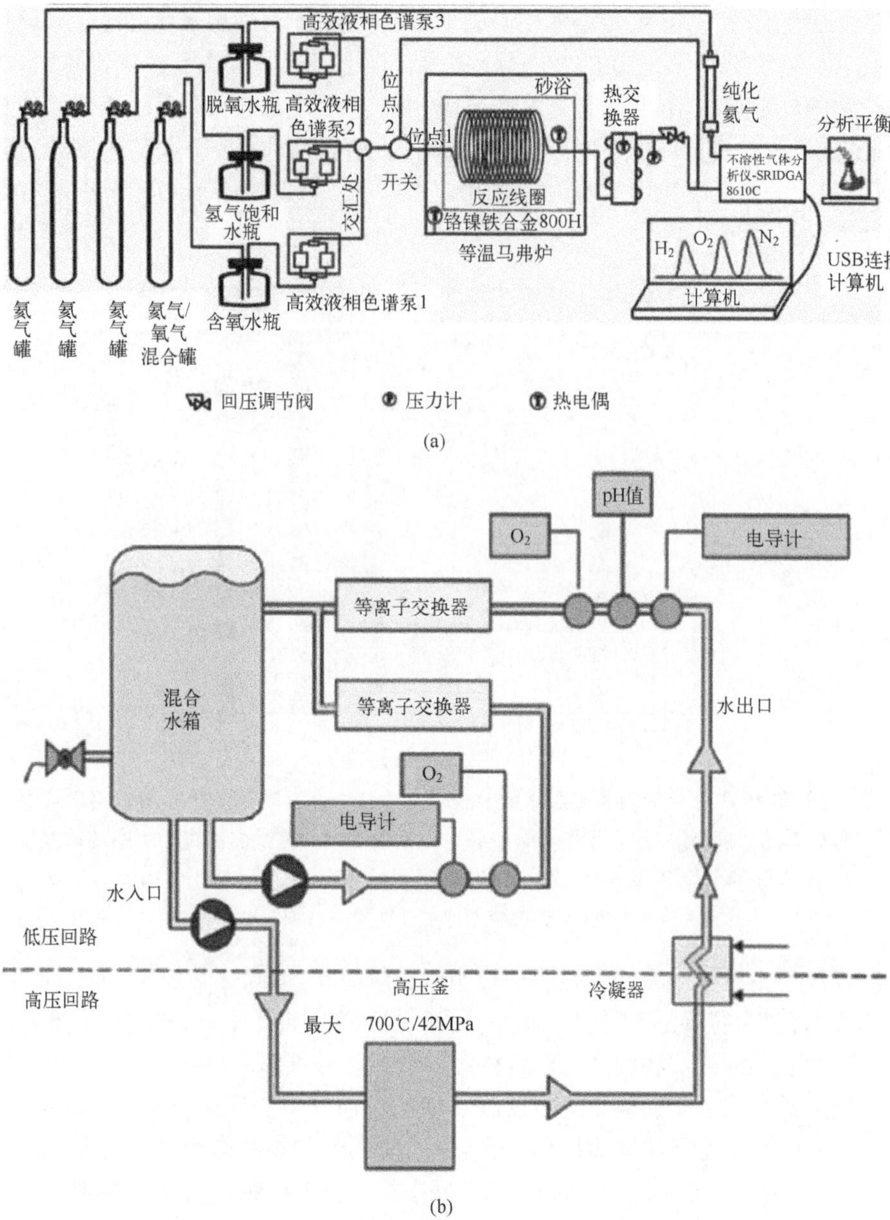

(a)

(b)

图 2.3 “直通式”循环中水的流动

(a) 实验室规模的一次性流通流动系统的原理图(Choudhry 等,2014)(所有通过试验区的水都用于取样或丢弃);(b) VTT“直通式”再循环回路原理图(Penttilä 等,2013)(经过试验区(高温高压容器)后,所有的水都被净化)。

在循环回路中无论净化与否,水都通过循环回路进行循环[①]。循环的主要缺点是系统复杂,通常需要进行很多操作和运行监督。

腐蚀试验装置的运行温度和压力受到容器和管道材料强度的限制。超过700℃时,测量在低压力下更容易开展,文献中描述了许多在高温蒸汽下的试验装置(如 Ruther 等,1966;Bsat 和 Huang,2015;Holom 等,2016)。此类装置,包括反应堆系统在内,在发展核过热蒸汽反应堆期间就被广泛使用(Fitzsimmons 等,1961;Spalaris,1963;Hazel 等,1965)。现存的试验装置能够在临界压力以上、800℃以下开展腐蚀试验研究(Behnamian 等,2016),尽管试验所能够持续的时间较短。Choudhry 等(2014)报道了在 750℃下实验室规模的 500h 连续试验测试结果,所采用的直通式装置如图 2.3(a)所示。

通过测定受热试验段前后的电导率可以评估水的纯度,需要注意的是:①电导率不受测量颗粒浓度和未带电荷溶解的金属水解物的影响(如 $Fe(OH)_2$);②在测试段下游采集的冷却后样品电导率不能准确地反映水的纯度,一方面可能是由于腐蚀产物溶解度通过临界点时急剧变化,另一方面可能是沉淀的胶体在热力学条件下不稳定,发生缓慢再溶解导致的。结果表明,试验段出口处的电导率几乎总是高于进口处电导率。Ampornrat 和 Was(2007)从 9 种 F/M 合金 T91、HCM12A 和 HT-9(进口电导率为 0.06μS/cm)的腐蚀试验中,报道了试验段出口电导率结果。在 400℃条件下,出口电导率为 0.075μS/cm,在600℃时增加到 0.12μS/cm,表明了腐蚀产物的释放。遗憾的是,由于更大的系统回路表面也会产生腐蚀产物(图 2.2),导致结果的定量分析相对困难,尽管研究人员试图利用电导率数据量化腐蚀产物的释放和质量的平衡,但基本上没有成功。

在 SCW 条件下,通过恒压下改变温度或通过恒温下改变压力都可以改变溶剂密度。虽然第一种情况与 SCWR 更相关,但这种试验会使数据解释变得复杂,因为温度(它影响被研究介质的动力学特性,如通过 Arrhenius 形式的相关性)和溶剂的密度(它影响溶剂中由溶解速率决定的物质浓度)在测试过程中都发生了改变。在与 SCWR 相关的低压力(低至 $P/P_c$ 接近于 1)下,最大的密度变化发生在略高于临界温度 $T_c$ 附近(图 2.4)。因此,在临界温度附近进行恒温变压(密度)试验,为研究 SCW 密度对腐蚀的影响提供了一种解决思路。在 25MPa、500℃以上时密度变化相对较小,在此温度以上进行的试验可以认为是在恒定密度(约 $80kg/m^3$)条件下进行的。堆芯内的温度状况类似于超临界化石燃料电厂(SCFPP)锅炉,可分为两部分:“蒸发器”是提高温度并达到

---

① 通常采用侧流净化系统来降低杂质浓度。

$T_c$的一个近临界温度区；"过热器"则是将温度提高到堆芯的出口温度。虽然"蒸发器"区域没有发生相变，但这期间确实为讨论温度/密度的状况提供了一个有用的依据。

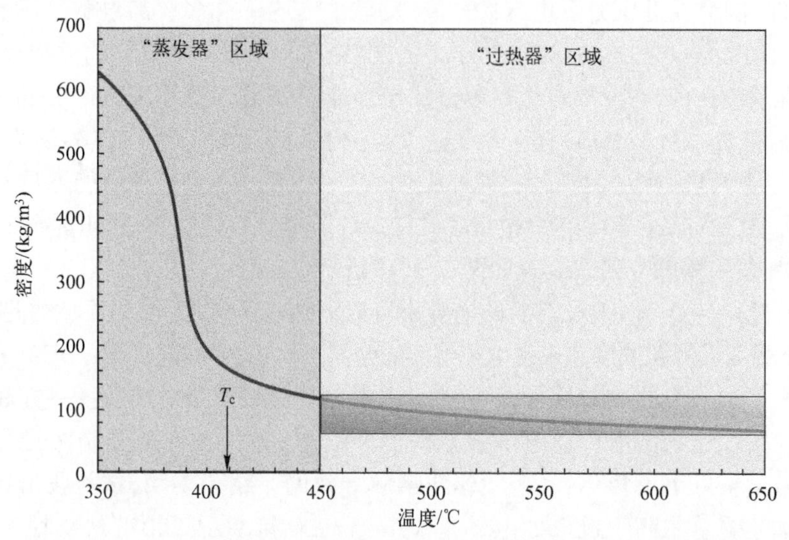


图 2.4　在 25MPa（概念上超临界水堆的设计运行压力）下密度随温度的变化（在接近 $T_c$（接近临界或蒸发器区域）温度时，密度随温度迅速变化，而在高于 450℃（过热器区域）的温度下，密度变化很小；在图中阴影区域，密度在 200℃ 的温度范围内仅降低了 30%）



水流量或高压反应釜更新时间是流动系统的一个关键参数，它决定了反应物向试件的输送速率和腐蚀产物远离试件的传输速率。根据 Novotny 等（2013b）的报告：在 500～550℃、25MPa 条件下，当流速为 15～17L/h 时（高温高压容器重新充水时间为 6min），进口的氧浓度为 1985～2020μg/kg，进口电导率小于 0.1μS/cm 时，测量发现出口电导率由开始的 0.4μS/cm 缓慢下降至 0.12μS/cm，容器出口氧浓度在试验初期从 0 增加到 1800μg/kg，此后连续加热期间基本保持（1800±10）μg/kg 不变。

作为国际合作实验室之间参比工作的一部分，Guzona 等（2016）仔细检查了高压反应釜质量增加与容器充水时间的关系（图 2.5）。在测试的几个试验回路中监测试验段前后的氧浓度，某实验室报告显示在进出口段之间有规律地损失约 600μg/kg，这一损失被猜想是由于试件的腐蚀和回路的高温部分腐蚀以及容器本身腐蚀造成的。通过对静态容器和流动回路中腐蚀产物脱除数据的比较，表明腐蚀产物在静态容器中的迁移并不是一个限制性因素。

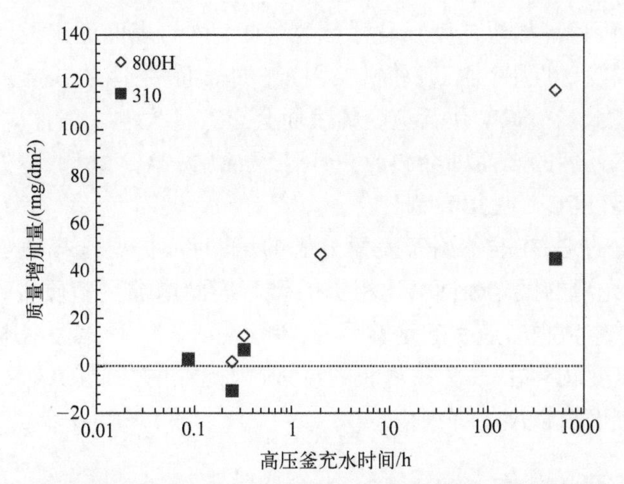


图 2.5　质量增加量与高温高压容器充水时间的函数关系(是针对 800H 合金和 310 SS,表面粗糙度为 1200 目,暴露于超临界水(550℃,25MPa,DO 浓度为 8mg/kg,室温下 pH 值为 7,室温下水的电阻率为 18MΩ,持续时间 500h))
(引自文献 Guzonas, D. , Penttilä, S. , Cook, W. , Zheng, W. , Novotny, R. , Sáez - Maderuelo, A. , Kaneda, J. , 2016. The reproducibility of corrosion testing in supercritical water e results of an international interlaboratory comparison exercise. Corros. Sci. 106 147-156 https://doi. org/10. 1016/j. corsci. 2016. 01. 034)



## 2.2 测试样品

### 2.2.1 电流化学效应

如同在水环境中进行的所有腐蚀试验一样,必须注意避免相邻腐蚀试验件之间以及试验件与容器之间形成电偶,试件的搁架应由绝缘材料建造。在 SCW 中这是非常具有挑战性的,因为包括 $ZrO_2$ 在内的许多陶瓷制品都可以在非常高的温度下溶解。在较高的温度下($T>500$℃),SCW 的离子电导率很低,在一定程度上减轻了对电流(化学)效应的担忧。

## 2.3 腐蚀速率测量

### 2.3.1 表面粗糙度

长期以来,人们已经认识到合金表面的粗糙度影响着高温水环境中金属的腐蚀速率(详见第 5 章)。对金属表面完整的描述应考虑 4 个因素(Berge,

1997)，即几何因素(如粗糙度)、化学因素(如主体与表面的组成成分差异)、结构因素(如晶粒尺寸和冷加工)和机械因素(如表面/主体的残余应力)。Ruther等(1966)注意到，在SCW中无应变的表面会导致最大的一般性腐蚀，而表面粗糙度经过严格处理的表面则会减少一般性腐蚀(见第5章)。冷处理也会影响应力腐蚀开裂(SCC)(见第6章)。

值得注意的是，在准备用于表面分析的试验件时，要选择适当的表面粗糙度，因为高度光滑的抛光表面会比粗糙的碾磨表面增加更多质量。在比较不同表面粗糙度增重数据时应引起足够重视，特别在 $T \geqslant T_c$ 时；因为不同表面粗糙度试验件的质量变化存在1~2数量级的差别，因此在腐蚀试验结果报告中，应该给出试验件表面粗糙度的详细信息。

## 2.3.2 质量增加与减少

整个设计寿命期内的金属材料损失(如管道壁厚的总减少量)是SCWR概念设计的关键性能要求。通常情况下，针对SCWR进行的腐蚀试验结果是按样品质量的变化(增重后-增重前)来表征的，例如，可以表征为置于试液中每平方分米表面的毫克质量变化($mg/dm^2$)。试验件的质量误差是按0.1mg或更小的精确度来确定。试验中建议对一个试验件进行5次称重(至少3次)，并以测量的平均值作为质量基准值。



虽然测量质量变化有助于不同合金之间的相对比较，但这并不是衡量金属损失最合理的标准，因为某些氧化金属可以溶解或脱落并被试液带走，或者从其他系统的表面释放的材料可能被运送到试验件表面并沉积在上面。Longton(1966)详细讨论了增重和减重测量之间的差异，强调只有当确定系统中没有氧化物流失时，增重测量才能获得可靠的信息，同时指出，报告中的氧化物损失高达50%。置于500℃、低氧浓度SWC中的304 SS的剩余质量数据(Guzona和Cook,2012)表明，试验中与试剂接触后质量变化估算值只是实际腐蚀值的1/6~1/5，即便计入氧化膜的金属也不足以解释金属质量的减少，这表明腐蚀产物释放到冷却剂中。虽然浸泡后质量变化基本不变，但腐蚀速率(这是由减重决定的)随着时间的推移而减小。Novotny等(2015)发现，在回路试验中，氧化物剥落对质量变化有显著影响，并指出对于所研究的合金(316L、347H、08Cr18Ni10Ti)，质量变化的测量结果始终低估了实际氧化层的生长。

图2.6显示了静置于SCW期间氧化物生长的简化横截面示意图，该图阐明了3种可能的情况(第4章和第5章分别详细讨论了氧化物溶解度和生长)。如果表面膜的溶解度高，则薄膜将完全溶解到溶液中，试件将失去质量，所测量的质量变化将直接相当于金属损失；对于中间的情况(氧化物溶解度低，但不可忽

略)，除非溶液的损失也可以被定量，否则测量的质量变化不能简单地与金属损失相关；如果表面膜的溶解度可以忽略不计，那么所有形成的氧化物都会停留在表面，导致试验件质量增加，增加的质量与薄膜形成导致的质量损失相抵。除了这些以溶解度为基础的现象外，氧化物还可能剥落导致质量减少，或者从试验容器表面释放的金属物质被运输到试验件表面并沉积导致质量增加。当且仅当腐蚀过程中所有被氧化的金属全部保留在表面氧化物层中，并且没有从溶液中沉积任何额外的物质时，那么质量的变化和真正的金属损失就会有直接的关系。测量金属损耗唯一可靠的方法就是将试验件置于试验环境后，去除氧化腐蚀产物并对试件除垢后(如 ASTM 61-03)获得的质量损失，或者直接测量横截面上剩余金属的厚度。在后一种情况下，必须知道试件的预测试厚度。

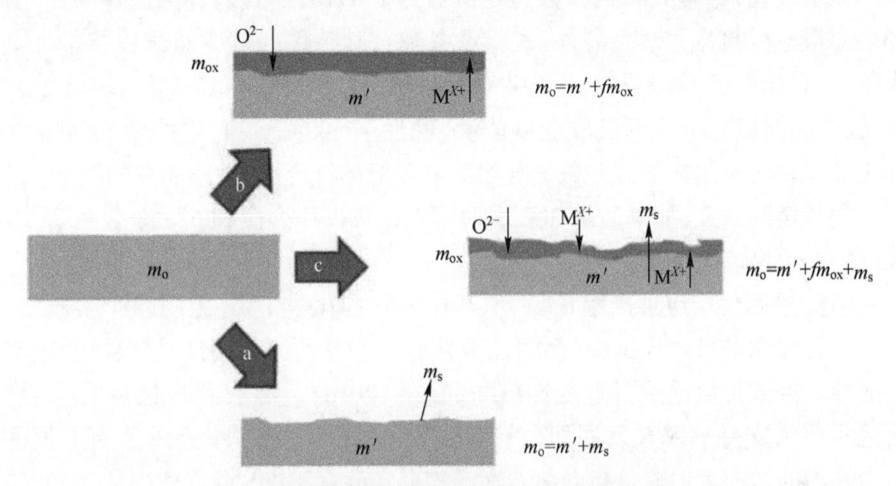


$m_o$—原始试件质量；$m'$—腐蚀后保留的本体材料质量；

$m_{ox}$—氧化物质量；$m_s$—溶液中的金属质量；$f$—将氧化物质量转化为金属质量的系数。

图 2.6　氧化物溶解度对质量变化的影响

(引自文献 Guzonas D A,2015. The physical chemistry of corrosion in a supercritical water-cooled reactor. In:Proceedings of the 16th Int. Conf. on the Properties of Water and Steam:Water,Steam and Aqueous Solutions Working for the Environment and Industry,September 1-5,2013,London,UK. Paper PWS-116)

目前，已有许多公认的成熟除垢程序，尽管对有些合金找到一种合适的除垢方法并不容易。不幸的是，除垢会破坏腐蚀膜，所以必须在除垢之前进行必要的表面表征，或者在测试中使用多个集片，并保留一个或多个集片用于表面分析。结合从表面分析中获得的表面氧化物结构信息和氧化物的质量数据，可以估算出腐蚀膜中铁、镍、铬和其他金属的质量，并将其与除垢过程中材料的表面损失质量进行比较。两者之间的差异意味着腐蚀产物要么在测试过程中被运输出去,要么从测试溶液中沉积在样品上(Guzona 和 Cook,2012)。

金属损耗也可以从截面的金相检查中推断出来,只要能找到曝光前的原始金属表面即可。通过金相检查可以确定是否发生了氧化物剥落,如果没有溶解,根据质量的变化就可以推断出氧化物的厚度,计算过程需要知道氧化物的组成和孔隙率。如果氧化物厚度不均匀,最好用最大穿透度来测量腐蚀,就像晶间腐蚀一样。

#### 2.3.2.1 可重复性

支撑 SCWR 发展的腐蚀试验研究数据重复性很难确定,因为数据图中的误差范围并不全面,而且试验中的不确定因素往往没有考虑。另外,测试(特别是范围界定试验)通常只包括少数特定的合金试验件,而且这些试验件采用不同的表面粗糙度制备方法。另外,试验设备和试验条件(温度、与试液接触时间、水化学、表面处理)的差异也可能使对比分析产生困难。Guzona 和 Cook(2012)发现,在相同试验条件下仅改变氧浓度,静置于 500℃ SWC 中的铁素体钢 T91 和 T92 在相同时间内的试验数据与实际有较好的符合性(见第 5 章的图 5.20)。其他类合金的研究,如普通的 300 系列不锈钢、800H 合金和 625 合金也有类似较好的一致性。

GIF(第四代反应堆国际论坛)SCWR 材料和化学 PMB 组织多个实验室开展一次国际性平行对比试验(循环试验),目的在于研究不同试验装置中获得的一般性腐蚀试验数据的可重复性(Guzona 等,2016)。考虑到全世界有大量的研究小组为支持 SCWR 发展而进行腐蚀测试,这种对比试验是非常有必要的。每个签约的参与者向其他参与者(最多 7 个实验室)提供经过适当加工的试验件(800H 合金、08Cr18Ni10Ti 不锈钢和 310 SS),每个参与者都遵循小组制定的标准程序,进行准备、记录试验件测试结果,数据报告仅限于试件质量变化。结果发现,质量增加的变化比预期值要大得多。在同一实验室内,一个给定合金的质量变化数据中的散度很小(通常小于平均值的±20%),但在不同的实验室之间却差别很大(通常大于平均值的 100%)。这种变化在很大程度上是由于没有考虑不同的测试装置在按测试程序进行过程中产生的差异,特别是高压釜的重新补水时间(图 2.5)。通过这些试验得到的一个关键结论是:在比较不同测试装置的数据时必须谨慎。试验报告中必须要记录试验台以试验前后对试验件处理的所有细节,并且在评估数据时对此要予以考虑,同时建议制定标准的腐蚀试验方法。

GIF SCWR M&C PMB 正在计划进行第二阶段国际多个实验室间的比较腐蚀试验,以解决第一阶段发现的一些缺陷。

### 2.3.3 电化学方法

目前,对于水冷堆腐蚀研究通常采用原位电化学方法来描述腐蚀特性(Lillard,2012)。正如在第4章中将讨论的那样,SCW在临界温度以上时的低电介质常数有利于离子缔合,从而导致离子电导率的下降。在临界温度附件各种电化学方法(Kritsunov和Macdonald,1995)、电化学噪声分析法(Liu等,1994)以及膜电阻膜阻抗法(Betova等,2007)被成功应用。Hettiarachchi等(1993a、b)介绍了超临界化石燃料电厂(SCFPP)SCW中的pH值和电位传感器的监测方法开发。Liu和Macdonald(1995)通过试验证明了在410℃、27.6MPa条件下,采用Pd/Pt相对电阻传感器监测SCW中溶解的氢浓度是可行的。在高温下使用非接触电化学方法的一个主要困难是缺乏稳定的参比电极,它的电位与环境无关,并且与一个合理的热力学标度有关。van Nieuwenhove(2012)介绍了一个参考电极,该电极由钇稳定氧化锆(YSZ)陶瓷管组成,管内的铁导体周围填充满铁和氧化铁粉末的混合物。在高温下,氧化锆是氧离子导体,允许氧离子在陶瓷/水和陶瓷/氧化铁界面上发生电化学反应,从而确定内部铁导体的电位。这种电极已成功应用于SCW回路,测量230~604℃(Novotny等,2017)的电化学电位和电化学阻抗。第4章和第5章讨论SCW中电化学测量的局限性。

### 2.3.4 原位分析的其他方法

在静置于SCW后,发展了一系列的非原位技术对试验件进行表征,从重量变化测量到各种表面分析方法等,如扫描电子显微镜(SEM)、X射线能量色散谱(EDX或EDS)、透射电子显微镜(TEM)、射频辉光放电光谱仪(GDOES)、X射线光电子能谱(XPS)、拉曼光谱分析和扫描螺旋显微镜(SAM)等。氧化膜的质量变化测量和表征通常是在样品冷却、漂洗和干燥后进行的原位测量,而SEM或TEM等技术则是在真空条件下进行的。氧化物膜在下面两种过程可能发生变化:①在冷却过程中,由于氧化物溶解度通过临界点发生较大变化,氧化物膜在冷却过程中可能发生变化;②在试验件运送过程中,必须考虑氧化层脱水的可能性。

原位表征方法虽然在试验上具有挑战性,但同时也避免了一些问题。Maslar等(2001、2002)利用现场拉曼光谱在25.2MPa和496℃下分别对铬和不锈钢上的氧化层进行测量。试验中最大的挑战是可视窗口材料的选择(2.4节进一步讨论)。测试区下游溶液中的金属离子浓度结果提供了有关金属释放速率的数据,这些数据可用于推断相对腐蚀速率。Han和Muroya(2009)报告了来自304 SS样品的$^{60}$Co放射性示踪剂在SCW中随时间和温度变化的释放数据。

他们收集了离子交换树脂释放的物质,并使用伽马能谱对释放量进行定量分析。Choudhry 等(2014)使用分级流动系统(图 2.3(a))监测了一些合金在 750℃以下腐蚀过程中的耗氧量、氢产生量和金属释放情况。试验溶液中溶解金属的浓度可通过标准分析方法测量,如电感耦合等离子体(ICP)、光学发射光谱(OES)、ICP 质谱法(MS)或吸附溶出伏安法,阴离子浓度可通过离子色谱仪(IC)测量。用气相色谱法可以测定氢和氧的浓度。

## 2.4 热力学性质的测量

构建 SCW 中溶解物种行为模型有助于预测腐蚀产物迁移产生的影响,构建这一模型则需要有关离子形态、平衡常数和其他热力学参数的温度和浓度依赖关系数据作为支撑。各类光谱方法可以定量地测量高温溶液中各种物质的浓度。上述试验中视窗材料的选择对于这些类型的试验也很重要。地球化学研究领域所用的高压可视腔体(HPOC)最新成果显示,已经能够在温度高达 500℃和压力高达 100MPa 的情况下利用拉曼光谱开展研究(Chou 等,2005)。拉曼光谱可以提供详细的结构信息,特别是当与现代计算方法相结合时,可以评估化合物的能量、化合物的高温结构和化合物的热力学稳定性(Pye 等,2014)。Applegarth 等(2015)介绍了熔融石英 HPOC[①] 在 30MPa(250~400℃)下对几种非络合阴离子热分解的研究(通常用拉曼的内部标准)。热液金刚石砧芯(HDAC)(Bassett 等,1993;Solferino 和 Anderson,2012)已经实现了在 600℃ SCW 中对金属氧化物行为的在线 X 射线荧光光谱仪(XRF)、X 射线吸收光谱(XAS)和拉曼光谱测量。Solferino 和 Anderson(2012)讨论了在 HDAC 中氧逸度的控制,包括使用过氧化氢(双氧水)以控制的方式改变逸度。Yan 等(2013)使用原位 Co K 层 XAS 测量,以研究在 500℃、约 220MPa 条件下 $Co^{2+}$ 在 $Fe_3O_4$ 纳米颗粒表面的吸附情况。Yan 等(2011)使用 X 射线光谱法对 SCW 中的 $Mo^{6+}$ 形成过程进行分析,试验过程中 SCW 被置于 HDAC 中。

在压力为 20MPa,温度为 295~625K,以及较宽的离子强度条件下,使用高温、高精度、高压导流式交流电导仪,已经测量了氢氧化锶和氯化锶水溶液的变频电导率(Arcis 等,2014)。所使用的流动模块由 Hnedkovsky 等(2005)建在特拉华大学,该模块是在 Zimmerman 等(1995)和 Sharygin 等(2002)的设计基础上改造而来的,可以测量强腐蚀性的溶液。该单元的使用温度达到 673K,压力为 28MPa,离子强度低至 $10^{-5}$mol/kg(Zimmerman 等,2012)。

① 一个需要考虑的问题是,酸性或碱性条件可能导致溶剂与窗口材料发生反应。

## 2.5 应力腐蚀开裂

目前,研究应力腐蚀开裂(SCC)和阻止裂纹扩展的方法很多,但各有优缺点。通常来讲,长期的高压试验被用来量化裂纹的萌生和扩展速度。试验件失效后,用显微镜观察断口表面。近年来,使用先进的样品制备方法和显微镜技术(聚焦离子束(FIB))(Li,2006;Li 等,2008),如分析透射显微镜(ATEM)和电子背向散射衍射(EBSD)等显微技术在裂纹尖端探测过程变得越来越普遍(Lozano-Perez 等,2014)。SCW 条件下的 SCC 测试比一般的腐蚀测试更具有挑战性,因为用于施加应变的设备要么完全安装在高温高压容器中,要么提供一种方法(使信号)能够穿透密封的容器壁,这样仪器体积要小、成本要高或两者兼而有之。Behnamian 等(2013)使用小型压力容器研究 SCC,根据加水量和温度计算容器所受应力。恒载试验中使用 C 形环或 O 形环(垫圈),以及固定在硬表面的一系列弹簧和紧固螺母实现;C 形环或 O 形环(垫圈)扩展到弹簧中,减少因膨胀引起的应力增长。Swift 等(2015)描述了用于 SCW 的恒负荷 C 形环组件,使用 Inconel(镍基合金) 718 Belleville 垫圈作为弹簧向试验件施加几乎恒定的负载。用力学模型计算弹簧在热膨胀作用下所受的力,从而可以预测 C 形环顶点处的应力。

Novotny 等(2013a)描述了一种用于断裂韧性、应力腐蚀裂纹、腐蚀疲劳、拉伸力和电化学测量的气动伺服控制系统,具有较高的灵敏度和准确性。该系统没有运动部件穿透压力边界,因此避免了密封件位置处的摩擦力,使负载控制比传统的伺服液压和步进电机驱动装置更精确。这同时也允许使用更小尺寸的样品,以便测试热室中的辐照材料、反应堆内测量以及裂纹扩展速率的研究。作为实验室放射性屏蔽室和未来反应堆内设备辐照促进应力腐蚀开裂(IASCC)测试装置的一部分,一个以微型波纹管为基体的装载设备原理样机(图 2.7)被开发出来。双波纹管(DB)装置弥补了以单波纹管为基体的加载装置的主要缺点,即波纹管内部初始压力很高(用于 SCWR 测试的压力约 25MPa)。伺服阀和波纹管不允许出现任何故障,尤其是针对放射性屏蔽室和反应堆内的操作环境。Pentilla 等(2017)介绍了一种以 DB 为基体的气动加载装置,该装置配有微型高温高压容器,能够在 SCW 中进行拉伸试验,用以测量裂纹扩展速率。在 25MPa、288℃和 550℃温度下,对气动加载装置进行了标定,同时采用 5mm 圆盘紧凑拉伸(DCT)的微型高压容器材料测试系统对 316 SS 试件进行了预裂验证试验。

应力腐蚀敏感性的定量分析技术值得重点关注,目前通用的测量方法包含两种,一种是基于破坏后断裂面晶间裂纹百分比(%IG)进行分析,另一种是基于试样表面的裂纹数量(密度、长度、长度/单位面积、深度)进行分析。正如Allen等(2012)所讨论的,这些测量结果并非完全一致。

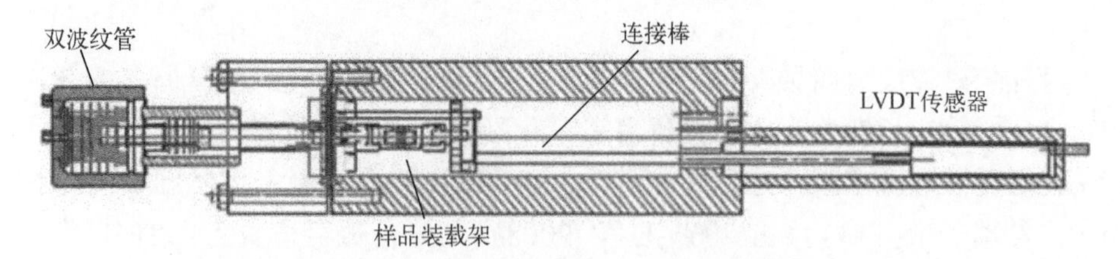


图2.7 微型高温高压容器双波纹管为基体的加载装置原理(Novotny等,2013a)

## 2.6 辐照条件下的试验

辐照对堆芯材料的损伤机理包括两种:①与材料直接作用导致原子位移和核反应(如中子活化);②通过水辐照分解的间接作用(图1.3)以及辐照分解产物与合金的相互作用。因此,在预期的水化学条件下进行SCW反应堆内测试必须是SCWR材料测试的最后一步(图2.1中第4级)。然而,在非SCWR条件下由辐照引起的力学性能变化可以用辐照来研究。上述这些研究,很重要的一点是要确保试验中能够再现预期的SCWR中子谱(快中子和热中子的正确平衡),并且在合理时间内达到充足的中子通量(Walters等,2017)。



### 2.6.1 材料测试

作为美国核蒸汽再热开发计划的一部分,一些合金材料在试验堆的过热蒸汽下进行了辐照测试。例如,在EVESR的SADE回路中,燃料包壳厚度为0.406mm的燃料组件在温度高达(738±83)℃的过热蒸汽中静置10292h(Comprelli等,1969)。目前,还无法实现在SCWR反应堆内开展SCW试验的装置。在CVR设计建造了用于研究反应堆内材料和水化学试验的SCW回路(图2.8)(Ruzickova等,2011)。该回路从功能上分为两部分,即辐照通道和辅助回路。辐照通道的设计能够在温度高达600℃、25MPa的条件下工作,计划将其放置在LVR-15测试反应堆的一个测试区。辅助回路将水调节到所需要的通道入口参数,并能够对水的参数进行监测和控制,同时根据需要向水中添加化学溶剂。另一个在概念上与之类似的回路被设计用于SCWR燃料鉴定试验(Vojacek等,2015)。活性通道的设计受到反应堆堆芯空间、测试条件下核材料的可用性以及核安全考虑的限制。因此,建议采用不锈钢08Cr18Ni10Ti作为活性通道的合金,将最高温度限制在450℃。截至2017年年中,反应堆内尚未安装材料试验和燃料鉴定回路。

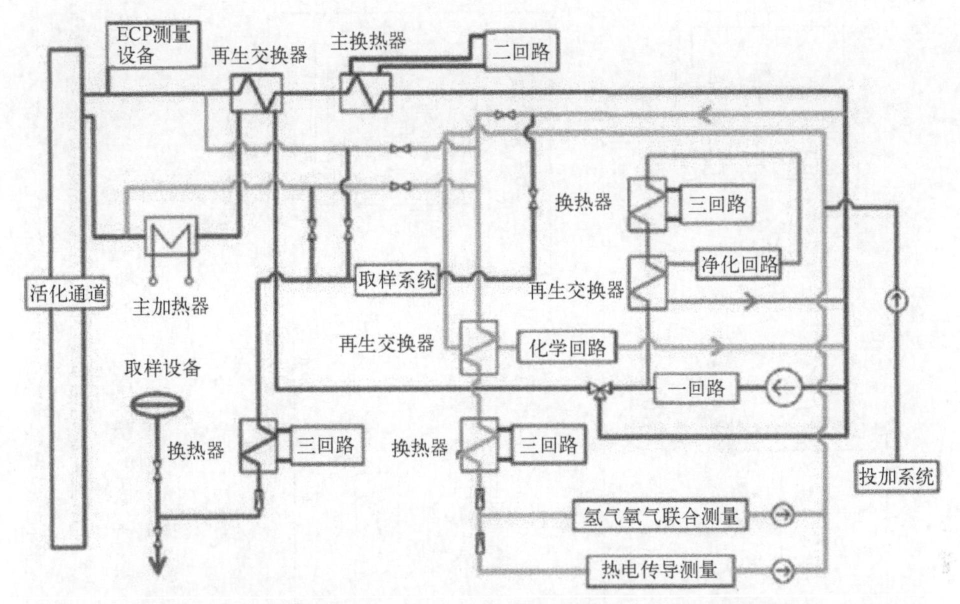


图 2.8　CVR 中 SCW 回路的流程图(Ruzickova 等,2011)。活性通道将被插入反应堆核芯,并包含试验部分。辅助回路包括主回路(红色)、溶剂添加系统(暗橙色)、测量系统(绿色)、净化系统(橙色)和冷却回路。溶剂添加系统可以添加化学物质(气体和溶液)。主回路中的化学监测(蓝色)由侧流回路提供。回路中的循环介质通过机械过滤器和离子交换器净化,以保持主回路中所需的化学条件(见彩插)

虽然反应堆内的辐照可能需要数年时间来积累目标损伤水平,并且需要专用的防护措施、设施和仪器来处理放射性样品,但在已明确规定的能量、剂量率和温度下进行离子辐照(Wa 和 Averback,2012)可很快达到 1~100dpa 范围内的损坏水平,且产生很少的或根本不残留放射性。这样可以简化样品处理流程,并大大减少周期和成本。通过量化对比分析中子辐照和离子辐照效应,可以设计出与中子辐照影响相当的离子辐照条件。高能($E$>1MeV)自相似离子辐照结合纳米力学测试技术,最近被用于研究某些备选 SCWR 合金的辐照硬化。

在 Kharkiv 物理与技术研究所建造了一个与 10MeV、10kW 直线电子加速器(LINAC)(Bakai 等,2011)相耦合的辐射单元 SCW 对流回路(图 2.9)。该回路配备了化学监测仪器(氧气浓度、电导率),这些监测仪器必须与辐照单元保持一定距离。Bakai 等(2016)在选定的样品上展示了辐照后表面氧化层的显微结构和硬度表征结果,其中包括从辐照细胞管道中切出的高度辐照样品 12Cr18Ni10T SS(辐照强度最高可达约 $10^{21}$ 电子/$cm^2$,$E_{dep}\approx16$keV/原子)。

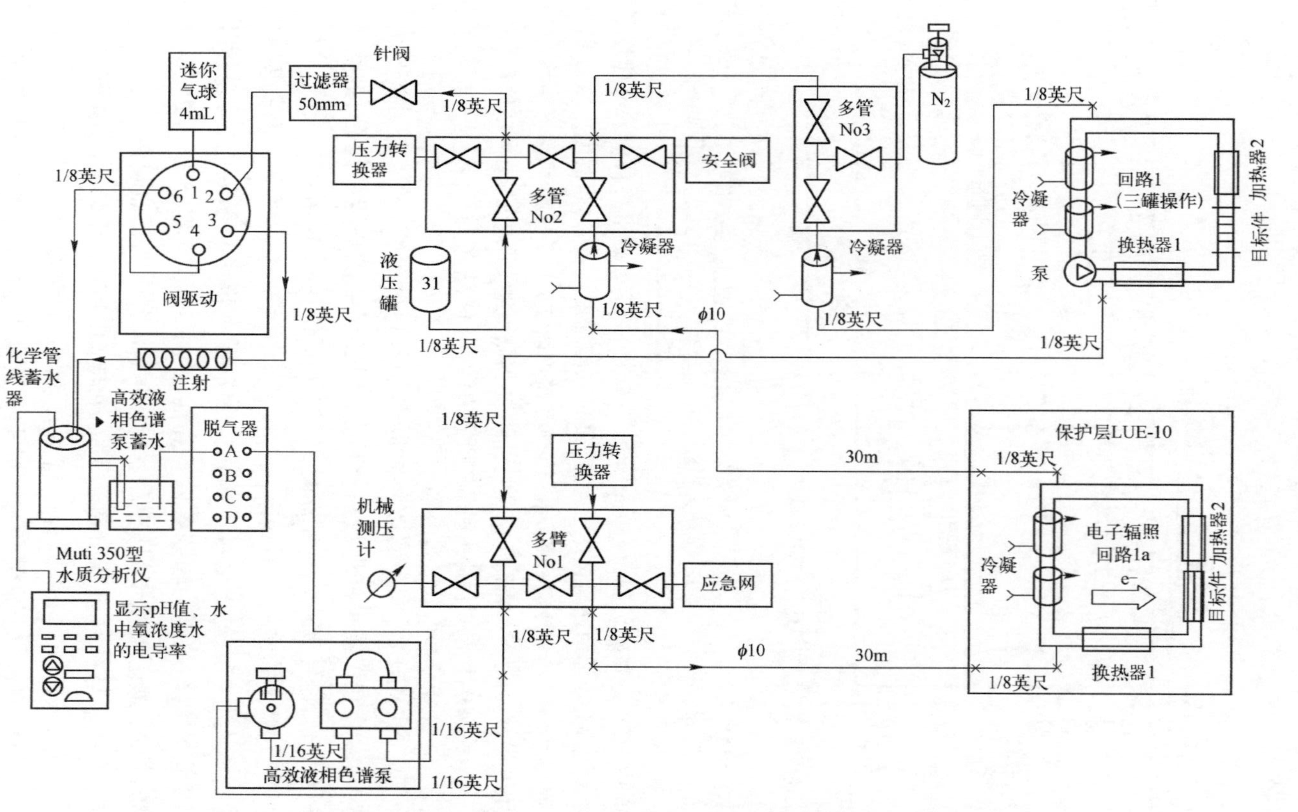

图 2.9 加拿大-乌克兰电子辐照试验装置原理图(Bakai 等,2011)。该回路配有盘管式冷却器,一个水平板式主加热器(加热器 1,≤6kW)和一个垂直启动加热器(加热器 2,≤3kW),用来控制自然循环方向。12Cr18Ni10Ti 不锈钢管道的内径为 32mm,壁厚为 4mm;加热器 1 的内管管壁厚度增加到 6mm

由于材料中能量为 10MeV 的电子射程很短(约 1cm),加之扫描通量在横向上的变形以及样件内较大能量的沉积,试验件表面在加热测试时不同区域有明显的变化差异。基于电子束空间分布数据和直线加速器能量分布数据,并利用蒙特卡罗模拟辐照区三维能量吸收图,再结合热场有限元法计算,分离了温度和辐照对这些物理量空间变化的综合影响。

### 2.6.2 水辐照分解研究

通过试验模拟 SCW 辐照分解特性,首先需要了解反应速率对流体温度和密度等参数的依赖关系(这部分内容将在第 4 章中详细讨论),这种依赖关系通常是通过脉冲辐照分解(Spinks 和 Woods,1976)试验获得,试验中使用电子或其他离子短时脉冲照射一个样品,并利用光谱法对产生的瞬态辐照分解物行为进行监测。一些用于高温、高压条件的光学元件已经设计成功(Fujisawa 等,2004;Takahashi 等,2000)。

Janik 等(2007)在一份研究报告中阐述了 Notre Dame 辐照实验室中的研究结果,试验中采用 3.0MeV 的 van de Graaff 加速器获得 2.5MeV 电子,并用其辐照一个定制的 SCW 样块,开展了 β 辐射分解试验。

威斯康星大学在其 1MW 的 TRIGA① 反应堆中嵌入了一个 SCW 回路,用于测量中子/伽马辐照条件下水的分解产率(Bartels 等,2006)。由于严格的安全要求,试验回路设计非常复杂,主要包括四部分:辐照体积/空体积和中子屏蔽、铅 γ 屏蔽、空的加热段、水屏蔽。辐照体积采用了多种不同的材料和设计方案,材料包括 Hastelloy C-276、800 合金和钛。为了了解反应堆全功率时辐照区的辐照环境,研究过程中首先采用蒙特卡罗粒子输运程序对反应堆进行建模,并基于计算结果对试验进行优化。当 CVR 安装 SCWR 回路时,它将成为研究水辐照分解的重要试验设备。

Pommeret(2007)描述了一个使用加速器实现脉冲辐照分解研究的小型($1.27cm^3$)Inconel 718 单元,操作温度和压力可以达到 500℃、100MPa。Katsumura 等(2010)和 Muroya 等(2012)描述了使用 S 波段直线加速器在高温条件下测定水合电子($e_{aq}$)的时间依赖性辐照分解率(图 2.10),间接利用甲基紫罗碱清除 $e_{aq}$ 和常规的纳秒脉冲辐照分解测量,并直接使用皮秒时间分辨率的脉冲辐照分解系统。为了克服 SCW 高温下信噪比的限制,研制了一种激光驱动的光电阴极 RF 枪和飞秒钛蓝宝石激光器。该系统产生了一个 11ps、22MeV 的电子脉冲(半高宽度(FWHM)),入射到水中以后通过飞秒激光脉冲测量所引起的瞬

① TRIGA:培训(Training),研究(Research),同位素(Isotopes),通用原子(General Atomic)。

态物质的吸收。整体时间分辨率约为 60ps,分析范围涵盖可见光到近红外波段,在 SCW 中具有较好的信噪比。

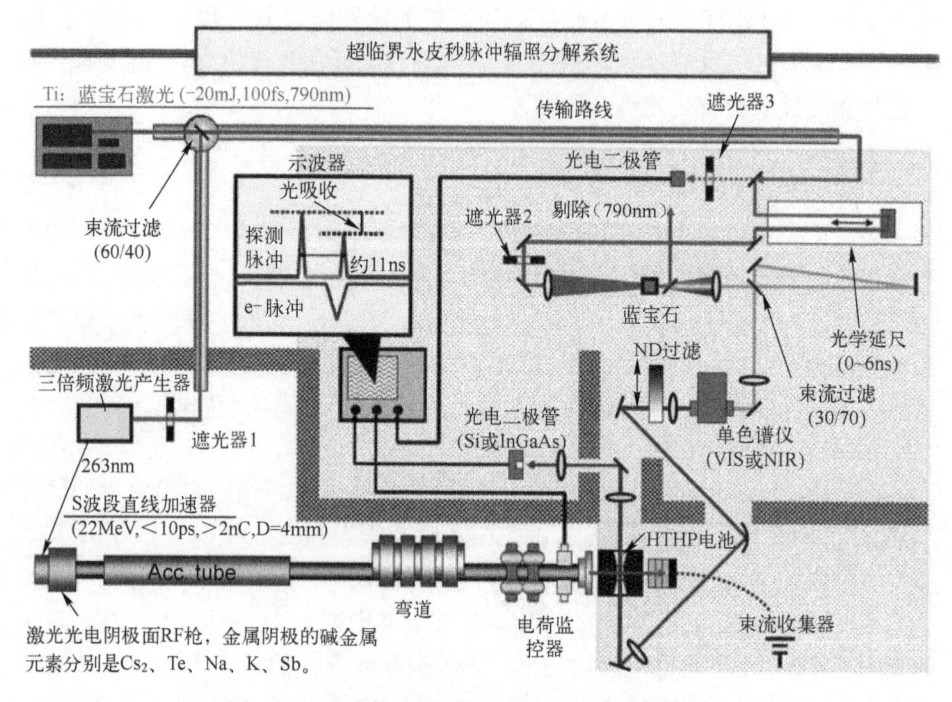


图 2.10 东京大学核学院(Katsumura 等,2010)皮秒时间分辨率的脉冲辐照分解系统原理(见彩插)

利用 μ 介子作为氢原子的类似物(Ghandi 和 Miyake,2010),μ 介子已被用于研究在 450℃以下 SCW 中的辐照分解特性(Alcorn 等,2014)①。在 μ 介子自旋光谱(μSR)试验中,一束正 μ 介子被定向发射到一个样品单元,通过调整光束动量,可以使 μ 介子穿过单元窗并存留在样品中。μ 介子光束是自旋极化的,在 μ 介子自旋方向横向施加一个弱磁场,就可以使停止的 μ 介子按其周围环境的频率特性产生旋动。磁场内的化学反应会导致穿过其间的 μ 介子发生分阶段自旋,而这种现象是可以测量的。Alcorn 等(2014)讨论了顺磁性物种可能产生的干扰(如试验单元腐蚀释放的金属离子),并指出这样的干扰通常发生在最初 24h,直到形成稳定的氧化层。

① μ 介子由一个正的 μ 介子($\mu^+$,寿命约 2.2μs,质量约 0.11u,$1u=1.66053886\times10^{-27}$kg)组成,它捕获一个电子,形成一个与氢原子类似的束缚原子状态(Ghandi 和 Miyake,2010)。由于 Mu 和 H 的还原质量几乎相同,Mu 可以被认为是氢的轻同位素,与氢反应相似,尽管速率常数可能因动力学同位素效应而有所不同(Walker,1981)。

Mayanovic 等(2012)报道了同步辐照条件下 HDAC 中 Fe 和 W 离子的 X 射线诱导辐照分解现象;这是样品辐照的意外结果,但确实为研究辐解效应提供了一种新的手段。例如,Men 等(2014)在 300~500℃条件下采用 XAS 来表征 Cr 标准溶液在 HDAC 中的辐射溶解氧化还原行为。稳态辐照分解产物中氢、氧和过氧化氢的浓度也可以通过置于伽马辐照器内的单元来监测。

## 参考文献


Alcorn, C. D., Brodovitch, J. -C., Percival, P. W., Smith, M., Ghandi, K., 2014. Kinetics of the reaction between H and superheated water probed with muonium. Chem. Phys. 435, 29-39.

Allen, T. R., Chen, Y., Ren, X., Sridharan, K., Tan, L., Was, G. S., West, E., Guzonas, D. A., 2012. Material performance in supercritical water. In: Konings, R. J. M. (Ed.), Comprehensive Nuclear Materials, vol. 5. Elsevier, Amsterdam, pp. 279-326.

Ampornrat, P., Was, G. S., 2007. Oxidation of ferritic-martensitic alloys T91, HCM12A and HT-9 in supercritical water. J. Nucl. Mater. 371, 1-17.

Applegarth, L., Alcorn, C., Bissonette, K., Noël, J., Tremaine, P. R., 2015. Non-complexing anions for quantitative speciation studies using Raman spectroscopy in fused silica high pressure optical cells under hydrothermal conditions. Appl. Spectrosc. 69, 972-983.

Arcis, H., Zimmerman, G. H., Tremaine, P. R., 2014. Ion-pair formation in aqueous strontium chloride and strontium hydroxide solutions under hydrothermal conditions by AC conductivity measurements. Phys. Chem. Chem. Phys. 16, 17688-17704.

Bakai, A. S., Boriskin, V. N., Dovbnya, A. N., Dyuldya, S. V., Guzonas, D. A., 2011. Supercritical water convection loop (NSC KIPT) for materials assessment for the next generation reactors. In: Proc. of the 5th Int. Symposium on Supercritical Water-Cooled Reactors. Vancouver, Canada, on CD-ROM, Paper 051.

Bakai, O. S., Guzonas, D. A., Boriskin, V. M., Dovbnya, A. M., Dyuldya, S. V., 2016. Combined effect of irradiation, temperature, and water coolant flow on corrosion of Zr-, Ni-Cr-, and Fe-Cr-based alloys. J. Nucl. Eng. Radiat. Sci. 2, 021007.

Bartels, D. M., Anderson, M., Wilson, P., Allen, T., Sridharan, K., 2006. Supercritical Water Radiolysis Chemistry Supercritical Water Corrosion. Progress Report Available at: http://nuclear.inl.gov/deliverables/docs/uwnd_scw_ii_sep_2006_v3.pdf.

Bassett, W. A., Shen, A. H., Bucknam, M. J., Chou, I. -M., 1993. A new diamond anvil cell for hydrothermal studies to 2.5 GPa and −190℃ to 1200℃. Rev. Sci. Instrum. 64, 2340-2345.

Behnamian, Y., Li, M., Zahiri, R., Kohandehghan, A., Mitlin, D., Chen, W., Luo, J. -L., Zheng, W., Guzonas, D., 2013. Stress corrosion cracking of austenitic alloys in supercritical water. In: The 6th International Symposium on Supercritical Water-Cooled Reactors, March 03-07, 2013, Shenzhen, Guangdong, China, Paper ISSCWR6-081.

Behnamian, Y., Mostafaei, A., Kohandehghan, A., Shalchi Amirkhiz, B., Serate, D., Sun, Y., Liu, S., Aghaie, E., Zeng, Y., Chmielus, M., Zheng, W., Guzonas, D., Chen, W., Luo, J. -L., 2016. A comparative study of oxide scales grown on stainless steel and nickel-based superalloys in ultra-high temperature supercritical water at 800℃. Corros. Sci. 106,188–207. https://doi.org/10.1016/j.corsci.2016.02.0040010-938X.

Berge,P.,1997. Importance of surface preparation for corrosion control in nuclear power stations. Mater. Perform 36,56–62.

Betova,I.,Bojinov,M.,Kinnunen,P.,Penttilä,S.,Saario,T.,2007. Surface film electrochemistry of austenitic stainless steel and its main constituents in supercritical water. J. Supercrit. Fluids 43,333–340.

Bsat,S.,Huang,X.,2015. Corrosion behavior of 310 stainless steel in superheated steam. Oxid. Metals 84, 621–631.

Chou,I. M.,Burruss,R. C.,Lu,W.,2005. A new optical capillary cell for spectroscopic studies of geologic fluids at pressures up to 100MPa. In: Chen, J., Wang, Y., Duffy, T. S., Shen, G., Dobrzhinetskaya, L. P. (Eds.),Advances in High Pressure Technology for Geophysical Applications. Elsevier,Amsterdam.

Choudhry,K. I.,Carvajal-Ortiz,R. A.,Kallikragas,D. T.,Svishchev,I. M.,2014. Hydrogen evolution rate during the corrosion of stainless steel in supercritical water. Corros. Sci. 83,226–233.

Comprelli,F. A.,Busboom,H. J.,Spalaris,C. N.,1969. Comparison of radiation damage studies and fuel cladding performance for Incoloy-800. In:Irradiation Effects in Structural Alloys for Thermal and Fast Reactors. Am. Soc. Test. Mater. 1969,400–413. ASTM STP 457.

Daigo,Y.,Watanabe,Y.,Sue,K.,2007a. Effect of chromium ion from autoclave material on corrosion behavior of nickel-based alloys in supercritical water. Corrosion 63,277–284.

Daigo,Y.,Watanabe,Y.,Sue,K.,2007b. Corrosion mitigation in supercritical water with chromium ion. Corrosion 63,1085–1093.

Fitzsimmons,M. D.,Pearl,W. L.,Siegler,M.,1961. A Simulated Superheat Reactor Corrosion Facility. General Electric Atomic Power Report GEAP-3778.



Fujisawa,T.,Maru,E.,Amita,F.,Harada,M.,Uruga,T.,Kimura,Y.,2004. Development and application of a multipurpose optical flow cell under supercritical condition of water. In:14th International Conference on the Properties of Water and Steam in Kyoto,445,p. 2004.

Ghandi,K.,Miyake,Y.,2010. Muon interactions with matter. In:Hatano,Y.,Katsumura,Y.,Mozumder, A. (Eds.),Charged Particle and Photon Interactions with Matter:Recent Advances,Applications and Interfaces. Taylor & Francis,pp. 169–208.

Guzonas,D.,2009. SCWR materials and chemistry-status of ongoing research. In:Proceedings of the GIF Symposium,Paris,France,September 9–10,2009.

Guzonas,D. A.,Cook,W. G.,2012. Cycle chemistry and its effect on materials in a supercritical water-cooled reactor:a synthesis of current understanding. Corros. Sci. 65,48–66.

Guzonas,D. A.,2015. The physical chemistry of corrosion in a supercritical water-cooled reactor. In:Proceedings of the 16th Int. Conf. on the Properties of Water and Steam:Water,Steam and Aqueous Solutions Working for the Environment and Industry,September 1–5,2013. London,UK. Paper PWS-116.

Guzonas,D.,Penttila,S.,Cook,W.,Zheng,W.,Novotny,R.,Sáez-Maderuelo,A.,Kaneda,J.,2016. The reproducibility of corrosion testing in supercritical water-results of an international interlaboratory comparison exercise. Corros. Sci. 106,147–156. https://doi.org/10.1016/j.corsci.2016.01.034.

Han,Z.,Muroya,Y.,2009. Development of a new method to study elution properties of stainless materials in subcritical and supercritical water. In:4th International Symposium on Supercritical Water-cooled Reactors,Heidelberg,Germany,March 8–11,2009. Paper No. 75.

Hazel, V. E., Boyle, R. F., Busboom, H. J., Murdock, T. B., Skarpelos, J. M., Spalaris, C. N., 1965. Fuel Irradiations in the ESADE-VBWR Nuclear Superheat Loop. General Electric Atomic Power report GEAP-4775.

Hettiarachchi, S., Song, H., Emerson, R., Macdonald, D. D., 1993a. Measurement of pH and Potential in Supercritical Water. Volume 1 Development of Sensors. Electric Power Research Institute. Report TR-102277/Volume 1.

Hettiarachchi, S., Makela, K., Song, H., Macdonald, D. D., 1993b. The viability of pH measurements in supercritical aqueous systems. J. Electrochem. Soc. 139, L3-L4.

Hnedkovsky, L., Wood, R. H., Balashov, V. N., 2005. Electrical conductances of aqueous $Na_2SO_4$, $H_2SO_4$, and their mixtures: limiting equivalent ion conductances, dissociation constants, and speciation to 673K and 28MPa. J. Phys. Chem. B 109, 9034-9046.

Holcomb, G. R., Carney, C., Doğan, Ö. N., 2016. Oxidation of alloys for energy applications in supercritical $CO_2$ and $H_2O$. Corros. Sci. 109, 22-35.

Janik, D., Janik, I., Bartels, D. M., 2007. Neutron and β/γ radiolysis of water up to supercritical conditions. 1. β/γ yields for $H_2$, H・atom, and hydrated electron. J. Phys. Chem. A 111, 7777-7786.

Katsumura, Y., Lin, M., Muroya, Y., Meesungnoen, J., Uchida, S., Mostafavi, M., 2010. Radiolysis of high temperature and supercritical water studied by pulse radiolysis up to 400℃. In: 8th Intl Radiolysis, Electrochemistry & Materials Performance Workshop, Quebec City, Canada, October 8, 2010.

Klassen, R. J., Rajakumar, H., 2015. Combined effect of irradiation and temperature on the mechanical strength of Inconel 800H and AISI 310 alloys for in-core components of a GEN-IV SCWR. In: The 7th International Symposium on Supercritical Water-Cooled Reactors (ISSCWR-7) March 15-18, 2015, Helsinki, Finland Paper ISSCWR7-2006.

Kriksunov, L. B., Macdonald, D. D., 1995. Corrosion testing and prediction in SCWO environments. In: Proceedings of the ASME Heat Transfer Division. 317-2, p. 281.

Li, J., 2006. Focussed ion beam microscope. J. Metal 58(3), 27-31.

Li, J., Elboujdaini, M., Gao, M., Revie, R. W., 2008. Investigation of plastic zones near SCC tips in a pipeline after hydrostatic testing. Mater. Sci. Eng. A 486, 496-502.

Lillard, S., 2012. Corrosion and compatibility. In: Konings, R. J. M. (Ed.), Comprehensive Nuclear Materials, vol. 5. Elsevier, Amsterdam, pp. 1-16.

Liu, C., Macdonald, D. D., Medina, E., Villa, J. J., Bueno, J. M., 1994. Probing corrosion activity in high subcritical and supercritical water through electrochemical noise analysis. Corrosion 50, 687-694.

Liu, C., Macdonald, D. D., 1995. An advanced Pd/Pt relative resistance sensor for the continuous monitoring of dissolved hydrogen in aqueous systems at high subcritical and supercritical temperatures. J. Supercrit. Fluids 8, 263-270.

Longton, P. B., 1966. The Oxidation of Iron- and Nickel-Based Alloys in Supercritical Steam: A Review of the Available Data. United Kingdom Atomic Energy Authority. TRG Report 1144(C).

Lozano-Perez, S., Dohr, J., Meisnar, M., Kruska, K., 2014. SCC in PWRs: learning from a bottom-up approach. Metallurg. Mater. Trans. E 1A, 194-210.

Maslar, J. E., Hurst, W. S., Bowers Jr., W. J., Hendricks, J. H., Aquino, M. I., Levin, I., 2001. In situ Raman spectroscopic investigation of chromium surfaces under hydrothermal conditions. Appl. Surf. Sci. 180, 102-118.

Maslar, J. E. , Hurst, W. S. , Bowers Jr. , W. J. , Hendricks, J. H. , 2002. In situ Raman spectroscopic investigation of stainless steel hydrothermal corrosion. Corrosion 58, 739-747.

Mayanovic, R. A. , Anderson, A. J. , Dharmagunawardhane, H. A. N. , Pascarelli, S. , Aquilanti, G. , 2012. Monitoring synchrotron X-ray-induced radiolysis effects on metal(Fe, W) ions in high-temperature aqueous fluids. J. Synchrotron Radiat. 19, 797-805.

Men, S. , Anderson, A. J. , Mayanovic, R. A. , 2014. In situ X-ray absorption spectroscopy study of radiolysis-induced redox in chromium nitrate aqueous solution under supercritical condition. In: 2014 Canada-China Conference on Advanced Reactor Development(CCCARD-2014), April 27-30, 2014. Niagara Falls, Ontario, Canada, Paper CCCARD2014-014.

Muroya, Y. , Sanguanmith, S. , Meesungnoen, J. , Lin, M. , Yan, Y. , Katsumura, Y. , Jay-Gerin, J. -P. , 2012. Time-dependent yield of the hydrated electron in subcritical and supercritical water studied by ultrafast pulse radiolysis and Monte-Carlo simulation. Phys. Chem. Chem. Phys. 14, 14325-14333.

Van Nieuwenhove, R. , 2012. Proceedings of a IAEA Technical Meeting Held in Halden, Norway, 21e24 August 2012, In-pile Testing and Instrumentation for Development of Generation-IV Fuels and Materials, Session 1 (Instrumentation Development), Development and Testing of Instruments for Generation-IV Materials Research at the Halden Reactor Project. IAEA TECDOC-CD-1726. http://www-pub. iaea. org/MTCD/Publications/PDF/TE-CD-1726/PDF/IAEA-TECDOC-CD-1726. pdf.

Novotny, R. , Moilanen, P. , Hahner, P. , Siegl, J. , Hausild, P. , 2013a. Development of crack growth rate SCC devices for testing in super critical water(SCW) environment. In: 6th International Symposium on Supercritical Water-Cooled Reactors, March 3-7, 2013, Shenzhen, Guangdong, China. Paper ISSCWR6-13070.

Novotny, R. , Janik, P. , Nilsson, K. F. , Siegl, J. , Hausild, P. , 2013b. Pre-qualification of cladding materials for SCWR fuel qualification testing facility. In: 6th International Symposium on Supercritical Water-Cooled Reactors, March 3-7, 2013, Shenzhen, Guangdong, China. Paper ISSCWR6-13069.

Novotny, R. , Janik, P. , Toivonen, A. , Ruiz, A. , Szaraz, Z. , Zhang, L. , Siegl, J. , Hau_sild, P. , Penttilä, S. , 2015. European Project 'Supercritical Water Reactor-Fuel Qualification Test' (SCWR-FQT): Summary of General Corrosion Tests. In: The 7th International Symposium on Supercritical Water-Cooled Reactors ISSCWR-7 March 15-18, 2015. Finland, Helsinki.

Novotny, R. , Macak, J. , Ruiz, A. , Timke, T. , 2017. Electrochemical impedance measurements in sub-critical and super-critical water. In: 8th International Symposium on Supercritical Water-Cooled Reactors(ISSCWR-8), March 13-15, 2017, Chengdu, China.

Penttilä, S. , Toivonen, A. , Li, J. , Zheng, W. , Novotny, R. , 2013. Effect of surface modification on the corrosion resistance of austenitic stainless steel 316L in supercritical water conditions. J. Supercrit. Fluids 81, 157-163.

Penttilä, S. , Moilanen, P. , Karlsen, W. , Toivonen, A. , 2017. Miniature autoclave and double bellows loading device for material testing in future reactor concept conditions-case supercritical water. ASME J. Nucl. Eng. Radiat. Sci. https://doi. org/10. 1115/1. 4037897.

Pommeret, S. , 2007. Radiolysis of supercritical water: an experimental and theoretical challenge. In: International Workshop on Supercritical Water Coolant Radiolysis, June 14-15, 2007. Řež, Czech Republic.

Pye, C. C. , Cheng, L. , Tremaine, P. R. , 2014. Overview on the investigation of metal speciation under supercritical water-cooled reactor coolant conditions by ab-initio calculations, spectroscopy and conductivity measurements. In: 19th Pacific Basin Nuclear Conference (PBNC 2014), Vancouver, Canada, August 24–28,2014.

Ruther, W. E., Schlueter, R. R., Lee, R. H., Hart, R. K., 1966. Corrosion behavior of steels and nickel alloys in superheated steam. Corrosion 22, 147–156.

Ruzickova, M., Vsolak, R., Hajek, P., Zychova, M., Fukac, R., 2011. Supercritical water loop for in-pile materials testing. In: Proc. of the 5th Int. Symposium on Supercritical Water-Cooled Reactors, Vancouver, Canada, on CD-ROM, vol. 11.

Sharygin, A. V., Wood, R. H., Zimmerman, G. H., Balashov, V. N., 2002. Multiple ion association versus redissociation in aqueous NaCl and KCl at high temperatures. J. Phys. Chem. B 106, 7121–7134.

Solferino, G., Anderson, A. J., 2012. Thermal reduction of molybdite and hematite in water and hydrogen peroxide-bearing solutions: insights on redox conditions in hydrothermal diamond anvil cell (HDAC) experiments. Chem. Geol. 322–323, 215–222.

Spalaris, C. N., 1963. Finding a corrosion resistant cladding for superheater fuels. Nucleonics 21, 41–49.

Spinks, J. W. T., Woods, R. J., 1976. An Introduction to Radiation Chemistry. John Wiley and Sons, Toronto.

Swift, R., Cook, W., Bradley, C., Newman, R. C., Zheng, W., 2015. Validation of constant load C-ring apex stresses for SCC testing in supercritical water. In: 7th International Symposium on Supercritical Water-Cooled Reactors ISSCWR-7, March 15–18, 2015, Helsinki, Finland Paper ISSCWR7-2088.

Takahashi, K., Cline, J. A., Bartels, D. M., Jonah, C. D., 2000. Design of an optical cell for pulse radiolysis of supercritical water. Rev. Sci. Instr. 71, 3345.

Vojacek, A., Ruzickova, M., Schulenberg, T., 2015. Design of an in-pile SCWR fuel qualification test loop. In: The 7th International Symposium on Supercritical Water-cooled Reactors, ISSCWR-7, March 15–18, 2015, Helsinki, Finland. Paper ISSCWR7-2065.

Walker, D. C., 1981. Muonium. A light isotope of hydrogen. J. Phys. Chem. 85, 3960.

Walters, L., Wright, M., Guzonas, D. A., 2017. Irradiation issues and material selection for Canadian SCWR components. In: 8th International Symposium on Supercritical Water-Cooled Reactors, ISSCWR-8, March 13–15, 2017, Chengdu, China.

Was, G. S., Averback, R. S., 2012. Radiation damage using ion beams. In: Konings, R. J. M. (Ed.), Comprehensive Nuclear Materials, vol. 5. Elsevier, Amsterdam, pp. 195–221.

Yan, H., Mayanovic, R. A., Anderson, A. J., Meredith, P. R., 2011. An in situ X-ray spectroscopic study of $Mo^{6+}$ speciation in supercritical aqueous solutions. Nucl. Instr. Methods Phys. Res. A 649, 207–209.

Yan, H., Mayanovic, R. A., Demster, J. W., Anderson, A. J., 2013. In-situ monitoring of the adsorption of $Co^{2+}$ on the surface of $Fe_3O_4$ nanoparticles in high-temperature aqueous fluids. J. Supercrit. Fluids 81, 175–182.

Zimmerman, G. H., Gruskiewicz, M. S., Wood, R. H., 1995. New apparatus for conductance measurements at high temperatures: conductance of aqueous solutions of LiCl, NaCl, NaBr, and CsBr at 28MPa and water densities from 700 to 260kg/$m^3$. J. Phys. Chem. 99, 11612–11625.

Zimmerman, G. H., Arcis, H., Tremaine, P. R., 2012. Limiting conductivities and ion association constants of aqueous NaCl under hydrothermal conditions: experimental data and correlations. J. Chem. Eng. Data 57, 2415–2429.

# 第3章
# 辐照效应和力学性能

所有堆内部件都将受到 α 粒子、β 粒子、中子和高能光子(γ 射线)的辐照(图 1.3),由于空位、间隙和空隙的形成,在原子水平上造成电离和微观结构退化的损伤。这些微观缺陷会引起物理和力学性能的变化,同时也是决定元件长期可靠性的主要因素。

有许多关于辐照损伤及其对材料(如合金、石墨、氧化物)影响的优秀参考文献(Boothby,2012;Garner,2012;Dai 等,2012;Stoller,2012;Zinkle,2012)。推荐读者从这一领域的早期研究开始阅读,以便了解在过去几十年里人们对这些现象认知的演变。由于超临界水冷反应堆(SCWR)的研究发展缺乏对堆内进行测试的设施,目前在堆内材料辐照方面的研究还很少。因此,这里只能对辐照损害的基本方面及其对与 SCWR 发展有关材料的影响作简要总结。对于穿透性强的辐照,如中子辐照对整个材料产生影响,从而改变整个材料的性质,但对于材料周围的流体(如 SCW)没有什么影响。因此,根据这些合金在其他流体中以及其他类似的温度、流场等条件下受到辐照产生的影响结果,可以对损伤模式的预估提供重要的参考。

## 3.1 一次辐射损伤

当一个原子被一个入射粒子撞击时,如果碰撞过程中传递的能量超过一定的阈值($E_d$),在晶格中正常位置的主碰撞原子(PKA)将被取代,这个阈值约为 25eV($1eV = 1.6 \times 10^{-19}J$),具体数值受金属的类型和弗伦克尔对方向([*hkl*])①

① 要获得关于 *hkl* 表示法、晶体结构和晶格缺陷的基本入门知识,请查阅任何固体物理学教科书,如 Kittel 著作 Kittel(1976)。

的影响。最初的损伤缺陷是成对形成的①。被取代的原子和反冲的原子通过弹性碰撞会撞击许多其他原子,最终处于间隙位置时就会产生空位。如果产生的空位和反冲原子没有充分分离,这两点缺陷就会相互湮灭。但是,如果它们之间的间隔超过空位俘获半径,就会形成一个弗伦克尔对。由于反冲原子本身会产生一系列的碰撞,PKA 通常会引起一连串的位移。一次碰撞和由此产生的一系列碰撞发生的时间大约为 $10^{-11}$ s,而由此产生的缺陷扩散到凹槽位置则需要几毫秒到几个月的时间(就像孔隙的形成和增长)。

Kinchin 和 Pease(1955)提出了一个线性位移理论,该理论将粒子和原子的二元碰撞模型视作两个坚硬小球的撞击过程。能量为 $E$ 的 PKA,其位移原子数($N_d$)为(Gittus,1978):

$$N_d = \frac{E}{2E_d} \tag{3.1}$$

式中:$E_d$为位移能阈值。

由于非弹性能量损失(由于原子势能和电子制动机制)和不稳定的邻近点缺陷对的非热复合,Norgett 等(1975)提出了一种估计位移剂量率的改进方法,即众所周知的 NRT 方法(以作者的首字母命名)。NRT 模型是标准评估位移级联损伤方法(ASTM,2009)的基础,则有

$$N_D = \frac{K_d E_D}{2E_d} \tag{3.2}$$

式中:$E_D$为损伤能,表示入射粒子向反冲原子传递的有效能;$E_d$为位移能阈值。

通常认为效率因子 $K_d$与化学和温度无关,在标准模型中为 0.8。然而,后续的研究表明它与能量有关,且随着损伤能量的大小而变化,最低为 0.3。不锈钢的位移能阈值为 40eV (ASTM,2009;Broeders 和 Konobeyev,2004)。也有人报道过纯镍的位移能阈值为 40eV (Averback 和 Diaz de la Rubia,1997)。

根据单位原子位移(dpa)的概念,材料特性的变化通常与位移损伤相关。虽然在反应堆堆芯中有足够的热激活扩散使许多空位和间隙重新组合(因此保留的位移损伤只是 dpa 值的一小部分),但 dpa 值已被证明是对辐照损伤进行定量描述的有效手段。图 3.1 显示了在 Au (OECD,2015)中一次损伤过程的分子动力学(MD)模拟结果,其中描述了一次损伤过程的许多关键特征。

① 一个稳定的空位-间隙对。

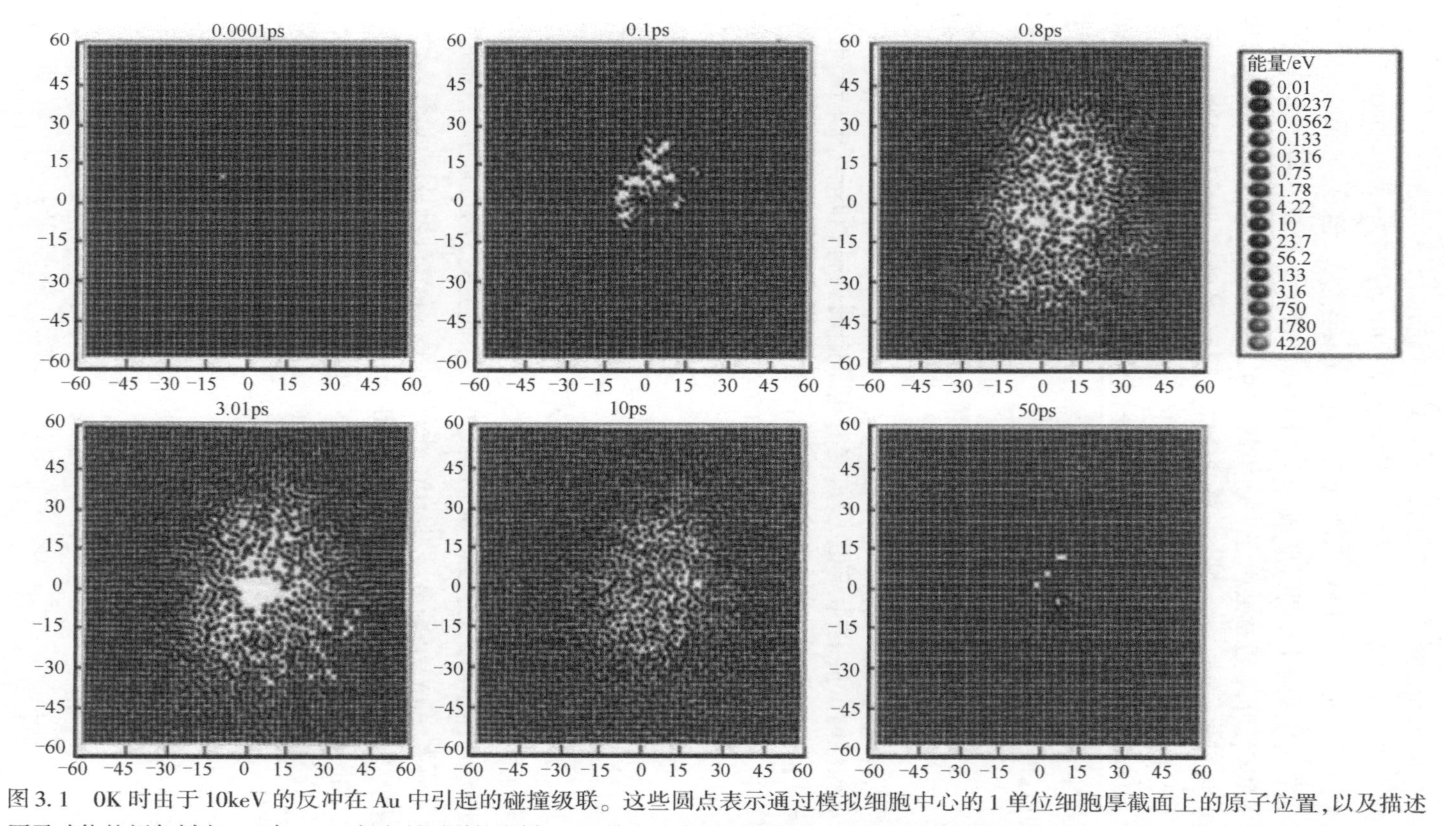


图 3.1　0K 时由于 10keV 的反冲在 Au 中引起的碰撞级联。这些圆点表示通过模拟细胞中心的 1 单位细胞厚截面上的原子位置，以及描述原子动能的颜色刻度。正如 NRT 方法所预测的那样，一开始会有大量原子移位，但是当级联冷却(大约 10ps 后)，几乎所有的原子都回到了正常的晶体位置(尽管许多原子没有回到它们原来的位置)，因此，原子位移的数目远远大于产生的稳定缺陷的数目(见彩插)

(引自文献 OECD, 2015. Primary Radiation Damage in Materials, NEA/NSC/DOC (2015) 9. https://www.oecd-nea.org/science/docs/2015/nsc-doc2015-9.pdf.)

随着损伤区(级联区)边缘间隙的形成,当级联中的原子从外部向中心重新结晶并回到正常的晶格排列时,空位簇被推入中心。这类似于冶金学家熟知的局部精炼过程。当这些缺陷重新排列成有利结构时,就形成了位错环。在面心立方(fcc)金属中,缺陷的聚集最初发生在{111}面上,形成弗兰克环①。Gittus(1978)很好地描述了缺陷簇的早期成核和生长。当流动性增加时也会形成空穴簇,而且当缺陷足够大时,在TEM显微照片上可以看到“黑点”。这些微观特征阻碍了位错运动,因此辐照金属会变硬。

在熔点温度的0.2~0.6倍($T_m$(K))之间,会出现更快的扩散过程并促进微观缺陷的更大范围聚集。因此,随着辐照剂量的积累,孔隙可以很容易地形成和生长。由于嬗变产生的小气泡(如He)也是孔隙成核的场所,这些孔隙的增长会导致孔隙膨胀,这不仅会改变金属构件的尺寸,还会降低其延展性。

所有堆内部件都要受到辐照,受到的总剂量取决于部件相对于燃料的位置、在核内的停留时间、中间材料的性质(提供屏蔽的材料)和中子能谱。到目前为止,关于SCWR核心部件辐照条件的文献报道甚少。Walters等(2017)估计了加拿大SCWR中的各种核心组件受到的通量、积分通量和温度数据。其中压力管接近燃料,且在堆芯停留的时间很长(75年),是吸收剂量最高的组件。然而,SCWR燃料包壳将经历最具挑战性的叠加条件(高温和高积分通量)。例如,加拿大SCWR设计中燃料包壳应用的条件是:温度为350~850℃(Nava-Domínguez等,2016)②;设计应力在正常工作条件下可忽略不计(Xu等, 2016b);设计寿命可达30000h③(Xu等,2014;Yetisir等,2017);总通量大约为$2.1\times10^{18}$个/($m^2\cdot s$);当$E>1MeV$时,总通量为$4.6-4.9\times10^{18}$个/($m^2\cdot s$);总积分通量④为$5.1\sim5.4\times10^{25}$个/$m^2$。Walters等(2017)报告表明,根据所选择的合金和在燃料组件中的位置,燃料包壳的总损伤将达到8.1~8.5dpa,并会产生19.3~63.9appm⑤的He,其他关于SCWR设计的细节较少。Ehrlich等(2004)报道了高性能轻水反应堆(HPLWR)核心材料在经历45000h后的平均损伤为27.6dpa,最大损伤为61.8dpa。他们还报道了经过3年后在1.4970 SS和在718合金中的与硼(n,α)反应生成He的产量分别为60~85appm和23.7appm。Matsui等(2007)列出了

① 弯曲的位错线两端固定于两个法向晶格平面之间的末端。弗兰克循环是有缺陷的,因为它们的价格矢量小于正常的晶间间距。

② 850℃表示估计的最大峰值包层温度。

③ 加拿大SCWR概念堆的燃料包壳被设计成坍塌在燃料上,使得冷却剂压力由燃料支撑。在事故发生期间,在没有冷却剂压力的情况下,燃料的内部压力将产生高达40MPa的环向应力,这取决于燃料的燃耗情况。

④ 通量和通量改变量随燃料芯块在堆芯中的空间位置而变化。

⑤ appm(atom parts per million)浓度单位。

日本 SCWR设计中考虑的材料辐照试验要求，其中中子通量包括 $2\times10^{21}$个/$m^2$的快中子和 $8\times10^{25}$个/$m^2$的热中子。

SCWR 核心材料所经历的温度范围将比目前的水冷反应堆大得多。由于热扩散效应，在燃料包壳的使用寿命期间，会根据包壳的温度产生一系列不同类型的辐照损伤，如图 3.2 所示（图 2.4 与压力容器设计中的位置大致相关）。

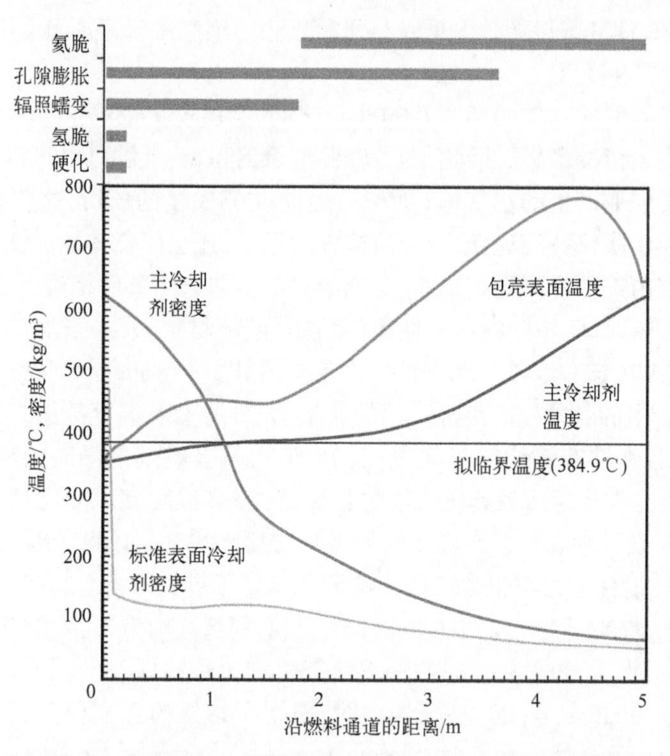


图 3.2　在加拿大 SCWR 堆芯内的特定温度范围内的辐照损伤的模型，显示了体积、表面温度和冷却剂密度沿燃料通道在理论上的分布（Zheng 等，2016）

加拿大 SCWR 设计使用了绝缘的 Zr 合金压力管，绝缘体材料（氧化钇-稳定的氧化锆）和压力管合金（Excel 合金）的辐照特性是验证这一设计的关键。Excel 压力管将在 120～230℃的重水慢化剂中工作。有学者提出了退火 Excel，因为这种材料可以使辐照蠕变和生长速率最小化（Chow 和 Khartabil，2008）。Walters、Donohue（2016）和 Walters 等（2017）总结了与加拿大 SCWR 设计相关的这种合金的现有数据。Excel 合金在 100～400℃温度范围内和高达 10dpa 的 Kr 离子辐照条件下显示出了 C 环的形成和 ω 相的析出，这表明该合金具有辐照生长的能力。虽然已证明退火后的 Excel 对延迟开裂（DHC）有相当大的作用，但众所周知，它具有很高的 DHC 速度，一旦开始裂纹就会迅速增长。

由于不锈钢和镍基合金是 SCWR 堆芯压力容器以及加拿大 SCWR 设计中燃料组件采用的主要合金,本章的其余部分将重点介绍这些合金中的代表性合金。

## 3.2 对力学性能的影响

### 3.2.1 硬化

关于材料辐照硬化的描述有数百种,包括面心立方合金(如奥氏体不锈钢、镍合金、铜及铜合金)、六方密堆积合金(hcp,如 Zr 和 Zr 合金)以及体心立方合金(bcc,如碳素钢、铁素体钢等)。从非常低的辐照剂量开始,硬化是辐照最早引起的合金微观结构变化的现象之一。随着屈服强度的增加,延展率降低,加工硬化率降低。图 3.3 是这种效应在不锈钢中的一个例子。这是位错环、微孔和气泡等由辐照引起的微观缺陷导致的塑性流动的典型结果(Garner,2012)。

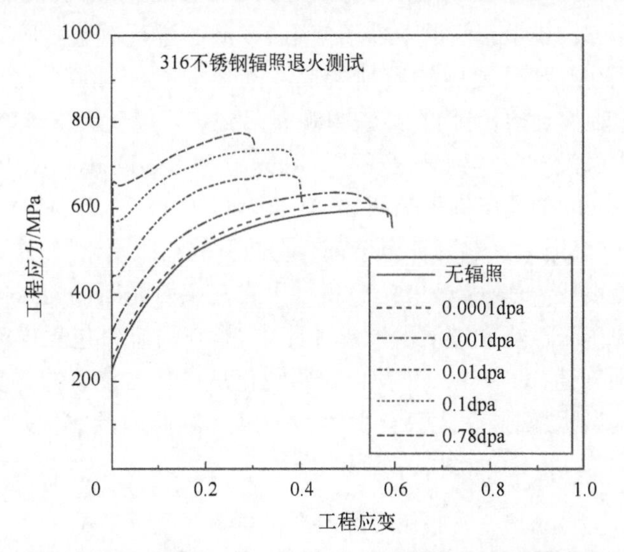


图 3.3 在 60~100℃高通量同位素堆(HFIR)混合谱堆中辐照退火的 316 不锈钢在 25℃条件下测试的工程应力应变曲线(Garner,2012)

剂量率对硬化有影响,对不同类型退火不锈钢(304、304L、316L、316LN、主要候选合金)的研究表明,这种硬化大概在 5dpa 左右条件下饱和(Pawel 等,1999)。然而,Xu 等(2016a)最近的研究结果表明,受 Xe 射线辐照后的试样硬度在 2dpa 左右条件下饱和,而受质子辐照时,其饱和剂量要高得多,可能超过 7dpa。这种差异被认为是质子和 Xe 辐照损伤分布的不同引起的。

Allen 等(2002)对从试验增殖反应堆(EBR-II)反射区回收的冷处理 316 SS(剂

量范围为1~56dpa)进行了研究,发现当剂量小于20dpa时硬化并未达到饱和。其中,辐照的温度为371~390℃,剂量为0.8~3.3×10$^{-7}$dpa/s。孔隙、沉淀和循环都有助于硬化。对其进行定量分析,由于环、空隙和环加上空隙的形成而导致的屈服强度测量值的增加可以描述为以下的总和(Allen 等,2002),即

$$\Delta\sigma_y = \Delta\sigma_y^{\text{dislocations}} + \sqrt{(\Delta\sigma_y^{\text{dislocations}})^2 + (\Delta\sigma_y^{\text{loops}})^2 + (\Delta\sigma_y^{\text{precipitates}})^2} \tag{3.3}$$

对于位错,屈服强度增量为

$$\Delta\sigma_y = M\cdot\alpha\cdot\mu\cdot b\cdot\sqrt{\rho_d} \tag{3.4}$$

式中:$\rho_d$为位错密度(单位体积线长);$M$为激活滑移面滑移所需的剪切应力;$\alpha$为阻挡强度;$\mu$为剪切模量;$b$为位错移动的伯格斯矢量。

Hardie 等(2013)最近对纯铁和含Cr的质量分数为5%、10%和14%的铁铬合金的辐照特性进行了研究,结果表明,高剂量率会使Fe-Cr合金产生更大的位错环密度和尺寸。同时,他们在低剂量率辐照的合金中发现了Cr型沉淀。他们在最高能量为2MeV,温度为300℃、400℃和500℃,剂量为0.6dpa且剂量率为6×10$^{-4}$dpa/s和3×10$^{-5}$dpa/s的条件下使用$Fe^+$离子进行试验。所有材料经受在300℃条件下的辐照后硬度均有所提高。在经受大于400℃的辐照后,只在Fe-Cr合金中观察到硬化,而在纯铁中观察不到。在经受大于500℃的辐照后,在任何材料中都没有观察到硬化现象。Klassen 和 Rajakumar (2015)使用8MeV的$Fe^{4+}$在15dpa的剂量条件下辐照800H合金和310 SS,然后使样品在400℃和500℃条件下等温退火。这两种合金都存在明显的辐照硬化,但作者发现其辐照损伤的热恢复很快。Hazel 等(1965)报道了在瓦莱西托斯(Vallecitos)沸水反应堆(VBWR)的膨胀过热先进演示试验回路中,暴露的燃料包壳样品在照前和辐照后的显微硬度值(表3.1),并发现辐照硬化是在其中较高的温度条件下退火的。这些数据表明,辐照硬化只会发生在SCWR堆芯内的“蒸发”区,在这里冷却剂的温度小于400℃,如图3.2所示。



表3.1　辐照后燃料包壳样品的硬度数据

| ESH 元素 | 材　料 | 暴露时间/天 | 显微硬度(维克斯硬度系数)/HV | |
|---|---|---|---|---|
| | | | 辐照前 | 辐照后 |
| 1.1-A-2 | 800 合金 | 106 | 164~165 | 162~168 |
| 1.1-BONUS | 348 SS | 106 | 150~153 | 147~150 |
| 1.1-F | 310 VM | 106 | 128~131 | 115~116 |
| 1.1-H | 800 合金 | 135 | 141~145 | 165~179 |
| 1.1-I | 600 合金 | 135 | 154~156 | 184~196 |
| 1-B | 800 合金 | 29 | 141~145 | 153~159 |

### 3.2.2 延展性

与硬化同时发生的是,辐照合金会随着剂量的增加而表现出局部流动的现象。在断裂面上,辐照不锈钢可以看到“沟道断裂”的平面,这是形变面上弗朗克环的位错清除的结果(图 3.4)。位错运动集中在这些狭窄的滑动带上,并抑制了均匀变形,在应力-应变曲线上过早地发生颈缩,导致均匀伸长大大降低。这些具有晶界的位错通道的切除可产生微空隙,并导致晶间裂纹(Stephenson 和 Was,2016)。

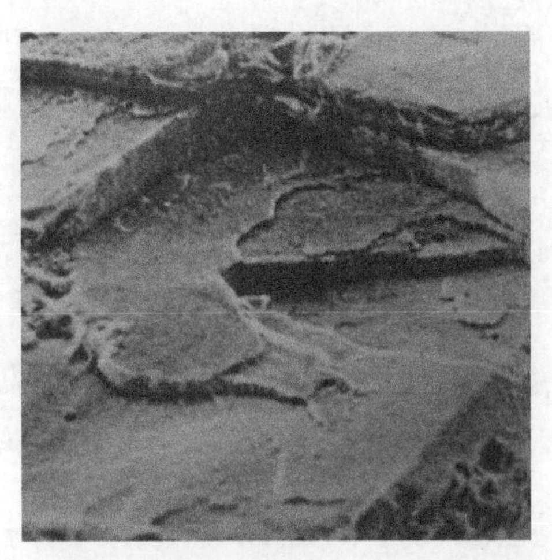

图 3.4　在铜中看到的通道断裂的一个例子(Gittus,1978)

在更微观的层面上,位错沿着有限数量的滑移面滑动,消除了它们路径上的小障碍,为进一步的塑性流动创造了一个通道。该通道的滑移带宽度通常仅为微米的几分之一。即使辐照合金的体积延展率只有几个百分点,其内部的局部应变也可能达到几百个百分点(Byun 和 Hashimoto,2006)。图 3.5 给出了在 316 SS、铁、钒和 Zr-4 合金中这种通道的 TEM 显微照片示例。在 316 SS 中,无缺陷通道之间的块体在没有通道的组织中几乎没有位错活动。活动性{111}<110>的导流滑移体系与变形均匀的导流滑移体系相同。

面心立方金属可以在高应变水平条件下发生双晶变形,辐照也会显著增加双晶变形形成的趋势(Lee 等,2001a、b)。Byun 和 Hashimoto(2006)将不锈钢的变形按等效应力范围进行了分类:

(1) 在低的等效应力下(小于 400MPa),位置缠结为主要形态;

(2) 小而孤立的堆垛断层(小于 1mm),应力范围为 400~600MPa;

(3) 大型叠加断层带(大于 1mm)和双叠错带,应力大于 600MPa。

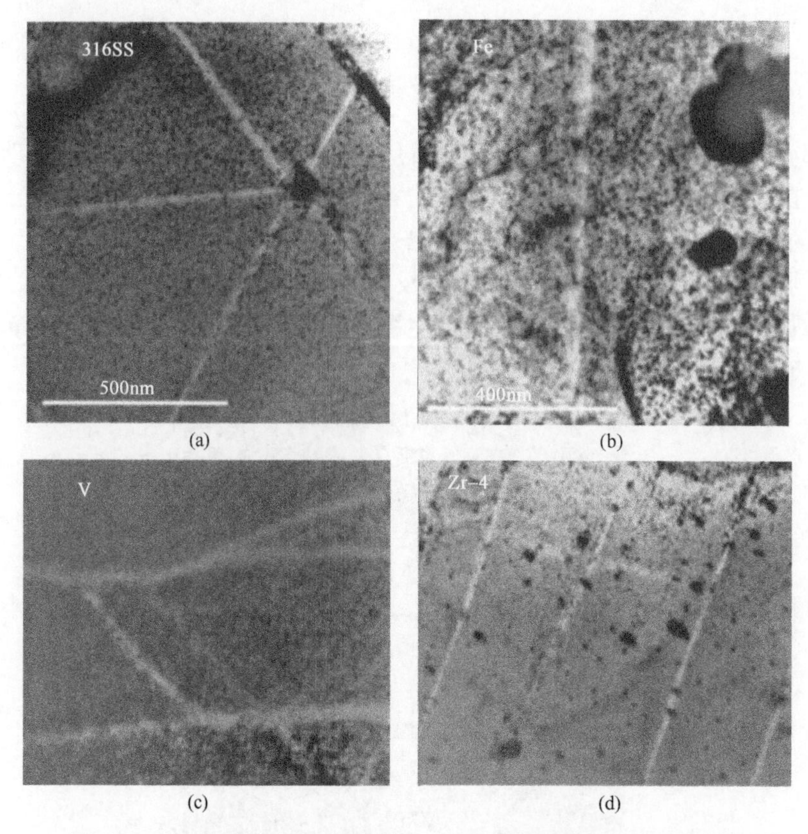

(a) (b)
(c) (d)



图 3.5 大约 80℃条件下中子辐照后的沟道微观结构在常温下变形

(a) 在 316 不锈钢中,0.78dpa、5%应变条件下的直通道及其截断;(b) 在 0.89dpa、0.2%应变条件下铁中的单通道;(c) 钒在 0.69dpa、10%应变条件下的河网通道;(d) Zr-4 合金在 0.8dpa、6%应变条件下的平行通道。

(引自文献 Byun,T.,Hashimoto,N.,2006. Strain localization in irradiated materials. Nucl. Eng. Technol. 38(7),619-638)

还有许多其他的机制,即辐照引起的微观结构或化学性质的变化可能降低合金的延展性,如 He 气泡的形成和孔隙的形成和扩大。

奥氏体不锈钢或镍基合金在 SCWR 堆芯中的应用面临的一个潜在的严重威胁是 He 的产生,这可能会导致氦气泡的形成和高温氦的脆化(Schroeder,1988;Dai 等,2012)。而且可能影响孔隙膨胀(Johnston 等,1974;Wolfer 等,1982;Garner,2012)。当含镍合金暴露于高热中子通量时,可以通过$^{58}$Ni(n,γ)$^{59}$Ni(n,α)$^{56}$Fe 连续反应产生 He(Greenwood,1983;Greenwood 和 Oliver,2006)。同样地,氢是由$^{58}$Ni(n,γ)$^{59}$Ni(n,α)$^{59}$Co 产生的。这些嬗变反应也会产生更多的反冲核,这可能会对累积的一次辐照损伤有严重影响。在这些反应过程中,自然产生的非放射性$^{58}$Ni同位素和中间同位素$^{59}$Ni最后会被消耗完。最终,He产量在热通量约为$10^{24}$个/cm$^2$时达到了饱和。在这种饱和通量下产生的He量是巨大的(大约为80000appm He)。

He的产量受合金中镍的含量以及中子能谱和通量的影响。加拿大SCWR燃料包壳的热通量在3个燃料循环(3.5年)后会小于$10^{21}$个/cm$^2$,这主要是由于在设计中使用了钍钚燃料的结果。Walters等(2017)报道,对在燃料包壳中He产量的范围进行计算发现,310 SS为19.3~23.7appm,625合金为50.8~63.9appm。这比日本研究报告中计算产量的数值大得多(表3.2)。Bysice等(2016)比较了加拿大SCWR设计中使用的PU-Th燃料和LEU-Th燃料的备选燃料包壳合金生成He的结果发现,相比于$^{238}$U,钚同位素的高热中子俘获截面可使PU-Th燃料产生的He量减少2倍。He产量的增加与合金中的Ni含量大致呈线性关系。

表3.2 日本超临界水冷反应堆概念中备选合金的He产量的计算值

| 材料 | 镍基合金中关键变形反应的He产量 | | |
|---|---|---|---|
| | $^{58}$Ni(n,γ)$^{59}$Ni(n,α)$^{56}$Fe | $^{58}$Ni(n,α)$^{55}$Fe | $^{60}$Ni(n,α)$^{57}$Fe |
| 310S SS | 105 | 17 | 2 |
| 316L SS | 68 | 11 | 1 |
| 690合金 | 338 | 56 | 5 |
| 纯Ni | 583 | 96 | 9 |

在某些合金中加入少量硼可以改善蠕变性能,且硼在许多合金中也以残余元素的形式存在。对于(n,α)反应,$^{10}$B有一个大的横截面,且硼被认为是在晶界分离的。因此,$^{10}$B向He的转化可能产生局部高浓度的He,从而促进晶界处气泡的形成。$^{10}$B向He的转化会在1~3年内完成,并取决于中子能谱和通量。$^{10}$B的损失(约20%的自然丰度)也会影响长期蠕变性能。因此,应尽可能减少硼在备选合金中的浓度。Erhlich等(2004)建议使用富含$^{11}$B的B,这会产生更低的(n,α)反应截面。

修理或维护过程中的焊接工艺会对处于He形成过程中的合金产生明显影响。He浓度低至1appm时会使焊接奥氏体不锈钢和镍基合金困难。

在预期的包壳温度峰值条件下辐照合金的延展性相关数据还很不足,特别是针对加拿大SCWR设计中高温条件更为缺少。Thiele并(1984)报告了在温度为375℃、热中子通量为1.1×$10^{21}$个/cm$^2$、剂量为1.4dpa条件下的HTRI中,经受辐照后的800合金和其他物质在不同温度下的延展性。从600℃到850℃,

随着温度的增加,所有合金的延展性都会快速下降。在700℃条件下,对于大多数合金的研究显示,延伸率大约为个位数或略高于10%。在850℃条件下,大多数材料的延伸率低于5%。

Comprelli 等(1969)在 VBWR、EVESR、通用电气试验反应堆 GETR 和试验增殖反应堆(EBR-II)中对合金 800 的辐照损伤和燃料性能进行了比较。这些试验中的条件维持在温度为 538~704℃,暴露时间为 200~500h,快中子通量为(0.17~67)×$10^{20}$个/$cm^2$,热中子通量为(0.3~2.5)×$10^{21}$个/$cm^2$;环境工况包括 Ar、Ar+He 以及钠和过热蒸汽(SHS)。与之特别相关的测试是将预切拉力试样插入 EVESR 的燃料束中并暴露于 SHS 环境。放置在速度助推器①中的样品在454℃的轴向峰值通量位置进行辐照。从峰值燃耗棒的外包壳分成两个试样,在538~621℃范围内辐照。表3.3表明,在热反应堆中辐照的主要作用是降低高温延展性。在加速器中的试样和从燃料包壳切割的试样之间延展性的差异是由于辐照温度或受到的应力-应变的差异造成的。

表3.3　704℃时在 EVESR Mark II 的过热蒸汽中的辐照样品的拉伸数据

| 情形 | 样品类型 | 样品数量 | 0.2%偏移屈服应力/(N/$m^2$) | 最终应力/(N/$m^2$) | 统一延长/% | 总延长/% |
|---|---|---|---|---|---|---|
| 退火 | 控制 | 1 | 1.75×$10^8$ | 3.47×$10^8$ | 17.7 | 34.4 |
| 冷态工作 | 控制 | 1 | 4.38 | 4.87 | 4.0 | 14.95 |
| 退火 | 速度增加 | 3 | 1.655~1.765 | 2.96~3.15 | 11.5~16.8 | 12.5~18.5 |
| 冷态工作 | 速度增加 | 2 | 2.85~3.90 | 3.18~4.25 | 3.4~3.6 | 4.3~6.5 |
| 退火 | 镀层 | 2 | 1.205~2.14 | 2.59~2.63 | 2.6~5.95 | 7.4~8.1 |



Rowcliffe 等(2009)指出,PE16 合金被广泛用作燃料包壳且效果良好。他们发现在 Dounreay 中的75%的 PFR 都装有 PE16 燃料元件,并在400~650℃和大约80dpa 的环境下工作,且没有发生一次故障。他们得出的结论是,通过合理的燃料设计,可以避免燃料包壳在发生晶间失效时受到温度-应力区域的影响。最大裂变气体压力估计小于70MPa,比引起晶界 He 气泡不稳定生长所需的应力小2~3倍。

## 3.2.3 辐照导致的应力腐蚀开裂

应力腐蚀开裂(SCC)的详细讨论内容在本书第6章中介绍;本节简要介绍与辐射效应有关的内容。

① 放置在每个燃料元件中央腔中的密封钢管,旨在增加蒸汽通过内部传热表面的速度。

除了在前面几节中讨论过的微观结构缺陷(点、线、位错和空隙),辐照还会在微观水平上引起材料的化学变化,如辐照诱导的偏析(RIS)和沉淀,两者都会影响 SCC 敏感性。这种损伤的程度取决于辐照剂量、通量、辐照光谱和辐照温度,其中辐照温度与扩散过程直接相关。例如,不锈钢中的 RIS 会消耗晶界中的 Cr,这是不锈钢中晶粒间开裂的已知原因之一(IAEA,2011),3.2.4 节将详细讨论。辐照不锈钢和镍基合金中形变通道的形成会在晶界处产生局部应变,这是与轻水反应堆(LWR)中辐照促进应力腐蚀开裂(IASCC)密切相关的因素(Jiao 和 Was,2011)。

在亚临界温度范围,Hojna (2013)对现有沸水反应堆(BWR)和压水反应堆(PWR)延长寿命期背景下的堆内 IASCC 的运行经验进行了总结。总的来说,随着损伤剂量的增加,辐照对 SCC 的影响越来越明显。辐照 315 不锈钢在 340℃下的恒载试验结果表明,随着损伤从 10dpa 增加到 30dpa 左右,IASCC 的应力阈值减小,其中 Cr 重量可达 19%,Ni 重量可达 9.5% (Freyer 等,2007)。这些影响对燃料包壳来说不那么重要,所以不锈钢燃料包壳在轻水堆中得到了广泛应用。拉克罗斯(La Crosse)BWR 使用 348 SS 燃料包壳运行了 20 年(Strasser 等,1982)。而在前 5 个操作周期中,约有 1.3%的棒件故障(几乎都高于堆芯平均燃耗)。通过改变最大燃耗限制和更严格的操作限制,降低了后续操作周期的故障率(如控制棒的移动)。失效是由于颗粒-包壳相互作用产生的内应力和包壳外表面(暴露于冷却剂中)的腐蚀共同作用的结果,这是由于水的辐射分解导致冷却剂中氧化剂浓度变高引起的。这一操作经验表明,优化设计(如设计和残余应力、使用寿命(辐照时间))可以使 IASCC 的风险最小化(见第 6 章)。

相比于亚临界水(使用 LWR 化学方法)中对 IASCC 的大量研究,在预期的 SCW 堆芯水化学中,IASCC 的数据很少(见第 4 章)。无论辐照与否,较高的温度都会对合金在机械载荷作用下的微观结构响应产生相当大的影响。Zhou 等(2009)将受质子辐照至 7dpa 的样品在 400℃和 500℃的 SCW 中进行恒定拉伸速率试验,比较了 316L 合金、D9、800H 合金和 690 合金的 SCC 敏感性。在所有情况下,辐照合金的裂纹深度都比未辐照的试样大。在不同温度条件下将 690 合金和 316L SS 合金辐照至 7dpa,它们在 500℃时的裂纹深度大于 400℃时的裂纹深度。应该指出的是,在这些试验中,SCC 样品被拉应力破坏,但不清楚裂纹是在应力-应变曲线的哪一点开始产生的;且在中子辐照下,不锈钢的延展性明显降低。

### 3.2.4 孔隙膨胀

孔隙膨胀是一种退化机制,可导致显著的尺寸不稳定和过早失效(Johnston

等,1974;Harries,1979;Wolfer 等,1982;Garner,2012)。因此,了解材料响应和冶金因素对孔隙膨胀产生的影响是至关重要的,包括奥氏体不锈钢在内的许多材料中的膨胀并不饱和。在没有外加应力或约束的情况下,体积膨胀是各向同性的;相反,在施加应力或约束情况下可以观察到各向异性的孔隙膨胀(Lauritzen 等,1987)。一般来说,必须存在以位移空位和空位簇形式出现的足够的位移损伤,才可以形成孔隙成核并促进其生长。高温条件下(大于 700℃),初始辐照损伤、密度和强度较低,空位簇稳定性较低,进而使膨胀驱动力降低。在低温条件(小于 300℃)下,缺陷扩散受限,因此孔隙膨胀的可能性较小。孔隙膨胀在 350~700℃之间最为明显。从图 3.2 可以看出,SCWR 堆芯中大部分材料的温度都在这个范围内。近期对传统奥氏体不锈钢重新研究评估表明,孔隙膨胀分为两个阶段,一个是体积变化极小的瞬态或潜伏期,另一个是稳定的膨胀期。在压水堆部件相关的剂量率下,钢中观察到的低孔隙膨胀是由于辐照产生沉淀(如 G 相、碳化物、γ′相)引起的(在 $T<370$℃ 时),这些沉淀充当了辐照引起的空位沉淀(Chung,2006)。

镍对孔隙膨胀的影响较大,但影响程度取决于辐照温度。在较低的辐照温度下,0%~24%(质量分数)的镍浓度对瞬时损伤剂量的影响不大。在较高的辐照温度下,Ni 的质量分数增加至 40%左右时,通常会延长稳定膨胀前的潜伏期(Johnston 等,1974;Johnston 等,1976;Garner,1984;Garner 和 Brager,1985)。铬对孔隙膨胀的影响不太明显;增加铬的含量可以降低瞬态损伤剂量(Garner 和 Kumar,1987)。虽然高镍合金对孔隙膨胀有一定的好处,但这些合金中 He 的产量也必须加以考虑。



对于奥氏体不锈钢,每 dpa 的稳态膨胀率约为 1%,并且与化学、温度和原子位移率无关。从图 3.6 可以看出,20%冷加工的 316 不锈钢的孔隙膨胀率随辐照温度和辐照剂量的变化是恒定的,在稳态下近似等于 1%/dpa。过渡阶段的时长受到多种因素的影响,包括主要合金和次要合金的添加(Garner 和 Brager,1985;Garner 等,1987;Allen 等,2000;Gessel 和 Rowcliffe,1977;Herschbach 等,1993)、氦浓度(Ayrault 等,1981;Singh 等,1981;Murphy,1987;Ohnuki 等,1988)、微观结构和相位稳定性、应力施加情况(Porter 等,1983)、先前的冷加工、辐照温度(Kozlov 等,2004;Okita 等,2000)和损伤剂量率(Kiritani,1989;Okita 等,2000;Okita 等,2002;Allen 等,2006)。

Zheng 等(2017)研究了在较低能量(100keV)的 $H^{2+}$ 条件下,氢的滞留对 AL-6XN SS 辐照试样中缺陷组织的影响。将高浓度的氢注入试件,位移损伤剂量可达 7dpa。用透射电镜对由辐照引起的位错环和位错空隙进行表征。在 290℃条件下,受辐照 7dpa 后,在其中发现了高密度的位错环和孔隙膨胀。在 380℃条

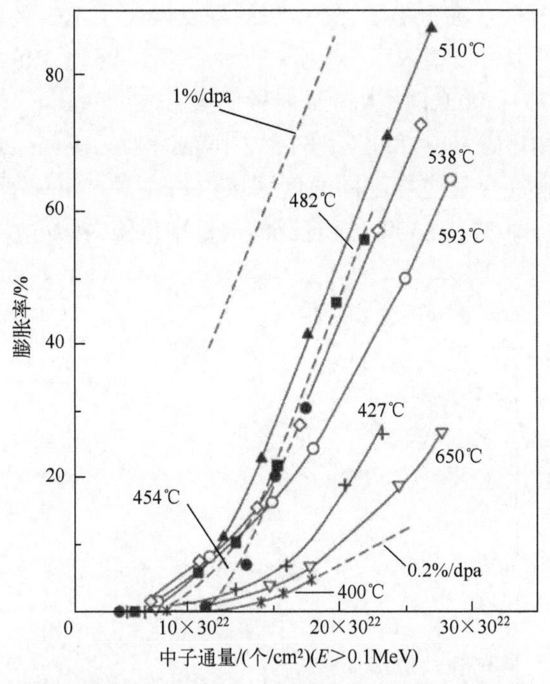


图 3.6 在试验增殖反应堆(EBR)快堆中,观察到辐照后 20%冷加工的 AISI 316 的膨胀,膨胀特征是通过密度变化来确定的,并与温度和剂量密切相关(所有测量都是在给定温度下同一样品上进行的。该方法最大限度地减少了标本间数据的分散,并有助于清晰地显示瞬态后的膨胀率)

(引自文献 Garner,F. A. ,2012. Radiation damage in austenitic steels. In:Konings,R. J. M. (Ed.),Comprehensive Nuclear Materials. Elsevier,Oxford,33-95,ISBN:9780 080560335,https://doi.org/10.1016/B978-0-08-056033-5.00065-3;originally by Garn -er,F. A. ,Gelles,D. S. ,1990. In:Proceedings of Symposium on Effects of Radiation on Materials:14th International Symposium,ASTM STP 1046,vol. II,pp. 673-683)

件下,受辐照 7dpa 后,大部分的位错环未断裂纠结,受到 5dpa 及以上损伤剂量时会出现孔隙膨胀。

## 3.3 对微化学的影响:辐照诱导的分离

空位和自间隙原子向晶界、错位或析出相表面的扩散,会引起合金在局部位置上化学性质的变化。与空位交换更快的 Cr、Mo 和 Fe 等元素在这些凹陷附近会消失,而通过间隙机制迁移的 Si、P 和 Ni 等元素会在点缺陷汇聚的区域附近富集(Bruemmer 等,1999;Scott,1994)。辐照温度和剂量率则会影响这些分离。

RIS 与温度相关。在低温下,迁移率的降低限制了 RIS 的范围。在非常高的温度下,反向扩散可以使浓度梯度分布更为平滑。因此,RIS 在中等温度下是最严重的。在 275~300℃时,当辐照剂量低至 0.1~5dpa 下,能观察到显著的 RIS (Scott,1994;Bruemmer,2001)。图 3.7 (Was 和 Andresen,2012)显示了缺陷的分离、反扩散和复合效应在不同温度区域内占主导。可以看出,在 SCWR 堆芯的温度范围内,即 375~850℃,很宽泛的剂量率范围内均可以产生 RIS。

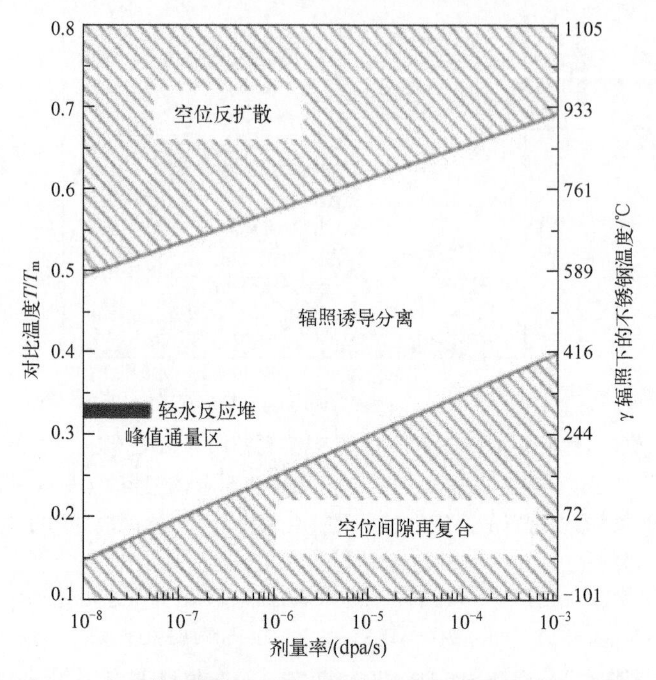


图 3.7 奥氏体不锈钢所处温度和剂量率对其辐照诱导分离的变化关系(Was 和 Andresen,2012)

RIS 在 IASCC 中的作用引起了广泛的研究兴趣。Busby 等(2002)报道了一项研究结果,该研究采用晶界化学测量和辐照后退火的方法,使受质子辐照后合金的晶界化学恢复到辐照前的状态。他们发现辐照导致的位错组织在退火过程中被优先去除。即使在 500℃条件下退火 45min 后,裂纹完全消除,主体的硬度仅略有恢复,说明辐照引起主体变形结构不是直接的相关参数。他们能够将 RIS 引起的化学变化通过在 600℃、90min 的退火过程中去除掉,因为这个退火过程使辐照引起的硬化完全恢复,但是 RIS 的轮廓没有改变。通过对比辐照和退火后的样品,他们发现 Cr、Ni、Si 或 P 本身的 RIS 不足以引起 IASCC。在辐照后的退火过程中,Ni 水平保持不变,而退火后裂纹迅速得到缓解,说明晶界处的 Ni 富集并不是 IASCC 的缓解因素。硅在晶界的富集是一个非控制因素,P 的富集对 IASCC 的影响最小。

在后续的研究中,Jiao 和 Was(2011)将他们的研究条件设定为晶界 Cr 含量为 12%~19%(质量分数),进一步排除了 RIS 的相关性。用 2~3.2 MeV 的质子将 Ni 含量为 8%~32%(质量分数)、Cr 含量为 15%~22%(质量分数)的奥氏体不锈钢在 360℃下辐照至 1~5dpa,最大应变为 3%,其中 SCC 试验是在 288℃条件下模拟 BWR 水环境进行的慢应变试验。通道高度的加权平均值被认为是用来表征局部变形的 IASCC 相关强度的最合适因素,包含了叠加的断层能量、硬度和 RIS。据推测,局部变形开裂是通过晶界滑移通道中位错间强烈的相互作用导致的。

Gupta 等(2016)在 450℃条件下用 Fe 辐照,对 304L 奥氏体不锈钢的 SCC 进行了研究。用 10MeV 的 Fe 辐照至 5dpa 后,用滑移线间距来量化微观结构的损伤,其滑移线间距是未辐照样品的 1.8 倍。在压水堆环境中,应变至 4%的铁辐照材料在裂纹尺寸和密度方面比未辐照的材料表现出更严重的 SCC。在惰性 Ar 环境中产生了 4%塑性应变的铁辐照材料没有出现任何裂纹,这也证实了局部变形本身不足以引发裂纹,SCC 的产生需要腐蚀性环境。

## 3.4 蠕　　变

### 3.4.1 介绍



蠕变性能是设计高温系统需要考虑的基本力学性能,蠕变断裂强度是选择 SCWR 燃料包壳材料的关键指标。设计要求通常以法规、标准和规范的形式给出。例如,美国机械工程师协会(ASME)规范第Ⅲ节第 NH 小节(ASME,2006)概述了高温应用中一类部件的设计应力要求,规范中(ASME,2006)要求设计应力小于极限抗拉强度的 1/3 或屈服强度的 2/3。但不锈钢和镍基合金除外,它们的屈服强度可达到 90%。随着超临界化石燃料电厂在超临界(USC)条件下的投入运行,抗蠕变合金的开发一直是超临界化石燃料电厂研发项目中的重点(Viswanathan 和 Bakker,2000)。在中子辐照条件下的蠕变比纯热蠕变发生得更早,其速度也更快。辐照蠕变与纯热蠕变的一个明显区别是,辐照蠕变在微观结构水平上是“无损伤”的,因为它的产生不会导致三点裂纹形成、晶界孔隙及马氏体的产生(Garner,2012)。从这个意义上说,辐照蠕变是有益的,因为它以一种类似于低温退火的方式缓解了外部施加的应力和内部产生的应力。

当使用温度 $T$ 高于 $0.3T_m$ 时,会产生纯热蠕变。典型的蠕变破坏可分为 3 个阶段(图 3.8),即初始阶段(第一阶段)、稳定阶段(第二阶段)和快速阶段(第三阶段)。在初始阶段,初始蠕变以较高的应变率发生,但随着时间的推移,由于加工硬化,使得变形速度变慢。当变形速率降至最低并最终趋于稳定时,就是第二阶段开始的地方。这个阶段在加工硬化和恢复之间存在一个平衡。高温元件的大部分使用寿命都在第二阶段。在使用后期,也就是第三阶段,微孔隙的积累(使用 TEM 经常可以观察到)和缩颈的发生导致应变率随时间呈指数增长。同时,合金的强度会很快降低。

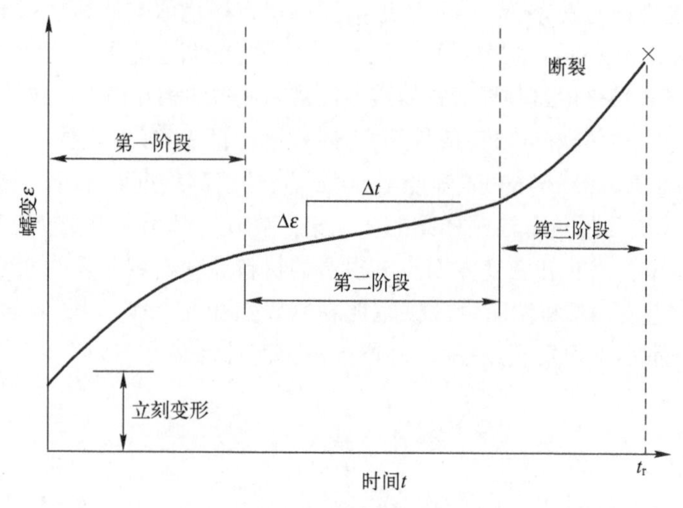


图 3.8　蠕变变形阶段



需要指出的是,尽管大多数金属和合金的蠕变是在高温条件下产生的现象,某些金属(如铅、镁、锡)即使在室温下也会发生蠕变。

根据温度范围和应力水平的不同,蠕变过程涉及不同的机制。其中包括以下几个。

(1) 位错蠕变。这是低温和高压力情况下的主要过程。滑移平面上的位置滑动在低温下很容易发生,但速率决定步骤是位错上升过程,而位错上升过程与受温度影响的空穴扩散运动有关。

(2) 晶界滑动。在剪切应力作用下,晶界会相互滑动,产生楔形裂纹和孔隙。认为晶界区域在 $0.6T_m$ 以上的抗剪强度低于晶内。

(3) 扩散流。随着原子沿合金扩散路径上的热激活扩散,原子从晶粒的侧面移动到顶部和底部,沿拉伸轴的晶粒变长。涉及晶界区原子的过程称为 Coble 蠕变(Coble,1963)。当扩散路径在晶粒内部时,这个过程称为 Nabarro-Herring 蠕变。后者通常存在于低应力、高温环境下的小晶粒合金中。

图 3.9 所示为 625℃ 下 P91 钢随时间变化的蠕变应变的测量实例。需要注意的是,当施加的应力大于 120MPa 时,应变率随应力迅速增大。这种合金通常

用于温度低于650℃的场所。

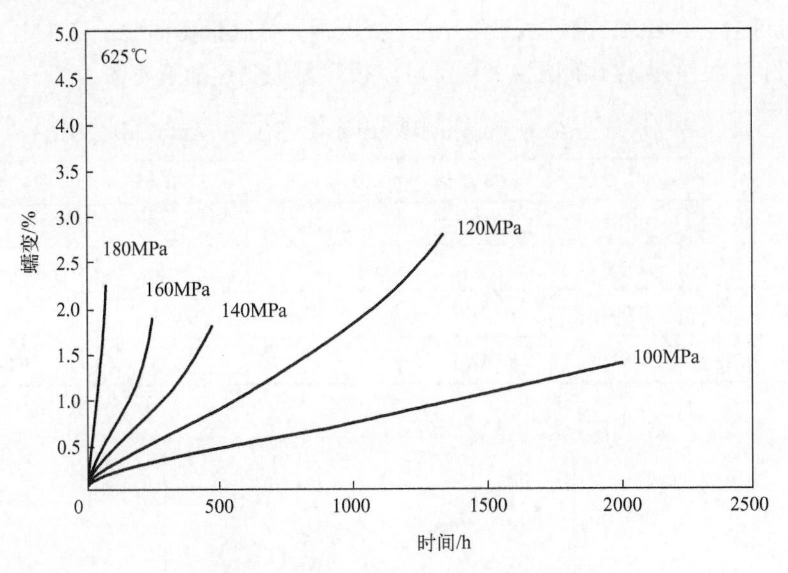


图 3.9　P91 钢在 625℃下的蠕变曲线

## 3.4.2 蠕变预测

Larson-Miller 参数(LMP)法(Larson 和 Miller,1952)是通过假定蠕变速率在给定应力下遵循 Arrhenius 温度关系外推蠕变数据的一种传统方法。将给定应力条件下的 LMP 定义为温度和断裂时间的函数,即

$$\mathrm{LMP}=T\cdot(\log t_r+C) \tag{3.5}$$

式中:$T$ 为温度(K);$t_r$为蠕变破裂寿命(h);$C$ 为 LMP 常数。

已经证明,可以用 LMP 方法较好地预测加拿大 SCWR 中考虑的备选不锈钢的 $t_r$(Xu 等, 2016b)。

Xu 和 Amirkhiz(2016)使用 Conway(1969)描述的拟合过程,获得了 5 种合金的 LMP 应力表达式,这些合金在加拿大 SCWR 中被广泛采用。试验数据拟合为关于 $\log\sigma$ 的多项式函数,即

$$\mathrm{LMP}=C_1+C_2\cdot\log\sigma+C_3\cdot(\log\sigma)^2 \tag{3.6}$$

式中:$\sigma$ 为应力(Pa);$C_1$、$C_2$、$C_3$为拟合常数。

这使得应力可以结合式(3.5)和式(3.6)转化为关于 LMP 的函数,并转化为关于特定使用时间内的温度的函数。

LMP 蠕变预测模型中的常数项(表 3.4)是通过对数据库中收集的数据进行拟合得到的,采用正态或对数正态分布的最小二乘回归法。以长期的试验数

据为基础,对试验数据的拟合优度进行了可视化评估。图 3.10 所示为 800H 合金和 625 合金的 LMP 图。对于给定的应力水平,可以根据曲线确定 LMP 的值。对于任何给定的温度(单位为 K),都可以很容易地确定断裂寿命。

表 3.4 所选合金的 Larson-Miller 参数(Xu 和 Amirkhiz,2016)

| 常　数 | 800H 合金 | 625 合金 | 310H SS | 370H SS | 214 合金 |
|---|---|---|---|---|---|
| $C$ | 14.99 | 23.9 | 9.74 | 11.05 | 22.71 |
| $C_1$ | 2589 | 3270 | 1252 | 1407 | 100.273 |
| $C_2$ | 10004 | 13402 | 7190 | 7699 | -12939 |
| $C_3$ | -1011 | -1271 | -727 | -763 | 483 |

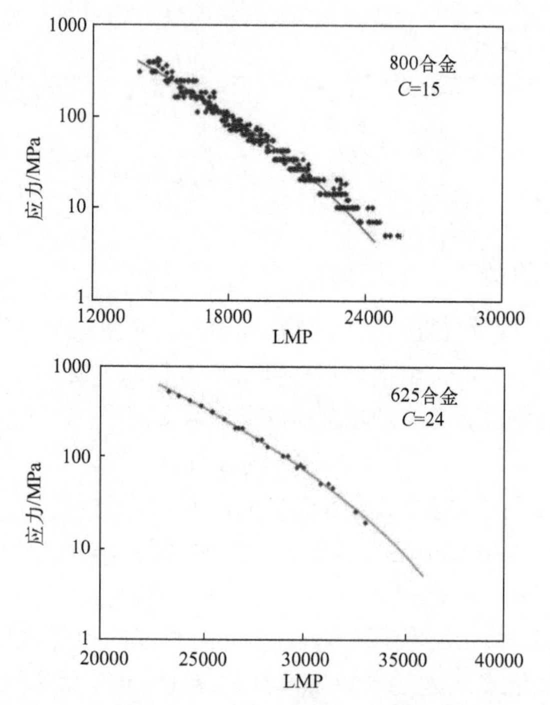


图 3.10 800H 合金和 625 合金的 Larson-Miller 参数(LMP)图($C$ 为 LMP 常数(式(3.5)))
(引自文献 Xu,S.,Zheng,W.,Yang,L.,2016b. A review of irradiation effects on mechanical properties of candidate SCWR fuel cladding alloys for design considera -tions. CNL Nucl. Rev. 5 (2),309-331. https://doi.org/10.12943/CNR.2016.00036)

利用 LMP 方法,Ukai 绘制了 20%冷加工的 316 SS 在反应堆内的蠕变断裂寿命,如图 3.11 所示(Ukai 等,2000)。其在反应器内受到的蠕变断裂强度低于在空气(大约为空气中的 70%~80%)和钠中的蠕变断裂强度。

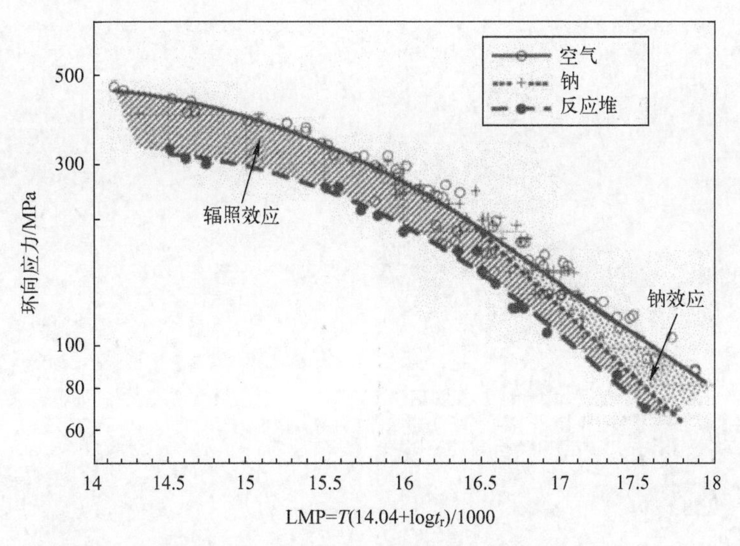


图 3.11 在空气、钠和反应堆条件下的蠕变破裂曲线（$20\times10^{26}$个/$m^2$($E$>0.1MeV)作为 Larson-Miller 参数的函数）（Ukai 等，2000）

对于燃料包壳外径为 8mm、壁厚约 0.6mm 的 SCWR 压力容器。Zhang 等（2012）估计，如果燃料棒没有内部加压，SCW 管壁在 25MPa 时受到的环应力约为 160MPa。与在 650℃条件下 SCWR 合金的屈服强度和蠕变强度相比，该应力是相当高的。

He 对燃料棒的内部加压显著降低了包壳的环应力，但是在冷却剂损失事故中，由于堆芯完全减压，燃料包壳必须承受来自内部氦气的压力和燃料释放的气体裂变产物的压力。Zhang 等（2012）提出了一种折中的方案，也就是假设初始 He 压力约为 4MPa。在正常工作条件下和完全减压（假设最大燃料棒温度不超过 650℃）时，此方案可将包壳上的环应力降低至约 85MPa。

TEM 研究了在 850℃条件下进行蠕变试验的 347H 试样，结果表明，合金中的 NbC 析出物对蠕变变形有一定的抑制作用（Shalchi Amirkhiz 和 Xu，2014）。在经过 1244h 蠕变的样品中，发现 NbC 粒子沿着[111]滑移线定向分布。此外，如图 3.12 所示，通过定住位错，可以防止位错的恢复和亚晶界的形成；否则将促进变形。通过这一机制，这些颗粒可以减缓蠕变变形，增强材料的蠕变寿命。在 1244h 的蠕变试样中没有观察到 σ 相或 $M_{23}C_6$ 碳化物相。在测试最低的蠕变应力水平（20.4MPa）下，观察到晶粒中有魏氏体形态的 $Cr_2N$ 相析出，晶界处有等轴 $Cr_2N$ 大颗粒。$Cr_2N$ 相在 NbC 沉淀上成核。

由于对堆内加载试样的蠕变进行测量存在困难，许多辐照蠕变研究是在空气中对未辐照样本进行辐照后进行的（在辐照期间没有施加应力）。这两种不

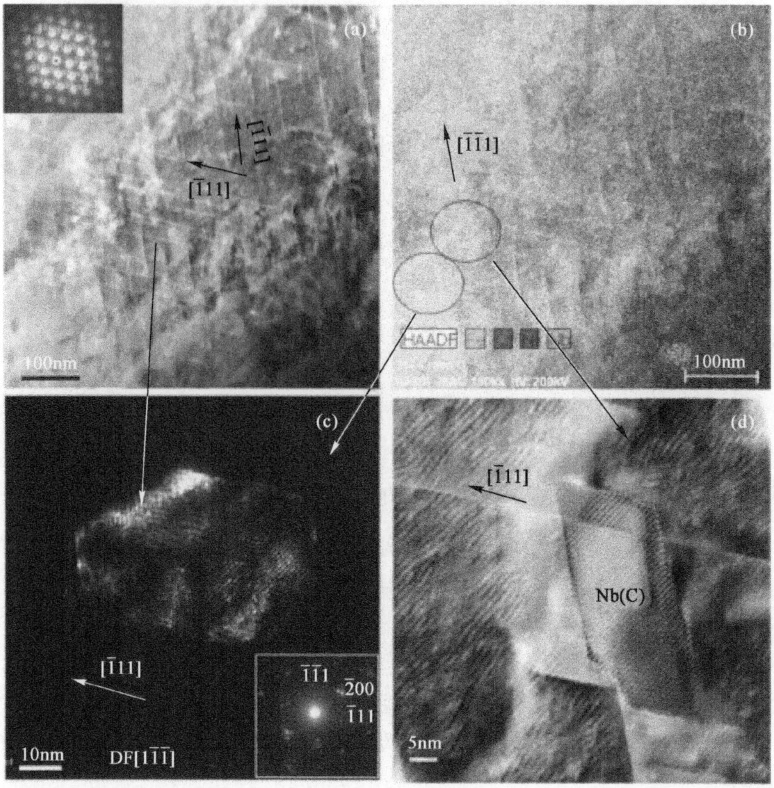

图 3.12 (a) 高角度环形暗场(HAADF)图像和会聚束电子衍射(CBED)插图;(b) 能谱图;(c) 反射暗场 TEM 图像;(d) 高分辨率 STEM-HAADF 图像(Shalchi Amirkhiz 和 Xu,2014)。

同方法测量的蠕变速率有很大差异。从文献中获得的辐照蠕变数据,通常代表了辐照对随时间变化的形变产生的整体效应,但也适用于针对设计目标的辐照效应的检验。值得注意的是,文献中的许多辐照数据都是在空气中辐照后的检测数据。在空气中辐照后的试验结果可能不同于堆内辐照蠕变试验结果(即在实际辐照、应力和温度条件下测量的蠕变应变),但在空气中辐照后的测试比堆内测试更有优势,因为可以对样本施加负荷和应变并进行准确测量,而且此测试比堆内测试更方便可用。

### 3.4.3 辐照蠕变

许多辐照蠕变机制已经被提出,并通过了试验和理论上的检验,且已经得到了广泛的论证(Simonen,1980;Garner,2006;Was,2007)。Matthews 和 Finnis(1988)总结了辐照蠕变机制,包括简单的辐照增强扩散模型,应力诱导的间隙环成核,应力诱导的位错点缺陷吸收以及攀升强化的滑移机制和强化恢复机制。

对于稳态辐照蠕变,其变形可以表示为辐照蠕变与膨胀的总和,下式为广义公式(Anderoglu 等,2012,Foster 等,1972),即

$$\frac{\varepsilon_{\text{creep}}}{\sigma^n}=B_0+D\dot{S} \tag{3.7}$$

式中:$\varepsilon_{\text{creep}}$为以蠕变率作为位移损伤(dpa)的函数;$\sigma$ 为压力;$n$ 为压力指数;$B_0$ 为蠕变柔量;$\dot{S}$ 为膨胀率随 dpa 的变化率;$D$ 为蠕变-膨胀耦合系数。

有时用蠕变柔量来比较辐照对蠕变的依赖关系。

在堆内的蠕变试验中,当累积的 dpa 达到一定水平时,孔隙膨胀、蠕变和沉淀相关的应变开始耦合在一起。在孔隙膨胀尺寸变化可测量之前,可以通过蠕变率的跳跃来检测试样的膨胀(Garner,2012)。Garner(2012)提出了一个观点,在热中子能谱的情况下,额外的蠕变变形可能与大量的 He 和 H 的转化有关,这可能导致气泡和孔隙的形成,并加速蠕变变形。

Hazel 等(1965)报告,燃料和反应堆功率等级的组合,对 ESH-1 和 ESH-1.1 燃料试验 ESADE-VBWR 过热蒸汽回路中使用的环形燃料元件施加了高应力,导致几乎所有测试包壳产生可测蠕变。得到主要的结论是,在测试条件下(540~700℃,106~135 天的辐照时间),没有一个元件可以被认为是有独立包壳的,并且燃料包壳会起皱。800 合金在尺寸上的稳定性优于 600 合金。

Brinkman 等(1973)报道了辐照处理后的 304 SS、304L(钛改性的)SS、316 SS 和 800 合金的蠕变疲劳研究,积分通量在(0.4~5)×$10^{22}$个/$\text{cm}^2$,温度为 700~750℃时 $E$>0.1MeV,试验的温度为 700℃。辐照合金的抗拉延展性和疲劳寿命较辐照前的显著降低,304 SS、304L SS 和 316 SS 降低为原来的$\frac{2}{3}$~$\frac{1}{5}$,800 合金降低为原来的$\frac{1}{35}$。日本 GIF 协会在日本的材料测试反应堆和 Joyo 试验快堆中,控制温度为 700℃对一些备选合金(316 SS、310 SS 的几个改进型和 690 合金)进行了辐照后的检测(Higuchi 等,2007)。他们指出,蠕变变形主要是热效应导致的,而不是 700℃时的辐照造成的。在 600℃条件下,Joyo 反应器辐照材料中可以观察到孔隙的形成,但认为其形成对构件变形的影响是可以接受的。

## 3.5 微观结构的不稳定

SCWR 设计的测试大多是在加工状态下使用已处理材料(通常是经过磨退火的)进行的,没有考虑在使用寿命期内同时暴露于高温和辐射中可能发生的

微观结构不稳定性。热老化损伤的影响直到最近才被确认是可能影响 SCWR 所用合金耐蚀性的一个因素(Li 等,2014;Jiao 等,2015)(5.3.6 小节)。Mo 等(2010)在开发超高温反应堆的背景下讨论了 617 合金和 230 合金的高温时效效应。Wang 等(2017)研究了一系列用于 SCWR 的 Fe-18Ni-12Cr-铝基-成形合金的长期时效效应。试样经受了 1000h、700℃条件,并测定了其组织稳定性和蠕变性能。随着时效时间的增加,合金的抗拉强度增加,这是由于析出相的强化作用。

由于缺陷汇处的 RIS,在这个温度范围内可能会形成各种辐照诱发的沉淀物。例如,一些扩散缓慢的元素(Ni、Si)通过间隙机制在金属内部移动,可以在晶界等下陷处富集,而在晶界等下陷处或其附近的辐照可以使移动更快的元素(Cr、Mn、Mo)耗尽。辐照温度、剂量和剂量率是影响 RIS 的主要变量。已经有一些关于奥氏体 Fe-Cr-Ni 合金在高温下微观组织不稳定性的综述(如 Weiss 和 Stickler,1972;Sourmail,2001)。对于辐照,主要是在 LWR 应用的背景下(Kenik 和 Busby,2012;Maziasz,1993)。Jin 等(2013)最近报道了离子辐照在奥氏体不锈钢中诱导沉淀 $Cr_{23}C_6$的研究结果。

## 3.5.1 高温引起的微观结构不稳定



高温下的微观结构不稳定性包括受温度/时间影响的第二相的成核和生长,通常首先发生在晶界上,然后发生在晶内。该过程通常以时间-温度沉淀图的形式表示。一般来说,因为会形成富含 Cr 的碳化物,如 $M_{23}C_6$,短期的暴露会使微观结构更敏感化。这些碳化物在长时间的暴露下可以分解形成 $M_6C$ 碳化物,这是一种由不同物质组成且稳定性与其他成分紧密相关的相。当加入 Ti、Nb 或 Zr 作为一种微量合金元素来“稳定”微观结构,以防止与晶间腐蚀有关的富铬碳化物的形成时,Ti、Nb、Zr 碳化物的形成优于富铬碳化物的形成。富含溶质合金元素(Cr、Ni 和 Mo)的更复杂的金属化合相,如 $\chi$、$\eta$ 和 $\sigma$ 相,需要更长的时间才能形成。$\chi$ 相和 $\eta$ 相中含有 Mo,含 Mo 的奥氏体 Fe-Cr-Ni 合金更受关注。$\chi$ 相需要的形成时间最短,而 $\sigma$ 相需要的形成时间最长。例如,在 700℃条件下,固溶退火的 316L SS 经过约 5h 后形成 $\chi$ 相,而 $\sigma$ 相的析出则需要超过 100h(Weiss 和 Stickler,1972)。$\chi$ 相的形成通常与碳化物的辐射效应和力学性能有关,但也可以独立形成。与 $\chi$ 相和 $\sigma$ 相比,$\eta$ 相主要在晶粒内形成,偶尔也在晶界上形成(Weiss 和 Stickler,1972)。

## 3.5.2 由于辐照形成的沉淀

在 3.3 节中介绍的 RIS 与受到辐照后材料中沉淀的形成有关。伴随着持续的辐照,缺陷处的 RIS 会随着剂量的增加而积累,从而达到溶解的程度,并在足够的驱动力作用下形成沉淀相。沉淀物一般分为 3 种,即辐照增强/阻碍、辐照修正或辐照诱导(Lee 等,1981)。前两种类型的沉淀只与高温有关,辐射会影响成核、生长速度和化学性质。第三种类型包括在高温下不常见的沉淀物(图 3.13)。

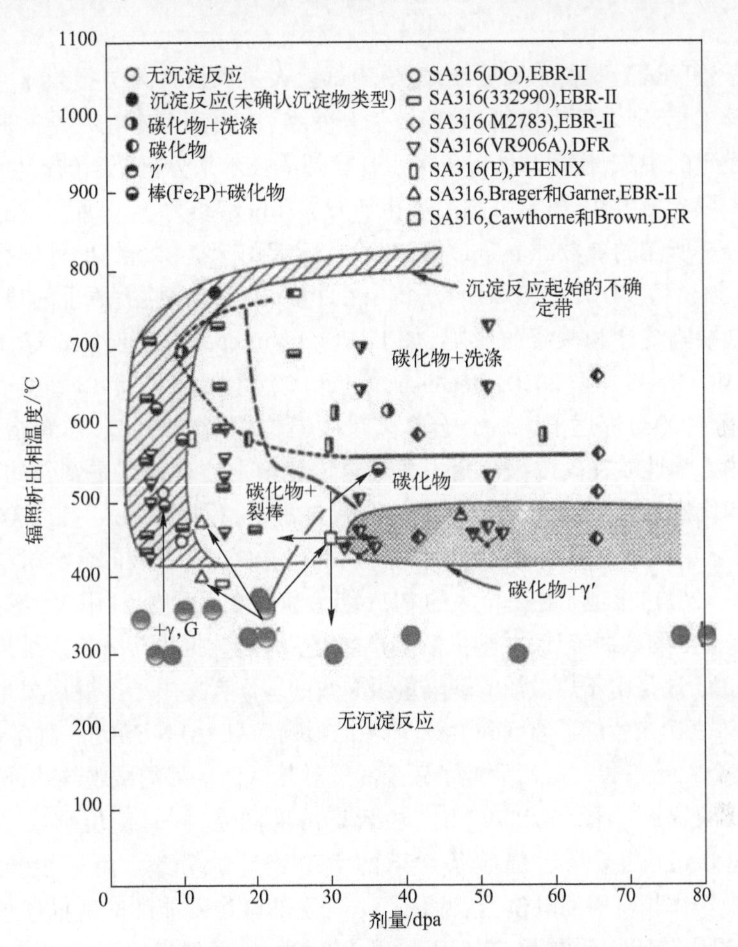


图 3.13　辐照后观察到的析出相随温度和剂量的变化(部分阴影数据点用于 γ′阶段,而固体数据点用于 G 及相关阶段或未知阶段)

(引自文献 Kenik, E. A., Busby, J. T., 2012. Radiation-induced degradation of stainless steel light water reactor internals. Mater. Sci. Eng. R. 73, 67, Kenik 等人首次把圆点添加到最初由以下作者制作的黑白图中, Maziasz, P. J., McHargue, C. J., 1987. Microstructural evolution in annealed austenitic steels by neutron irradiation. Int. Mater. Rev. 2, 190)

Kenik 和 Busby(2012)总结了受中子辐照后 316 SS 中可以形成的析出物类型。不锈钢的冷处理会影响其析出过程。例如,冷处理产生的位错最初是主要的缺陷，它吸引来自 RIS 过程的 Ni 和 Si，随后这些元素的扩散会导致 $\gamma'$ 相的形成。

## 3.6 模　型

由于辐照试验既费时又费钱，建模为试验人员和设计人员提供了一个有价值的指导工具，即多尺度模型，它正处于开发中，该模型考虑在一定时间和长度范围内发生的现象，使用从头计算和 MD 模拟等技术来分析缺陷的产生，还有利用蒙特卡罗模拟方法，研究缺陷的演化动力学和位错动力学，进而计算连续体和有限元分析所需的参数（Samaras 等，2009），后者用于提供组件的性能数据。对于 α-Fe 损伤的 MD 位移级联研究，集中在 PKA 能量与温度、原子间势、溶质添加对合金缺陷统计和聚集的影响，如 Fe-Cr（Virtler 等，2008；Deo 等，2007）和 Fe-Ni（Masterakos 等，2010；Beamish 等，2010）对缺陷统计和聚集的影响。Malerba 等（2008）综述了 Fe-Cr 合金多尺度模拟的最新进展。作者指出，辐照响应和热力学性质在高浓缩合金中尤其难以建模。这种模型是倒叙的，因为几乎所有四代概念反应堆中，辐照和温度条件基本是相同的（Fazio 等，2009）。

很少有专门针对合金的建模工作，特别是针对 SCWR 发展的合金材料。Boyle 和 Shabib（2011）报道了在一个大的 PKA 能量和温度范围内，利用 MD 模拟 α-Fe 位移级联缺陷的数据统计，并根据缺陷产生率（$\eta$）对数据进行分析。结果表明，$\eta$ 与温度有关，这是由于热激活的缺陷重组导致的。作者提出了扩展后关于 $\eta$ 的幂律形式，该表达式考虑了温度和 PKA 能量的影响。AL-6XN SS 是一种备选的 SCWR 硅合金，Zheng 等（2017）研究了残余的氢对其试件缺陷的微观结构的影响，通过将较低的辐照能量（100keV）$H^{2+}$ 注入高浓度的氢，同时使位移损伤剂量达 7dpa。利用 MD 模拟揭示辐照条件下的大位错环的分裂。由于没有面心立方（fcc）奥氏体相可用的电位，且为 bcc Fe 开发的电位不能准确地描述包含 1/3 <111>断层位错环的系统，作者使用了 fcc Ni 的电位，并发现这并不理想。选择电位并进行测量，发现在 10ps 之后 1/3 <111>环的结构被分割成与测量值一致的小位错环。在一项并行研究中，Yu 等（2017）提出了一种速率理论来计算这些位错环的平均尺寸和密度演化。在 550～900K 的辐照温度和 15dpa 的辐照剂量下，计算缺陷簇的尺寸和密度，模拟结果与 563K 条件下的试验结果比较吻合。



## 参考文献

Allen, T. R. , Busby, J. T. , Gan, J. , Kenik, E. A. , Was, G. S. , 2000. Correlation between swelling and radiation-induced segregation in iron-chromium-nickel alloys. In: Hamilton, M. L., Kumar, A. S., Rosinski, S. T., Grossbeck, M. L. (Eds.), 19th International Symposium: Effects of Radiation on Materials, No. 1366. ASTM Special Technical Publication, pp. 739-755.

Allen, T. R., Tsai, H., Cole, J. I., Ohta, J., Dohi, K., Kusanagi, H., 2002. Properties of 20% coldworked 316 stainless steel irradiated at low dose rate. In: ASTM Radiation Effects on Materials Conferences, Tucson Arizona, June 18-20, 2002.

Allen, T. R., Cole, J. I., Trybus, C. L., Porter, D. L., Tsai, H., Garner, F., Kenik, E. A., Yoshitake, T., Ohta, J., 2006. The effect of dose rate on the response of austenitic stainless steels to neutron radiation. J. Nucl. Mater. 348 (1-2), 148-164.

Anderoglu, O., Byun, T. - S., Toloczko, M., Maloy, S. A., 2012. Mechanical performance of ferritic martensitic steels for high dose applications in advanced nuclear reactors. Metall. Mater. Trans. 44A, S70-S83.

ASME, 2006. Class 1 Components in Elevated Temperature Service. ASME Boiler & Pressure Vessel Code, Section III, Subsection NH, New York, NY, USA.

ASTM, 2009. Standard Practice for Neutron Radiation Damage Simulation by Charged-Particle Irradiation. American Society for Testing and Materials, Philadelphia, 1996, reapproved 2009.

Averback, R. S., de la Rubia, T., 1997. Displacement damage in irradiated metals and semiconductors. Solid State Phys. 51, 281-402.

Ayrault, G., Hoff, H. A., Nolfi Jr., F. V., Turner, A. P. L., 1981. Influence of helium injection rate on the microstructure of dual-ion irradiated type 316 stainless steel. J. Nucl. Mater. 104, 1035-1039.

Beamish, E., Campa, C., Woo, T. K., 2010. Grain boundary sliding in irradiated stressed Fe-Ni bicrystals: a molecular dynamics study. J. Phys. Condens. Matter 22 (34), 345006.

Boothby, R. M., 2012. Radiation effects in nickel-based alloys. In: Konings, R. J. M. (Ed.), Comprehensive Nuclear Materials. Elsevier, Oxford, ISBN 9780080560335, pp. 123-150. https://doi.org/10.1016/B978-0-08-056033-5.00090-2.

Boyle, K. P., Shabib, I., 2011. Temperature dependence of defect production efficiency in α-Fe. In: 5th International Symposium on Supercritical Water Cooled Reactors, Canadian Nuclear Society. Paper P116.

Brinkman, C., Korth, G. E., Beeston, J. M., 1973. Influence of irradiation on the creep/fatigue behavior of several austenitic stainless steels and Incoloy 800 at 700 C. In: Effects of Radiation on Substructure and Mechanical Properties of Metals and Alloys, ASTM STP 529, American Society for Testing and Materials, pp. 473-492.

Broeders, C. H. M., Konobeyev, A. Y., 2004. Defect production efficiency in metals under neutron irradiation. J. Nucl. Mater. 328, 197-214.

Bruemmer, S. M., Simonen, E. P., Scott, P. M., Andresen, P. L., Was, G. S., Nelson, J. L., 1999. Radiation-induced material changes and susceptibility to intergranular failure of lightwater-reactor core internals. J. Nucl. Mater. 274, 299-314.

Bruemmer, S. M., 2001. In: Proc. 10th Intl. Symp. on Environmental Degradation of Materials in Nuclear Power Systems-Water Reactor. NACE, Houston, TX. Paper No. 0008V.

Busby, J. T., Was, G. S., Kenik, E. A., 2002. Isolating the effect of radiation-induced segregation in irradiation-assisted stress corrosion cracking of austenitic stainless steels. J. Nucl. Mater. 302, 20-40.

Bysice, S., Pencer, J., Walters, L., Bromley, B., 2016. Comparison of dpa and helium production in candidate fule cladding materials for the Canadian SCWR. CNL Nuc. Rev. 5, 269-275.

Byun, T., Hashimoto, N., 2006. Strain localization in irradiated materials. Nucl. Eng. Technol. 38 (7), 619-638.

Chow, C. -K., Khartabil, H. F., 2008. Conceptual fuel channel designs for CANDU-SCWR. Nucl. Eng. Technol. 40, 139-146.

Chung, H. M., 2006. Assessment of Void Swelling in Austenitic Stainless Steel PWR Core Internals. Argonne National Laboratory Technical report ANL-04/28 TRN: US1006107.

Coble, R. L., 1963. A model for boundary diffusion controlled creep in polycrystalline materials. J. Appl. Phys. 34, 1679.

Comprelli, F. A., Busboom, H. J., Spalrais, C. N., 1969. Comparison of radiation damage studies and fuel cladding performance for Incoly-800. In: Irradiation Effects in Structural Alloys for Thermal and Fast Reactors, ASTM STP 457, American Society for Testing and Materials, pp. 400-413.

Conway, J. B., 1969. Stress-Rupture Parameters: Origin, Calculation and Use. Gordon and Breach, New York.

Dai, Y., Odette, G., Yamamoto, T., 2012. The effects of helium in irradiated structural alloys. In: Konings, R. J. M. (Ed.), Comprehensive Nuclear Materials. Elsevier, Oxford, pp. 141-193.

Deo, C. S., Okuniewski, M. A., Srivilliputhur, S. G., Maloy, S. A., Baskes, M. I., James, M. R., Stubbins, J. F., 2007. The effects of helium on irradiation damage in single crystal iron. J. Nucl. Mater. 367-370A, 451-456.

Erlich, K., Konys, J., Heikinheimo, L., 2004. Materials for high performance light water reactors. J. Nucl. Mater. 327, 140.



Fazio, C., Alamo, A., Almazouzi, A., De Grandis, S., Gomez-Briceno, D., Henry, J., Malerba, L., Rieth, M., 2009. European cross-cutting research on structural materials for generation IV and transmutation systems. J. Nucl. Mater. 392, 316-323.

Foster, J. P., Wolfer, W. G., Biancheria, A., Boltax, A., 1972. Analysis of irradiation-induced creep of stainless steel in fast spectrum reactors. In: Proc. European Conference on Irradiation Embrittlement and Creep in Fuel Cladding, British Nuclear Society, London, pp. 273-281.

Freyer, P. D., Mager, T. R., Burke, M. A., 2007. Hot cell crack initiation testing of various heats of highly irradiated 315 stainless steel component obtained from three commercial PWRs. In: Proceedings of the 13th International Conference on Environmental Degradation of Materials in Nuclear Power Systems, Whistler, British Columbia, Canada, April 19-23, 2007.

Garner, F. A., Brager, H. R., 1985. Dependence of neutron-induced swelling on composition in iron-based austenitic alloys. In: Garner, F. A., Gelles, D. S., Witten, F. W. (Eds.), Optimizing Materials for Nuclear Applications. The Metallurgical Society of AIME, pp. 87-109.

Garner, F. A., Brager, H. R., 1988. The influence of Mo, Si, P, C, Ti, Cr, Zr and various trace elements on the neutron-induced swelling of AISI 316 stainless steel. J. Nucl. Mater. 155-157, 833-837.

Garner, F. A., Kumar, A. S., 1987. The influence of both major and minor element composition on void swelling in austenitic steels. In: Garner, F. A., Packan, N. H., Kumar, A. S. (Eds.), Radiation-Induced Changes in Microstructure: 13th International Symposium (Part I). ASTM Special Technical Publication, pp. 289-314.

Garner, F. A., Brager, H. R., Gelles, D. S., McCarthy, J. M., 1987. Neutron irradiation of Fe-Mn, Fe-Cr-

Mn and Fe-Cr-Ni alloys and an explanation of their differences in swelling behavior. J. Nucl. Mater. 148, 294-301.

Garner, F. A., 1984. Recent insights on the swelling and creep of irradiated austenitic alloys. J. Nucl. Mater. 122 (1-3), 459-471.

Garner, F. A., Gelles, D. S., 1990. Proceedings of Symposium on Effects of Radiation on Materials: 14th International Symposium, ASTM STP 1046, vol. II, pp. 673-683.

Garner, F. A., 2006. Irradiation performance of cladding and structural steels in liquid metal reactors. In: Frost, B. R. T. (Ed.), Materials Science and Technology: A Comprehensive Treatment. VCH Publishers, pp. 419-543 (Chapter: 10A).

Garner, F. A., 2012. Radiation damage in austenitic steels. In: Konings, R. J. M. (Ed.), Comprehensive Nuclear Materials. Elsevier, Oxford, ISBN 9780080560335, pp. 33-95. https://doi.org/10.1016/B978-0-08-056033-5.00065-3.

Gessel, G. R., Rowcliffe, A. F., 1977. The effect of solute additions on the swelling of an Fe-7.5Cr-20Ni alloy. In: Bleiberg, L., Bennett, J. W. (Eds.), Radiation Effects in Breeder Reactor Structural Materials. Metallurgical Society of AIME, pp. 431-442.

Gittus, J., 1978. Irradiation Effects in Crystalline Solids. Elsevier Applied Science Publishers, Limited.

Greenwood, L., Oliver, B. M., 2006. Comparison of predicted and measured helium productionin U. S. BWR reactors. J. ASTM Int. 3 (3), 1-9.

Greenwood, L. R., 1983. A new calculation of thermal neutron damage and helium production in nickel. J. Nucl. Mater. 115, 137-142.

Gupta, J., Hure, J., Tanguy, B., Laffont, L., Lafont, M.-C., Andrieu, E., 2016. Evaluation of stress corrosion cracking of irradiated 304L stainless steel in PWR environment using heavy ion irradiation. J. Nucl. Mater. 476, 82-92.

Hardie, C. D., Williams, C. A., Xu, S., Roberts, S. G., 2013. Effects of irradiation temperature and dose rate on the mechanical properties of self-ion implanted Fe and Fe-Cr alloys. J. Nucl. Mater. 439 (1-3), 33-40.

Harries, D. R., 1979. Neutron irradiation-induced embrittlement in type 316 and other austenitic steels and alloys. J. Nucl. Mater. 82, 2-21.

Hazel, V. E., Boyle, R. F., Busboom, H. J., Murdock, T. B., Skarpelos, J. M., Spalaris, C. N., 1965. Fuel Irradiation in the ESADA-VBWR Nuclear Superheat Loop. General Electric Atomic Power Report GEAP-4775.

Herschbach, K., Schneider, W., Ehrlich, K., 1993. Effects of minor alloying elements upon swelling and in-pile creep in model plain Fe-15Cr-15Ni stainless steels and in commercial DIN 1.4970 alloys. J. Nucl. Mater. 203, 233-248.

Higuchi, S., Sakurai, S., Ishida, T., 2007. A study of fuel behavior in an SCWR core with high power density. In: Proc. ICAPP'07, Nice, France, May 13-18, 2007. Paper 7206.

Hojná, A., 2013. Irradiation-assisted stress corrosion cracking and impact on life extension. Corrosion 69(10), 964-974.

IAEA, 2011. Stress Corrosion Cracking in Light Water Reactors: Good Practices and Lessons Learned. IAEA Nuclear Energy Series No. NP-T-3.13. International Atomic Energy Agency, Vienna.

Idrees, Y., Yao, Z., Sattari, M., Kirk, M. A., Daymond, M. R., 2013. Irradiation induced microstructural changes in Zr-Excel alloy. J. Nucl. Mater. 441, 138-151.

Jiao, Z., Was, G. S., 2011. Impact of localized deformation on IASCC in austenitic stainless steels. J. Nucl. Mater. 408 (3), 246-256.

Jiao, Y., Zheng, W., Guzonas, D. A., Cook, W. G., Kish, J. R., 2015. Effect of thermal treatment on the corrosion resistance of Type 316L stainless steel exposed in supercritical water. J. Nucl. Mater. 464 (2), 356.

Jin, S., Guo, L., Luo, F., Yao, Z., Ma, S., Tang, R., 2013. Ion irradiation-induced precipitation of $Cr_{23}C_6$ at dislocation loops in austenitic steel. Scr. Mater. 68 (2), 138-141.

Johnston, W. G., Rosolowski, J. H., Turkalo, A. M., Lauritzen, T., 1974. An experimental survey of swelling in commercial Fe-Cr-Ni alloys bombarded with 5 MeV Ni ions. J. Nucl. Mater. 54 (1), 24-40.

Johnston, W. G., Lauritzen, T., Rosolowski, J. H., Turkalo, A. M., 1976. Void swelling in fast reactor materials: a metallurgical problem. J. Met. 28 (6), 20-24.

Kenik, E. A., Busby, J. T., 2012. Radiation-induced degradation of stainless steel light water reactor internals. Mater. Sci. Eng. R. 73, 67.

Kinchin, G. H., Pease, R. S., 1955. The displacement of atoms in solids by radiation. Rep. Prog. Phys. 18 (1), 1.

Kiritani, M., 1989. Radiation rate dependence of microstructure evolution. J. Nucl. Mater. 169, 89-94.

Kittel, 1976. Introduction to Solid State Physics. John Wiley and Sons, New York.

Klassen, R. J., Rajakumar, H., 2015. Combined effect of irradiation and temperature on the mechanical strength of Inconel 800H and AISI 310 alloys for in-core components of a GEN-IV SCWR. In: 7th International Symposium on Supercritical Water-cooled Reactors (ISSCWR7), Helsinki, Finland, March 15-18, 2015. Paper ISSCWR7-2006.

Kozlov, A. V., Portnykh, I. A., Bryushkova, S. V., Kinev, E. A., 2004. Dependence of maximum swelling temperature on damage dose in cold worked 16Cr-15Ni-2Mo-1Mn cladding irradiated in BN-600. In: Grossbeck, M. L., Allen, T. R., Lott, R. G., Kumar, A. S. (Eds.), Radiation-induced Changes in Microstructure: 21th International Symposium, No. 1447, pp. 446-453.

Larson, F. R., Miller, J., 1952. A time-temperature relationship for rupture and creep stresses. Trans. ASME 74, 765-775.

Lauritzen, T., Vaidyanathan, S., Bell, W. L., Yang, W. J. S., 1987. Irradiation-induced swelling in AISI 316 steel: effect of tensile and compressive stresses. In: Garner, F. A., Packan, N. H., Kumar, A. S. (Eds.), Radiation-Induced Changes in Microstructure: 13th International Symposium (Part I). ASTM Special Technical Publication, pp. 101-113.

Lee, E. H., Maziasz, P. J., Rowcliffe, A. F., Holland, J. R., et al., 1981. The Metallurgical Society of AIME, Warrendale, PA, pp. 191-218.

Lee, E. H., Byun, T. S., Hunn, J. D., Yoo, M. H., Farrell, K., Mansur, L. K., 2001a. On the origin of deformation microstructures in austenitic stainless steel: Part I- Microstructures. Acta Mater. 49, 3269.

Lee, E. H., Yoo, M. H., Byun, T. S., Hunn, J. D., Farrell, K., Mansur, L. K., 2001b. On the origin of deformation microstructures in austenitic stainless steel: Part II-Mechanisms. Acta Mater. 49, 3277.

Li, J., Zheng, W., Penttila, S., Liu, P., Woo, O. T., Guzonas, D., 2014. Microstructure stability of candidate stainless steels for GEN-IV SCWR fuel cladding application. J. Nucl. Mater. 454, 7.

Malerba, L., Caro, A., Wallnius, J., 2008. Multiscale modelling of radiation damage and phase transformations: the challenge of FeCr alloys. J. Nucl. Mater. 382, 112-125.

Mastorakos, I. N., Le, N., Zeine, M., Zbib, H. M., Khaleel, M., 2010. Multiscale modeling of irradiation induced hardening in α-Fe, Fe-Cr and Fe-Ni systems. Mater. Res. Soc. Symp. Proc. 1264, 259-264.

Matsui, H., Sato, Y., Saito, N., Kano, F., Ooshima, K., Kaneda, J., Moriya, K., Ohtsuka, S., Oka, Y., 2007. Material development for supercritical water-cooled reactors. In: Proc. ICAPP 2007, Nice, France, May 13-18, 2007. Paper 7447.

Matthews, J. R., Finnis, M. W., 1988. Irradiation creep models - an overview. J. Nucl. Mater. 159, 257-285.

Maziasz, P. J., McHargue, C. J., 1987. Microstructural evolution in annealed austenitic steels by neutron irradiation. Int. Mater. Rev. 2, 190.

Maziasz, P. J., 1993. Overview of microstructural evolution in neutron-irradiated austenitic stainless steels. J. Nucl. Mater. 205, 118-145.

Maziasz, P. J., 1993. Swelling and swelling resistance possibilities of austenitic stainless steels in fusion reactors. J. Nucl. Mater. 122 (1-3), 472-486.

Mo, K., Lovicu, G., Tung, H. -M., Chen, X., Stubbins, J. F., 2010. High temperature aging and corrosion study on alloy 617 and alloy 230. In: Proc. Int. Conf. on Nuclear Engineering (ICONE18), May 17-21, 2010, Xi'an, China.

Murphy, S. M., 1987. The influence of helium trapping by vacancies on the behavior of metals under irradiation. In: Garner, F. A., Packan, N. H., Kumar, A. S. (Eds.), Radiation-Induced Changes in Microstructure: 13th International Symposium (Part I). ASTM Special Technical Publication, pp. 330-344.

Nava-Domínguez, A., Onder, N., Rao, Y., Leung, L., 2016. Evolution of the Canadian SCWR fuel-assembly concept and assessment of the 64 element assembly for thermal hydraulic performance. CNL Nuc. Rev. 56, 221-238.

Norgett, M. J., Robinson, M. T., Torrens, I. M., 1975. A proposed method of calculating displacement dose rates. Nucl. Eng. Des. 33, 50-54.

OECD, 2015. Primary Radiation Damage in Materials. NEA/NSC/DOC(2015)9. https://www.oecd-nea.org/science/docs/2015/nsc-doc2015-9.pdf.

Ohnuki, S., Takahashi, H., Nagasaki, R., 1988. Effect of helium on radiation-induced segregation in stainless steel. J. Nucl. Mater. 155, 823-827.

Okita, T., Kamada, T., Sekimura, N., 2000. Effects of dose rate on microstructural evolution and swelling in austenitic steels under irradiation. J. Nucl. Mater. 283-287, 220-223.

Okita, T., Sato, T., Sekimura, N., Garner, F. A., Greenwood, L. R., 2002. The primary origin of dose rate effects on microstructural evolution of austenitic alloys during neutron irradiation. J. Nucl. Mater. 307-311, 322-326.

Pawel, J. P., Ioka, I., Rowcliffe, A. F., Grossbeck, M. L., Jitsukawa, S., 1999. In: In Effects of Radiation on Materials: 18th International Symposium; ASTM STP 1325, pp. 671-688.

Porter, D. L., Takata, M. L., Wood, E. L., 1983. Direct evidence for stress-enhanced swelling in type 316 stainless steel. J. Nucl. Mater. 116, 272-276.

Rowcliffe, A. F., Mansur, L. K., Hoelzer, D. T., Nanstad, R. K., 2009. Perspectives on radiation effects in nickel-base alloys for applications in advanced reactors. J. Nucl. Mater. 392, 341-352.

Samaras, M., Victoria, M., Hoffelner, W., 2009. Advanced materials modeling - E. U. perspectives. J.

Nucl. Mater. 392, 286-291.

Schroeder, H., 1988. High temperature helium embrittlement in austenitic stainless steels-correlations between microstructure and mechanical properties. J. Nucl. Mater. 155-157, 1032-1037.

Scott, P., 1994. A review of irradiation assisted stress corrosion cracking. J. Nucl. Mater. 211, 101-122.

Shalchi Amirkhiz, B., Xu, S., 2014. NbC precipitation and deformation of SS 347H crept at 850℃. Microsc. Microanal. 20 (Suppl. 3). Paper 1494 (Proceeding of Microscopy and Microanalysis Conference, August 3-7, 2014, Hartford, CT, USA).

Simonen, E. P., 1980. Irradiation creep in transient irradiation environments. J. Nucl. Mater 90, 282-289.

Singh, B. N., Leffers, T., Makin, M. J., Walters, G. P., Foreman, A. J. E., 1981. Effects of implanted helium on void nucleation during Hvem irradiation of stainless steel containing silicon. J. Nucl. Mater. 104, 1041-1045.

Sourmail, T., 2001. Precipitation in creep resistant austenitic stainless steels. Mater. Sci. Technol. 17, 1-14.

Stephenson, K., Was, G. S., 2016. The role of dislocation channeling in IASCC initiation of neutron irradiated stainless steel. J. Nucl. Mater 481, 214-225.

Stoller, R. E., 2012. Primary radiation damage formation. In: Konings, R. J. M. (Ed.), Comprehensive Nuclear Materials. Elsevier, Oxford, ISBN 9780080560335, pp. 293-332.

Strasser, A., Santucci, J., Lindquist, K., Yario, W., Stern, G., Goldstein, L., Joseph, L., 1982. An Evaluation of Stainless Steel Cladding for Use in Current Design LWRs. Electric Power Research Institute Report NP-2642.

Thiele, B. A., Diehl, H., Ohly, W., Weber, H., 1984. Investigation into the irradiation behavior of high temperature alloys for high-temperature gas-cooled reactor applications. Nucl. Technol. 66, 597-606.



Ukai, S., Mizuta, S., Kaito, T., Okada, H., 2000. In-reactor creep rupture properties of 20% CW modified 316 stainless steel. J. Nucl. Mater. 278, 320-327.

Virtler, K., Bjorkas, C., Terentyev, D., Malerba, L., Nordlund, K., 2008. The effect of Crconcentration on radiation damage in Fe-Cr alloys. J. Nucl. Mater. 382, 24-30.

Viswanathan, R., Bakker, W. T., 2000. Materials for boilers in ultra supercritical power plants. In: Proc. of 2000 Int. Joint Power Generation Conference, Miami Beach, Florida, July 23-26, 2000. Paper JPGC2000-15049.

Walters, L., Donohue, S., 2016. Development of high performance pressure tube material for Canadian SCWR concept. JOM 68 (2), 490-495.

Walters, L., Wright, M., Guzonas, D., 2017. Irradiation issues and material selection for Canadian SCWR components. In: 8th Int. Symp. on Super-critical Water-cooled Reactors, March 13-15, 2017, Chengdu, China.

Wang, M., Zhou, Z., Jang, J., Sun, S., 2017. Microstructural stability and mechanical properties of AFA steels for SCWR core application. In: 8th Int. Symp. on Super-critical Water-cooled Reactors, March 13-15, 2017, Chengdu, China.

Was, G. S., 2007. Irradiation creep and growth. Fundam. Radiat. Mater. Sci. 71, 17-63.

Was, G. S., Andresen, P. L., 2012. Irradiation Assisted Corrosion and Stress Corrosion Cracking (IAC/IASCC) in Nuclear Reactor Systems and Components. Woodhead Publishing.

Weiss, B., Stickler, R., 1972. Phase instabilities during high temperature exposure of 316 austenitic stainless steel. Metall. Trans. 3, 851-866.

Wolfer, W. G., Garner, F. A., Thomas, L. E., 1982. On radiation-induced segregation and the compositional dependence of swelling in Fe-Ni-Cr alloys. In: Brager, H. R., Perrin, J. S. (Eds.), 11th International Symposium: Effects of Radiation on Materials, 782. ASTM Special Technical Publication, pp. 1023-1041.

Xu, R., Yetisir, M., Hamilton, H., 2014. Thermal-mechanical behavior of fuel element in SCWR design. In: Canada - China Conference on Advanced Reactor Development, Niagara Falls, Ontario Canada, April 27-30, 2014. Paper CCCARD2014-005.

Xu, S., Shalchi Amirkhiz, B., 2016. Mechanical properties of fuel cladding candidate alloys for Canadian SCWR concept. J. Met. 68 (2), 469-474.

Xu, C., Zhang, L., Qian, W., Mei, X. L. J., 2016a. The studies of irradiation hardening of stainless steel reactor internals under proton and xenon irradiation. Nucl. Eng. Technol. 48 (3), 758-764.

Xu, S., Zheng, W., Yang, L., 2016b. A review of irradiation effects on mechanical properties of candidate SCWR fuel cladding alloys for design considerations. CNL Nucl. Rev. 5 (2), 309-331. https://doi.org/10.12943/CNR.2016.00036.

Yetisir, M., Hamilton, H., Xu, R., Gaudet, M., Rhodes, D., King, M., Kittmer, A., Benson, B., 2017. Fuel assembly concept of the Canadian SCWR. In: 8th Int. Symp. on Super-critical Water-cooled Reactors, March 13-15, 2017, Chengdu, China.

Yu, Y., Guo, L., Zheng, Z., Tang, R., Yi, W., 2017. Rate theory studies on dislocation loops evolution in AL-6XN austenitic stainless steel under proton irradiation. In: 8th Int. Symp. on Super-critical Water-cooled Reactors, March 13-15, 2017, Chengdu, China.

Zhang, L., Bao, Y., Tang, R., 2012. Selection and corrosion evaluation tests of candidate SCWR fuel cladding materials. Nucl. Eng. Des. 249, 180-187.

Zheng, Z., Tang, R., Gao, N., Yu, Y., Zhang, W., Shen, Z., Long, Y., Wei, Y., Guo, L., 2017. Evolution of dislocation loops in austenitic stainless steels implanted with high concentration of hydrogen. In: 8th Int. Symp. on Super-critical Water-cooled Reactors, March 13-15, 2017, Chengdu, China.

Zheng, W., Guzonas, D., Boyle, K. P., Li, J., Xu, S., 2016. Materials assessment for the Canadian SCWR core concept. J. Met. 68 (2), 456-462.

Zhou, R., West, E. A., Jiao, Z., Was, G. S., 2009. Irradiation-assisted stress corrosion cracking of austenitic alloys in supercritical water. J. Nucl. Mater. 395, 11-22.

Zinkle, S. J., 2012. Radiation-induced effects on microstructure. In: Konings, R. J. M. (Ed.), Comprehensive Nuclear Materials. Elsevier, Oxford, ISBN 9780080560335, pp. 65-98.

# 第4章 水化学

## 4.1 引言

火力发电厂需要包含以下材料:

(1) 将热量从热源(如化石燃料、核裂变)转移的传热介质;

(2) 传热介质的管道。

水是一种常用的传热介质,它有丰富、便宜、无毒、耐高温、热性能好的优点。然而,水可以与很多材料发生反应,这种反应发生的临界温度条件恰好在高热效率下的最佳运行温度范围内。在任何使用高温水作为传热介质的动力循环中,水化学对材料的腐蚀有重要影响,因此有必要了解在材料腐蚀过程中,水的性质所产生的影响。需要注意的是,对于任何一个超临界水冷反应堆(SCWR),超临界水(SCW)的性质直接影响系统材料的一般腐蚀和局部腐蚀(如腐蚀产生的裂纹)以及腐蚀产物进出堆芯的输运过程。腐蚀产物在管道等构件表面的沉积会影响冷却剂的传热,从而导致金属温度升高,这也会影响材料腐蚀机制,如蠕变。Cohen(1985)和Kritsky(1999)的著作是当代水冷反应堆(WCR)水化学方法基础的优秀参考文献。

### 4.1.1 什么是超临界水

所有的流体①在 $P-T-V$ 的图中都存在一个临界点,该临界点由 $T_c$、$P_c$、$V_c$来定义,在该临界点以上,是不可能通过加压将气体液化的。对于水,$T_c$ = 647.096K、$P_c$ = 22.064MPa、$\rho_c$ = 322kg/m$^3$。在超临界区域,没有导致液体和蒸汽共存的相分离(没有沸腾)。临界过渡区的物理、化学变化一直是人们非常感兴趣的课题,用无量纲热力学量的临界功率定律和临界标度律描述了近临界区流

① 在本章中,流体指的是物质的一种状态,如果不受限制,则不会保持其形状。

体的热力学行为，用 $T_c$、$P_c$、$V_c$ 定义无量纲热力学量。在很大程度上，SCWR① 不需要对临界转变有如此详细的了解，临界标度定律只适用于临界点附近的小范围温度和密度。对这一主题感兴趣的读者会发现，Anisimov 等(2004)的评论是一个很好的介绍。

当 $k_B T$ 超过 $\Delta\varepsilon$ 时，分子总是从势阱中逃出，不可能形成持久的液相。通过考虑流体分子之间的相互作用能($\Delta\varepsilon$)，对临界转变在分子尺度上有了一些深入的了解(图 4.1)。当 $T>\Delta\varepsilon/k_B$ 时，流体分子的平均热能大于分子间的相互作用能($\Delta\varepsilon$)，这时不可能形成持久的液相；当温度 $T_c=T=\Delta\varepsilon/k_B$ 时，短寿命的分子团簇仍然会成形和分裂。而超临界流体由高密度分子簇区域和单个分子的低密度区域组成(图 4.2)。这种分子分布的不均匀性是超临界流体的一个特性。在临界点附近，流体经历了明显的密度波动。

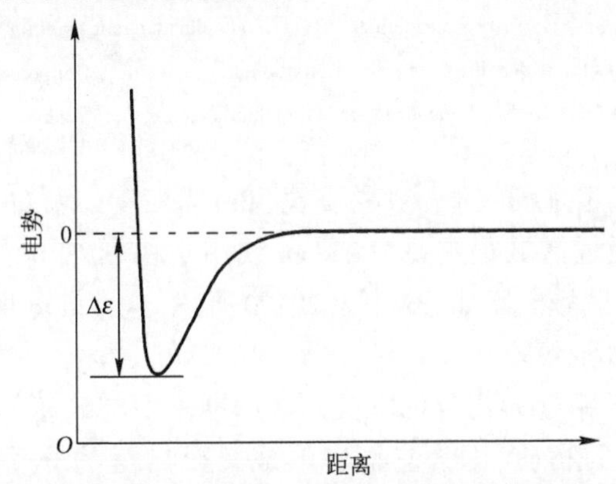


图 4.1　两个相互作用的液体分子的距离与电势关系的简化示意图(当 $k_B T$ 超过 $\Delta\varepsilon$ 时分子总是逃避势阱且没有永久的液相形式)

(引自文献 Tabor, D., 1969. Gases, Liquids and Solids; And Other States of Matter. Cambridge University Press)

然而 SCW 这一术语有时被限制在 $T$–$P$ 相图中 $T>T_c$、$P>P_c$ 满足的区域。考虑将整个区域中 $T>T_c$ 的区域作为 SCW，这更好地强调了高温蒸汽($T>T_c$、$P<P_c$)和高压下 SCW 中的腐蚀现象($T>T_c$、$P>P_c$)之间的联系，因此本书采用了这种划分方法，即认为整个区域中 $T>T_c$ 的区域作为 SCW。这里需要注意的是，各种物理性质(如等温压缩性、热容、声速、热扩散率)都经过临界点附近的极值，这一现象扩展到单相区域，是汽-液相平衡的"延续"②。这些 Widom 线③(图 1.1)随

① 对模型进行开发的例外情况发生在临界点附近条件下。

② 例如，当压力大于 $P_c$ 时，流体的等压热容量达到最大值的温度称为伪临界点(384.87℃、25MPa)。

③ 热力学响应函数(或相矢量)在 $P$–$T$ 空间中的极值轨迹(Imre 等，2012)。这些极值在 $P$–$T$ 空间中很接近但不重合，并随着临界点距离的增加而发散，因此存在一组 Widom 线。

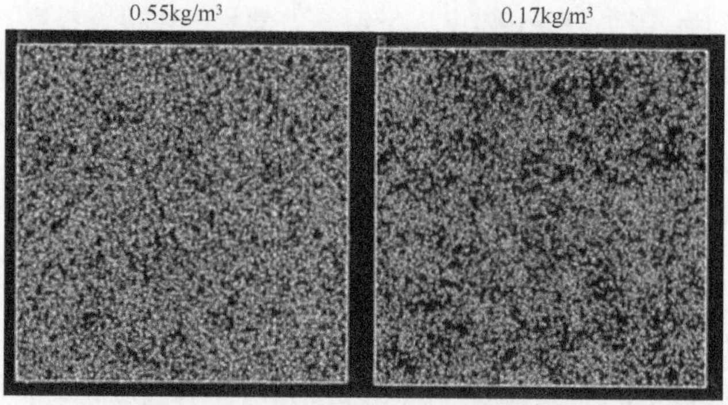

0.55kg/m³　　　0.17kg/m³

图 4.2　密度为 55kg/m³ 和 17kg/m³ 超临界水的分子动力学模拟(显示氧(红色)和氢(白色)原子的三维分布,$T$=400℃)(见彩插)

(引自文献 Metatla N,Jay-Gerin J-P,Soldera A,2011. Molecular dynamics simulation of sub-critical and supercritical water at different densities. In:5th International Symposium on Supercritical Water-cooled Reactors (ISSCWR-5),Vancouver,British Columbia,Canada,March 13-16,2011)

着离临界点的距离增加而幅值减小并变宽(Brazhkin 等,2011;Imre 等,2012),因此只有在临界点附近区域很重要。Widom 线用于将 SCW 流体区域划分为"类液体"和"类气体"区域。正如 Imre 等(2012)所述,这种划分比仅仅基于临界压力的划分更具有物理意义。



与其他分子量相近的化合物以及元素周期表中与氧相邻的氢的化合物相比,水是一种独特的物质,因为它能够形成一个广泛的分子间氢键的三维网络,这导致其性质(沸点、介电常数)与非氢键液体的差异。图 4.3 所示为元素周期表第 2 周期 p 区元素的氢化合物的沸点;随着分子氢键能力的增加,沸点达到最大值。

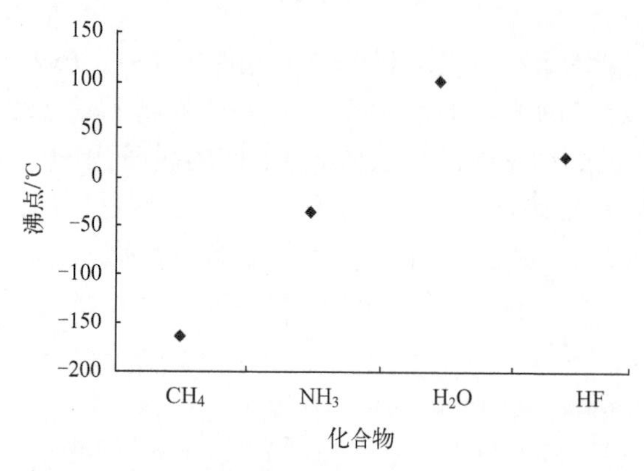


图 4.3　氢化合物和元素周期表第 2 周期 p 区元素(甲烷、氨、水和氟化氢)的沸点

氢键键能不大，随着温度的升高，热运动使氢键键能减小(图 4.4)，同时，介电常数从室温下的 80 降至温度为 $T_c$时的 10 或温度为 425℃时的 2。低介电常数显著地改变了离子生成量($K_w$)和 pH 值(图 4.5)，并且与亚临界水相比，超临

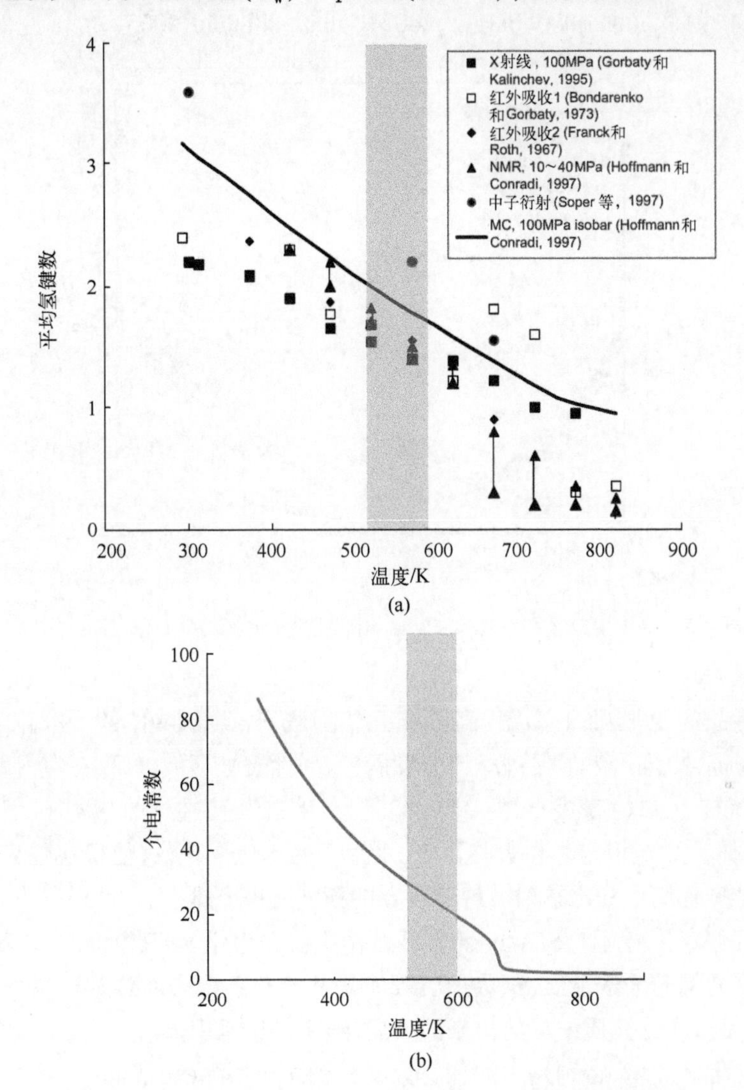


图 4.4 热运动使氢键键能减小灰色条带表示当前水冷堆主冷却剂工作温度范围

(a) 每个分子的平均氢键数与温度的关系；(b) 水的介电常数。

(图(a) 引自文献 Kalinichev, A. G., Bass, J. D., 1997. Hydrogen bonding in supercritical water. 2. Computer simulations. J. Phys. Chem. A 101, 9720-9727。图(b) 引自文献 Guzonas, D., 2014. Extreme water chemistry-how Gen IV water chemistry research improves Gen III water-cooled reactors. In: 19th Pacific Basin Nuclear Conference (PBNC 2014), Vancouver, B. C., Canada, August 24-28, 2014)

界水只有轻微的解离。而水在低温下是一种不可压缩的流体，其等温压缩系数[①] $\kappa_T$ 在临界点附近迅速增大，在临界点处发散。这些巨大的性能变化使 SCW 成为理想的化学介质，因为这些性能（如介电常数）可以通过温度的微小变化在大范围内"调整"（Akiya 和 Savage，2002；Galkin 和 Lunin，2005）。

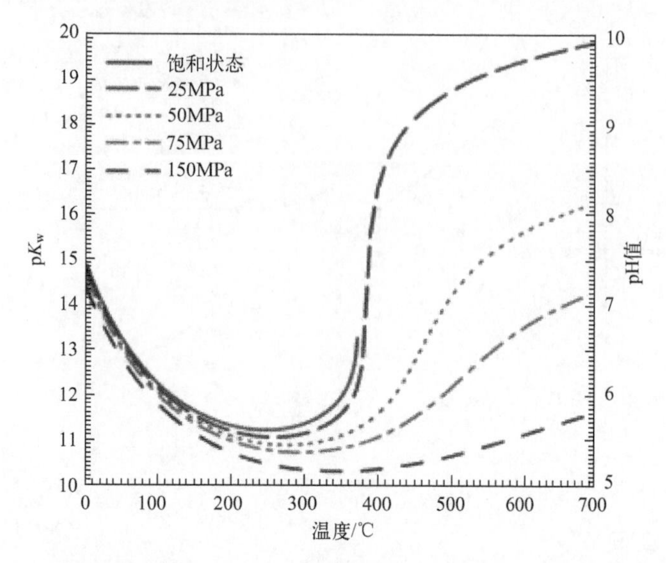


图 4.5　离子生成量和中性水的 pH 值随温度变化（Guzonas 和 Cook，2012）



水性质的变化，加上高温、高压和强辐射场，导致 SCWR 的"极端"堆芯环境（Guzonas，2014），为了促进材料的发展，水的性质必须加以研究。如图 2.4 所示，水在堆芯入口（"蒸发器"区域），密度为"液态"，而在"过热器"区域的密度更像"气体"，相对恒定。在近临界区，水性质的迅速变化，这使得在临界点附近建立腐蚀产物沉积和水辐射分解等现象的模型变得困难。

在任何直接循环的 SCWR 概念中，3 个主要的水化学问题如图 4.6 所示。

（1）向堆芯输送杂质，这些杂质可能是从给水系统表面释放出来的腐蚀产物，或者是从净化介质中释放出来的杂质，再由补给水引入。

（2）堆芯环境，特别是水的辐射分解和腐蚀产物沉积的影响。

（3）杂质从堆芯输送到汽轮机和低压区。SCWR 中主要关注的是氧化、辐射分解产物和放射性同位素的运输（活性运输）。

① 等温可压缩性是对流体或固体在压力（或平均压力）变化情况下相对体积变化的响应的度量：$k_T=\rho^{-1}\left(\frac{\delta\rho}{\delta p}\right)_T \equiv -V^{-1}\left(\frac{\delta V}{\delta p}\right)_T$。

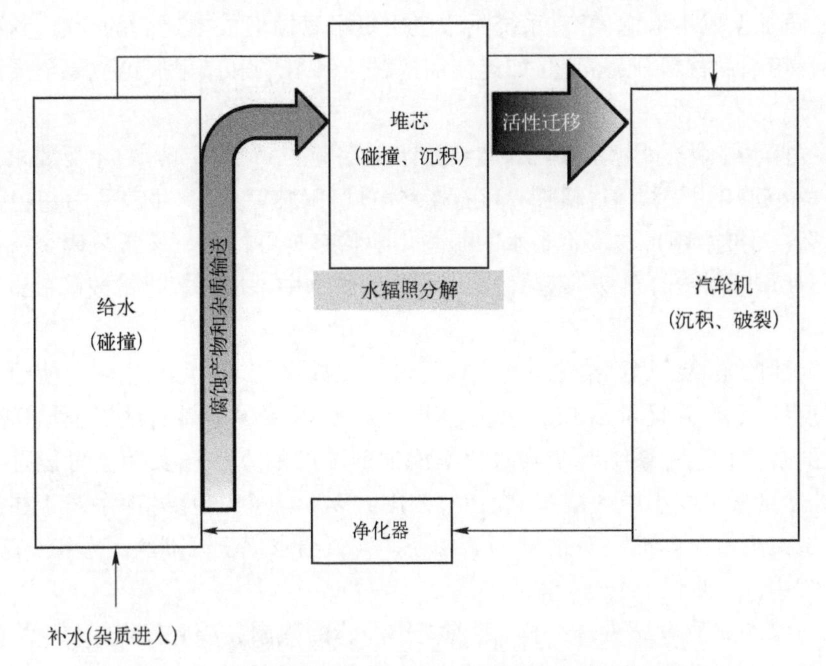


图 4.6　直接循环超临界水冷反应堆的水化学研究区域

(引自文献 Guzonas, S., Cook, W., 2015. Water chemistry specifications for the Canadian supercritical water-cooled reactor concept. In: 7th International Symposium on Supercritical Water-cooled Reactors (ISSCWR-7), March 15-18, 2015, Helsinki, Finland, Paper ISSCWR7-2089)

类似的问题也存在于间接循环 SCWR 的一次回路中,其中主回路取代了给水系统,蒸汽发生器取代了汽轮机,净化系统是附在主回路上的侧流系统。

本章大致介绍了 SCWR“回路”的水化学,认识到这 3 个关注点不是独立的,而是从一个区域到下一个区域的叠加。例如,给水系统受水化学的影响,而产生的腐蚀产物的行为会受到堆内化学物质、水辐照分解等因素的影响。腐蚀产物可以通过中子活化而变成放射性物质,然后被运送出堆芯,沉积在下游部件表面,如汽轮机叶片。

## 4.2 给水系统

水冷式热电厂的电厂配套设施(BOP)是一个庞大而复杂的化工/机械工程系统,通过几十年的运营经验,它的设计及运营得到了优化。核电站的维护和维修费用中有很大一部分都与核电厂配套设施有关。例如,由于材料腐蚀而更换压水堆蒸汽发生器的费用估计约为 3500 万美元。BOP 中的温度和压力变化范围大,再加上成本考虑,所有系统无法使用更耐腐蚀但成本更高的合金,这可能导致环境辅助材料开裂问题(如流体加速腐蚀(FAC) 和接触腐蚀),最终导致管道故障。

因此,控制给水化学①是工厂经营者“保存资产、节约资源”的主要要求。与目前的 SCWR 概念类似,超临界化石燃料电厂(SCFPP)锅炉通常采用“直流”方式运行(无再循环),这决定了水的哪些化学状态是可行的。通常采用全流程净化,不使用吗啉或环己胺等有机胺,也不能向给水中添加氢氧化钠或磷酸钠等固体试剂。

重要的是要认识到给水化学并不总是与锅炉(堆芯)化学相同。使给水系统温度低于 $T_C$ 并且对 SCFPP 水化学进行优化,以减少进料腐蚀以及输送到汽轮机的杂质。通过系统改变给水化学添加剂浓度的方法在实际上可能是可取的,如尽量减少锅炉中的溶解氧(DO)浓度。Woolsey(1989)讨论了对于在过热器区域使用奥氏体不锈钢的 SCFPP 中使用含氧给水化学试剂的问题和经验(在第 6 章中进一步讨论)。

与水冷式热电厂设计一样,随着操作经验的积累,SCFPP 对给水化学控制进行了精心优化。EPRI、VGB 和 IAPWS 等组织以推荐的操作规范或最佳方法的形式提供指导,这些规范或最佳方法随着操作经验的发展而不断更新。Woolsey (1989)、Dooley (2004)、Gabrielli 和 Schwevers(2008)及其他人也提供了 SCFPP 中化学控制方法的优秀综述。鼓励读者查阅这些文档,以便更深入地了解当前最佳方法背后的基本原理。本节只关注那些影响 SCWR 堆内材料降解过程的给水化学,以便理解所提出 SCWR 的给水化学规范背后的基本原理。



### 4.2.1 腐蚀产物及其他杂质的输运

在任何直接循环运行的电厂中,BOP 管道的腐蚀是不可避免的。在腐蚀过程中,部分金属氧化物会溶解在流动的水中。此外,由于冷凝管泄漏、过滤器的浸出以及除盐装置和补充水中的离子交换树脂,可能会发生微量杂质(如氯化物、硫酸盐和二氧化硅)的渗入。虽然它们的浓度通常很低(μg/kg 或 ng/kg 量级),但存在的各种浓度机制可导致已知存在杂质的局部高浓度。此外,即使短时间不合格操作(由于管道泄漏而无法进行净化)可能会导致 $10^{-6}$ 浓度的杂质对材料的损害。因此,了解这些杂质在堆芯中的行为和影响是至关重要的,这样

① SCFPP 和 BWR 的给水化学与 PWR 和 PHWR 的二次侧化学相似。后两种 WCR 设计也严格控制了一次侧的化学反应,但由于一回路是一个封闭的加压回路,故化学方面的考虑有些不同。

才能对它们在堆芯入口的浓度进行限制。一旦这些杂质进入堆芯，就很难降低它们的影响。

1960 年，人们认识到在 SCWR 中腐蚀产物的输运和沉积的重要性（Machaterre 和 Petrick，1960），他们指出，与反应堆系统有关的 SCW 技术的主要缺陷在于缺乏关于外部系统中放射性物质沉积和内部放射性物质①在辐照下积聚的信息。自 WCR 发展以来，一回路的腐蚀产物沉积一直是一个难题（Cohen，1985），并引起相关的运行问题，特别是当运行条件发生变化时（如功率上升、水化学的变化）。最近的问题包括 PWR 中的轴向偏移异常（AOA）（Henshaw 等，2006）和污垢导致的沸水反应堆（BWR）局部腐蚀（Cowan 等，2011、2012）。沉积的腐蚀产物在芯材中输送会导致：①燃料包壳表面的污垢，导致传热减少和燃料失效的可能性增加；②流量限制导致燃料冷却不足；③中子活化增加了放射性物质的产量，导致堆芯外辐射场和工人剂量的增加。系统管道内沉积物的形成也是超临界水氧化工艺发展的一个问题（Hodes 等，2004）。

尽管 Pencer 等（2011）模拟了堆内腐蚀产物在堆芯物理堆积的影响，但腐蚀产物对 SCWR 堆芯的热工水力的影响尚未得到评估。Edwards 等（2014）假设氧化物热导率为 3W/（m·K），热通量为 40kW/m，氧化物的孔隙率为 0%~75%，估算了由于表面氧化物的存在而导致的包壳表面温度的升高。致密的（无孔的）20μm 厚的氧化物层的存在导致包壳温度大约升高 5K，而相似厚度的孔隙率为 75%氧化物会导致包壳温度大约升高 60K。在 SCW 中形成的氧化层往往是多孔的，其孔隙率随试验条件的不同而变化很大（第 6 章）。Solberg 和 Brown（1968）比较了铁素体钢和奥氏体钢在过热蒸汽（SHS）下的长时间传热。铁素管（2.25Cr1Mo、3Cr1Mo、5Cr0.5Mo、5Cr0.5Ti、5Cr1.25Si、9Cr 及 GeCrMoVa）被放置在 593℃、649℃，压力为 13.8MPa 的环境中，为期 30 个月。奥氏体钢（304、310、316、321、347、16-25-6、17-14CuMo、15-15N）被放置在 649℃、732℃、816℃的环境中，为期 18 个月。作者报告说，对于铁素体合金，氧化物减少了传热，这是由于氧化物外层的多孔性造成的。高铬钢的传热减少率较低，由于奥氏体合金表面粗糙度的增加弥补了氧化物热阻的增加，奥氏体合金在氧化过程中表现出更好的传热性能。这些结果强调了用氧化物覆盖的样品进行 SCW 传热研究的必要性，以及将工作的 SCWR 堆芯中形成多孔氧化物层最小化的必要性。

关于 SCWR 中腐蚀产物的行为我们知之甚少，但是通过对 SCFPP 数据的分

---

① 用于表示沉积和/或循环的颗粒腐蚀产物。据说是白垩河不明沉积物（Chalk River Unidertified Deposits）。

析可以发现,从燃料包壳表面释放出的腐蚀产物很有可能沉积在燃料包壳表面。SCFPP 通常使用低成本合金(碳素钢、低合金钢)作为原料,但模型表明,即使使用低腐蚀速率的原料,也会在 SCWR 堆芯中沉积。锅炉(使用 SCW)管内氧化铁沉积物的工厂数据显示,最大的沉积物出现在温度 $T_c$ 附近,并且表面热通量对沉积峰的位置有显著影响。Dickinson 等 (1958) 用试验工厂的化石燃料锅炉研究了 SCW 中的沉积和管道过热问题。锅炉管由碳钼(A-209)钢(外径 1.43cm、内径 0.80cm、长 150m),以及 33m 长的 A-321 型不锈钢过热器管(其他尺寸与锅炉管相同)组成。水的 pH 值为 8.5~9.5 并含氨,每千克水中包含 Cu 5μg、Fe 10μg、Si 20μg。冷却剂的温度在 379~482℃范围内随距离的增加大致呈线性上升。沉积发生在 304~416℃范围内,最大沉积(4.84mg/cm$^2$)产生于 416℃,压力为 34MPa。当锅炉压力为 23.9MPa 时,从壁面温度的变化可以推断出 $T_c$ 处的沉积。基于 Zenkevitch 和 Sekretar(1976)描述的俄罗斯超临界锅炉的数据,Woolsey(1989)得出结论,氧化物沉积场在很大程度上取决于 $T_c$ 处的伪相变化,而不是氧化物溶解度的降低。

图 4.7 显示了 SCW 在 $pH_{25}$值为 8~8.5(用联氨调节)条件下运行 8400h 后,锅炉内管沉积物作为锅炉低辐射段温度函数的数据(Chudnovskaya 等,1988)。3 层沉积物的外层是棕色到黑色,这取决于它的表面是在火侧还是在水侧。内沉积层①呈黑色,颗粒细小。在冷却剂中只加入联氨时,沉积物才会附着在管道表面,而加入联氨和氨时则不会。外层沉积在 350℃时达到峰值,为 180g/m$^2$,



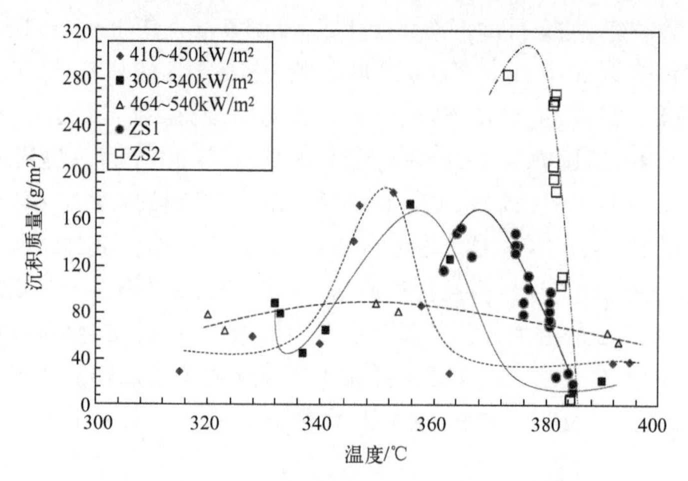


图 4.7 超临界化石燃料电厂锅炉管沉积数据

① 内沉积层可能在疏松的外层沉积物下,但与腐蚀膜不同。

内层沉积在320~340℃时达到峰值,为80~150g/m$^2$。沉积物的质量是表面热流密度、水温和管材的复杂函数,Zenkevitch 和 Sekretar(1976)描述的来自俄罗斯超临界锅炉的数据(Woolsey,1989)也包括在图4.7内,以进一步说明沉积物质量和可能的峰值温度。Vasilenko (1976)报告的沉积数据来源于俄罗斯的SCFPP(用联氨和氨同时调节 $pH_{25}$ 值为8.8~9.8,并工作10000h),指出沉积物质量在温度为370~390℃内达到峰值。尽管最大的沉积质量是在很大的温度范围内被观察到,最大沉积质量总是出现在 $T_c$±20℃的温度范围内。

使用氨或氨-联氨水化学方法导致的腐蚀产物沉积通常比使用含氧水化学方法的沉积高。Kontorovich 和 Rogal'skaya(1987)指出,在SCFPP中,当给水系统中使用氨-联氨化学,沉积物的形成导致需要对锅炉受热面进行化学清洗,根据热负荷的不同,清洗间隔为4000~15000h。虽然化学清洗① SCWR 中的沉积物是可能的,在换料期间,可以进行超声波燃料清洗,清除燃料组件上的沉积物,但是,由于沉积物的活化而产生的额外困难大大增加了技术和财政上的挑战。

很少有实验室研究在SCWR条件和无辐照条件下的腐蚀产物沉积。Karakama 等(2012)报道了在23MPa、温度约为 $T_c$,$pH_{25}$ 分别为3和9的条件下,316L SS上氧化铁的沉积厚度,结果,$Fe^{2+}$ 的浓度为5mmol/L。试验结束后,用破坏分析法测定沉积层厚度和氧化层附着力。测试期间的在线温度测量能使测试期间的沉积变化得到监测,温度沿测试段长度保持恒定。当温度高于 $T_c$ 时,在研究的两个pH值中,均观察到壁温的升高,其中pH值为9时最高。值得注意的是,在pH值为9时,在线温度数据表明氧化剥落很容易发生。作者还报道了 $T_c$ 以上形成的沉积物比 $T_c$ 以下形成的沉积物更难去除。由于在使用的模型中忽略了表面附着物和去除率,建立沉积模型的尝试没有成功。

使用SCFPP数据,可获得(但有限的)磁铁矿在超临界水中的溶解度数据并能对亚临界水中测量数据进行外推。Burrill(2000)预测,SCWR堆芯的沉积物质量峰值大约是俄罗斯SCFPP的10倍(图4.7),比从PHWR中取出的燃料的典型观测值高 $10^4$ 倍。Saito 等 (2006)通过构建Fe、Cr、Ni的氧化物在温度分别400和450℃、压力分别为50MPa和100MPa条件下的电势-pH值图,指出腐蚀产物很可能沉积在SCWR的燃料包壳上。Woolsey(1989)指出,如果用孔口控制锅炉管流量,腐蚀产物沉积会导致堵塞或流量限制,导致冷却剂流量不足,这可能是SCWR燃料组件的一个复杂几何布置问题,也是在使用脱氧给水化学体系的SCFPP中的一个大问题。

---

① 核反应堆一回路的化学清洗被核工业称为化学去污(IAEA,1994),包括注入能够溶解金属氧化物而不破坏结构材料的化学物质。

大量的氧化物溶解度数据是从对水热地质过程和矿物形成过程感兴趣的地球化学家获取的，而且，这些数据通常是在比 SCWR 中使用的压力（如 100MPa）高得多的情况下测量的。在 SCW 条件下，溶解度与压力（密度）有很强的相关性，因此从 SCWR 发展的角度来看，地球化学数据只能用于定性分析。Walther（1986）总结了一些现有的地球化学数据，Qiu 和 Guzonas（2010）对 SCWR 发展的相关数据进行了简要的总结。对超临界水处理技术发展最有意义的是 Fe、Cr、Ni 的氧化物和氢氧化物的溶解度随温度（高于 $T_c$）的变化。尽管它们很重要，但由于涉及的浓度极低，以及在高温和高压下测量溶解度的实验十分困难，这些数值的试验数据有限（图 4.8）（Wesolowski 等，2004）。Adschiri 等（2001）将这些溶解度作为 pH 值的函数，在温度为 400℃，压力为 30MPa 和 60MPa 的条件下进行了估计。金属氧化物的溶解度在临界点附近一般会急剧下降，但有些不会，如 $WO_3$ 的溶解度随着温度的增加而增加。

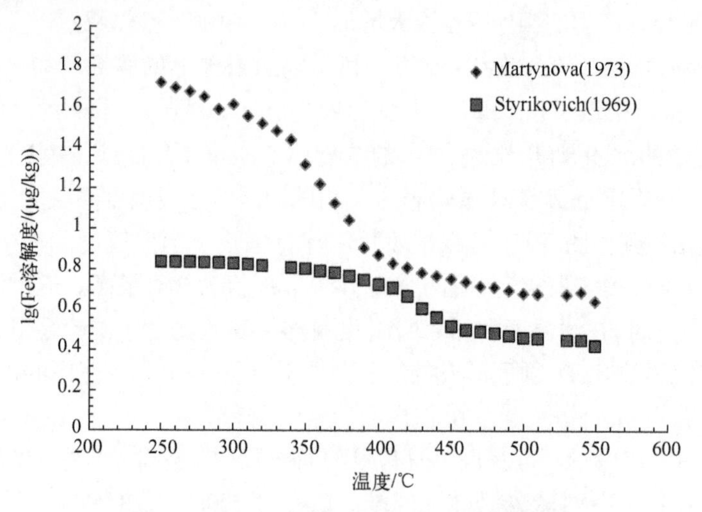


图 4.8　测量了磁铁矿在温度 $T>T_c$ 时的溶解度

目前关于磁铁矿在超临界水中的溶解度的测量报道较少。Holster 和 Schneer（1961）的早期研究报告称磁铁矿在纯水中的溶解度低于其方法检测极限（MDL）的测量范围（压力 48.5MPa，温度 400~500℃，检测极限为 20μg/kg）。Martynova（1973）在压力为 25MPa 的纯水条件下，把磁铁矿的溶解度作为温度（250~550℃）的函数，但是论文没有报道测量技术的细节。在临界点附近，随着温度的升高，溶解度迅速降低；当 $T>400$℃时，溶解度下降的速度减慢，温度为 550℃时，溶解度为 5.4μg/kg。Styrikovich（1969）报道了磁铁矿在蒸汽中的溶解度随过热区温度的变化而变化（图 4.8），在 550℃时，其值（2.1μg/kg）与 Martynova（1973）在同等强度下报道的值相符。

在可以开展有效试验之前,在 SCWR 堆芯中建立腐蚀产物沉积过程的模型是预测可能出现状况的唯一途径。SCWR 中主要的杂质是电解质(如 NaCl),而对非电解质(如气体)的研究也有一定的价值。非电解质在临界点附近的行为模型比电解质的模型发展得更好。标准理论研究了在温度低于 $T_c$时,电解液在临界点附近的行为。电解质溶液中含有电荷载体,随着温度的升高,水的压缩性更高,而介电常数的降低导致离子缔合。Majer 等(2004)对用标准热力学性质作为温度和压力函数的各种模型进行了总结。Anisimov 等(2004)讨论了非电解质和电解质系统的近临界行为,总结了在临界点附近模拟带电系统存在的困难。Tremaine 等(2004)以及 Tremaine 和 Arcis(2014)总结了在高温水中模拟热力学性质的方法。

建立水溶液模型要着重考虑标准状态和浓度变量的选择。虽然现在有几个半经验模型,但迄今为止,只有 Helgesone-Kirkhame-Flowers (HKF)方程及其改进方程才能对 SCWR 中腐蚀产物的运输进行建模。Guzonas(2012)、Brosseau(2010)和 Olive(2012)等描述了该模型在 SCWR 腐蚀产物沉积中的应用。所有半经验方法的一个主要限制是用于开发模型的相关性和验证模型预测的可靠性所需的高温试验数据有限。值得关注的是,可能存在不稳定的高温物质,而这些高温不稳定物质在较低的温度下不会被发现,因此仅从低温数据得出的相关关系不能预测这些物质。Guzonas 等(2012)强调了确保用于模型开发的任何数据库的内部自洽的重要性。

磁铁矿的溶解度建模特别困难,因为磁铁矿是 $Fe^{2+}$ 和 $Fe^{+}$ 氧化物的混合。因此,它的溶解涉及许多氧化还原和水解反应(图 4.9),这些反应的重要程度取决于 pH 值以及氧化剂和还原剂的浓度。虽然这些参数可以在实验室环境中小心地控制,但它们的值在发电厂的运行中往往不为人所知。关于磁铁矿在亚临界水中溶解度随 pH 值和温度变化的数据集已经发表多次((Sweeton 和 Baes,1970;Tremaine 和 LeBlanc,1980 年;Bohnsack,1987;Ziemniak 等,1995;Wesolowski 等,2000),结果显示,考虑到试验的不确定性,关于 Fe(Ⅱ)和 Fe(Ⅲ)在高温下水解物的相对重要性以及

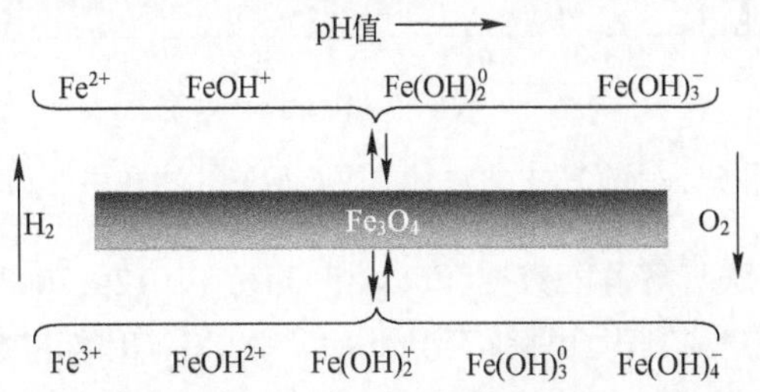


图 4.9 水中与磁铁矿溶解有关的各种反应及其对 pH 值和氧化还原条件的依赖性

在碱性溶液中使用的模型仍存在争议,需要预测腐蚀产物在PWR和PHWR中沉积的情况(如Guzonas和Qiu,2004)。由于SCW中关于水辐射分解的知识有限,目前SCWR堆芯中的氧化还原条件基本上完全未知(4.4节),在这种情况下,对SCWR堆芯腐蚀产物沉积的预测只能是定性的。Wesolowski等(2004)、Brosseau等(2010)、Brosseau等(2010)和Olive等(2012)对有关高温磁铁矿溶解度测量和建模的问题进行了分析。

图4.10显示了使用修正HKF方程得到的用于SCWR的合金中若干金属氧化物溶解度的计算(Guzonas和Cook,2012;Olive,2012)。溶解度在临界点附近迅速下降,但一些金属($NiO$、$Cr_2O_3$)在500℃左右的溶解度略有上升。400℃的计算数据与Adschiri等(2001)所预测的相符合,但是当$T>T_c$时,计算出的磁铁矿溶解度比图4.8所示的测量值低几个数量级,这可能是因为使用的试验技术或氧化还原条件的选择所造成的,或者因为模型中使用的Fe种类不同。Burrill(2000)根据对SCFPP锅炉中氧化物沉积速率的评估,认为Martynova(1973)报告的溶解度值高了100倍。由于缺乏在严格控制条件下获得的SCW中磁铁矿溶解度的试验数据,因此不可能进一步细化模型。最后,需要注意的是,虽然图4.10所预测的腐蚀产物浓度可能较低,但如果沉积是局部的,并且由于在WCR(SCWR)中大量冷却剂的循环,仍然会导致大量腐蚀产物的沉积。



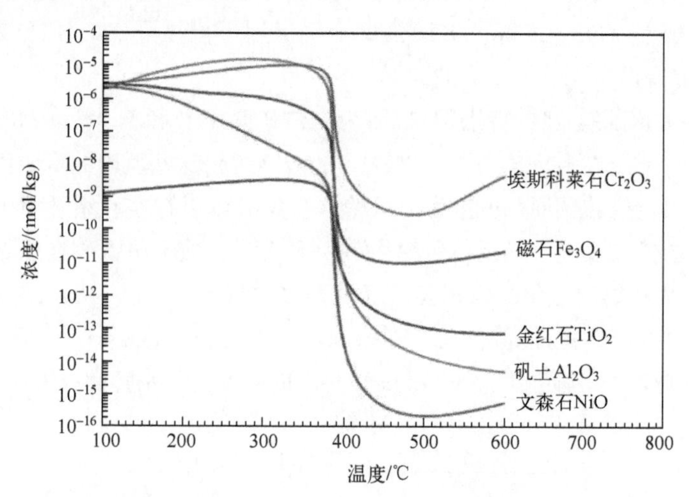


图4.10 用修正后的HKF方法计算了金属氧化物在中性水中的溶解度

使用最新的磁铁矿溶解度数据,Cook和Olive(2012a,2013)以及Olive(2012)使用修正的HKF方程模拟了两种工况下加拿大SCWR堆芯中的铁和镍的沉积:①堆芯入口的冷却剂对相应的金属溶解达到饱和;②堆芯入口的冷却剂对相应金属未达到饱和(图4.11)。加拿大SCWR中有一个向下的中央流管,冷却剂在通道底部反转方向之前，以最小的热量通过该流管，在向上流动时被燃料加热(Yetisir 等,2013)。当冷却剂中金属溶解达饱和时，在堆芯入口开始沉积，并在堆芯内部约 1m 处达到最大值，然后一直持续到堆芯出口。对于金属溶解不饱和的冷却剂(溶解度为 1μg/kg 的 Fe)，沉积从堆芯大约 1m 处开始，一直持续到堆芯出口。通过参数分析，研究了沉积速率常数(SCW 中从未测量过)对铁载荷的影响。当沉积速率常数改变时，从 0.05m/s 到 0.1m/s，铁载荷的差异很小，这表明，在 0.05m/s 或更高的值时，导致沉积速度被限制的因素是流体通过堆芯时被加热而导致溶解度下降。当沉积速率常数分别为 0.01m/s 和 0.001~0.05m/s 时，Olive(2012)指出，由于最大铁负荷下降，导致向上通道的负荷增加；在 0.001m/s 时，载荷几乎与沉积同时开始。研究发现，对于溶解金属不饱和的冷却剂(1μg/kg)，在堆芯入口溶解的铁，根据沉积速率常数和腐蚀速率进行计算，一年后磁铁矿的峰值厚度可以在 10~65μm 之间。

图 4.11 和图 4.7 的对比表明，该模型未能预测 SCFPP 数据中观测到的峰值。然而，Cook 和 Olive 的模型不包含热通量的相关影响，从图 4.7 可以看出，在 SCFPP 中，热通量对沉积量和沉积峰值温度都有显著影响，这可能是造成差异的原因。并且沉积速率常数的值也有相当大的不确定性。因此，HKF 方程也可能没有恰当地描述通过接近临界区域的过渡现象。目前建模工作的另一个不足之处是忽略了辐照可能对铁氧化还原状态的影响；Mayanovic 等(2012)发现，在辐照条件下，铁的种类可以根据压力、温度和辐照条件的不同而被氧化、还原或沉淀。



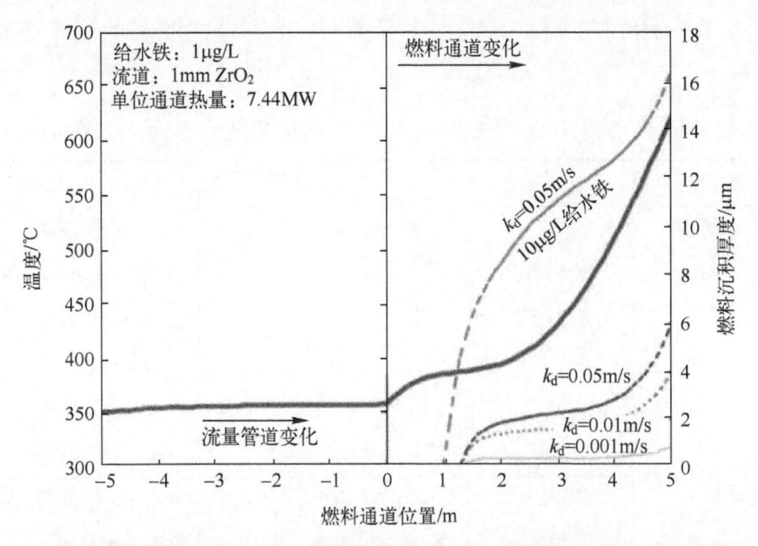


图 4.11　加拿大超临界水冷反应堆(SCWR)的中央流管使用一年后的氧化物厚度

模拟结果表明,如果能最大限度地降低堆芯进口处的铁浓度,沉积的腐蚀产物可以维持在一个可接受的水平。Guzonas 等(2007)的一项早期预测是,如果给水系统使用腐蚀速率为 0.1mg · $dm^2$/天(mdd)的材料,将使堆芯内沉积最小化。Cook 和 Olive (2012a)预测的简单外推表明,对于在堆芯入口加入溶解度为 0.1μg/kg 的 Fe 能够优化给水系统,一个燃料循环(425 天)周期后的沉积厚度约为 15μm,与 Hazelton(1987)报道的压水堆乏燃料的平均氧化层厚度 5.4μm 相比,没有显著提高。这表明,如果对给水系统和化学控制系统进行优化, SCWR 中的堆芯内沉积是可以控制的,堆芯入口处 Fe 的溶解度将小于 0.1μg/kg。为了在随后的燃料循环过程中减少沉积,可以对重新装载的燃料棒束进行超声波清洗,这是一些 PWR 处理 AOA 问题的做法(EPRI,2004)。更精确的预测需要在 SCWR 条件下的沉积试验数据,特别是沉积速率常数以及水的辐照分解和热流的影响,后两个方面可以在堆内回路试验中进行研究。

WCR 中,溶解的 Fe 和 Ni 在冷却剂中产生的电流主要是材料的腐蚀速率和在合金表面形成的氧化物溶解度的函数。人们早就认识到,在酸性条件下,磁铁矿和混合铁氧体的溶解度随温度的升高而降低。由于能发生水解的金属存在,将 pH 值提高到弱碱性会导致这些氧化物的溶解度随温度的升高而增加。对于低合金钢,氧化还原电位也起重要作用,还原条件有利于磁铁矿和混合铁氧体,氧化条件有利于溶解度较低的赤铁矿。SCFPP 中已经采用了许多化学控制策略来控制腐蚀速率和腐蚀产物的溶解度,包括 pH 值控制添加剂的使用,如 $NH_3$,氧化还原添加剂的使用,如 $N_2H_4$ 或 $O_2$(表 4.1)。然而,中性的 $Fe(OH)_2^0$、$Fe(OH)_3^0$ 在超临界区占主导地位,这使得在较高的温度下控制 pH 值会失败。



表 4.1　超临界化石燃料电厂全挥发和含氧水处理技术总结(Guzomas 等,2012)

| 水化学 | $pH_{25}$值 | 注解 | 参考 |
|---|---|---|---|
| 氨+联氨 | — | 0.7~1mg/kg 胺 | Larsen 等(1996) |
| | 9.1~9.4 | $NH_3+N_2H_4$ | Chudnovskaya 等(1998) |
| | 8.5~9.5 | $NH_3$ | Dickinson 等(1958) |
| | 9.1~9.6 | — | Margulova(1978) |
| | 9.1 | $NH_3$<0.8mg/kg 避免铜转运 | Vasilcuko(1976) |
| | >9 | $NH_3$ 或环己胺+联氨 | Cialone 等(1986) |
| 仅联氨 | 8~8.5 | $N_2H_4$:60~100μg/kg | Chudnovskaya 等(1988) |
| | 7.7 | $N_2H_4$:60~100μg/kg | Deeva 等(1986) |

续表

| 水化学 | $pH_{25}$值 | 注解 | 参考 |
| --- | --- | --- | --- |
| 螯合物+$NH_3$+$N_2H_4$ | — | 80μg/kg 螯合物,0.8mg/kg $NH_3$,0.2mg/kg $N_2H_4$ | Oliker 和 Armor(1992) |
| | — | 增加氧化铁沉积的热导率 | Vasilenko(1978) |
| 有氧环境下 pH=7 | — | $O_2$ 含量 50~200μg/kg,电导率<0.1μS/cm | Larsen 等(1996) |
| | 6.5~7.3 | — | Margulova(1978) |
| | — | 在沸腾炉入口处,$Fe(OH)_2$ 与 $O_2$ 的反应速率比与 $H_2O_2$ 的反应速率慢,$H_2O_2$ 的电导率小于 0.15μs/cm,$O_2$ 含量为 0.2~0.4mg/kg。 | Kontorovich 等(1982) |
| 组合模式 | 8~8.5 | $NH_3$+$O_2$——$NH_3$ 提供了一个缓性调节的 pH 值缓冲以至于杂质不能引起 pH 值更宽范围的波动,尤其是在酸性边界腐蚀层可能增加 | Kontorovich 和 Rogal'skava(1987) |

在 PWR 和 PHWR 中常用 LiOH(或 KOH)来控制主回路的 pH 值。但是它不用于 BWR 或化石发电厂,因为它是非挥发性的,因此不能有效地转变成蒸汽态(从 pH 值控制的角度来看)。必须在蒸汽中加入 pH 值控制剂,以减轻堆芯或锅炉下游电厂配套设施(BOP)部件的腐蚀。然而,在超临界压力下,LiOH 的溶解度要比在沸水堆高,Guzonas 等(2012),Brosseau 等(2010),Plugatyr(2011a)等、Carvajal-Ortiz(2012a、b)等和 Cook 和 Olive (2012a)研究了在 SCWR 中使用 LiOH 控制 pH 值,且所有这些作者都认为 LiOH 浓度高于 1mmol/kg。温度大约为 400℃时,LiOH 可以保证磁铁矿在 SCWR 堆芯中有一个正的溶解度梯度;当温度高于 400℃时, LiOH 未发生显著解离,随着 LiOH 离子对成为溶液中 LiOH 的主要形态(Carvajal-Ortiz 等,2012b),pH 值控制逐渐失效(图 4.12),当 SCW 密度小于 86kg/m$^3$(25MPa、$T>514$℃)时,使用 LiOH 控制 pH 是不可行的。Svishchev等(2013a)对此进行了论证,他们在温度范围为 350~650℃ 的中性($pH_{25}$值为 7±0.3)条件下以及碱性($pH_{25}$值为 10±0.3,以 NaOH 为 pH 控制剂的水化学反应)条件下测量 316 不锈钢的金属释放量(间接测量腐蚀速率)。如图 4.13所示,当温度低于 500~550℃时,两种释放速率是不同的;在这个温度范围以上,两种释放速率是相同的。这与在图 4.12 中观察到的 LiOH 溶解度的降低相一致。由于这个温度是在对接近堆芯入口处的燃料包壳表面温度,LiOH 将不能有效地通过堆芯的大部分区域,进而无法控制 pH 值。此外,随着冷却剂密

度沿堆芯方向减小,LiOH 溶解度降低,可能导致燃料包壳、主蒸汽管道和高压汽轮机上的 LiOH 析出,使其无法使用。

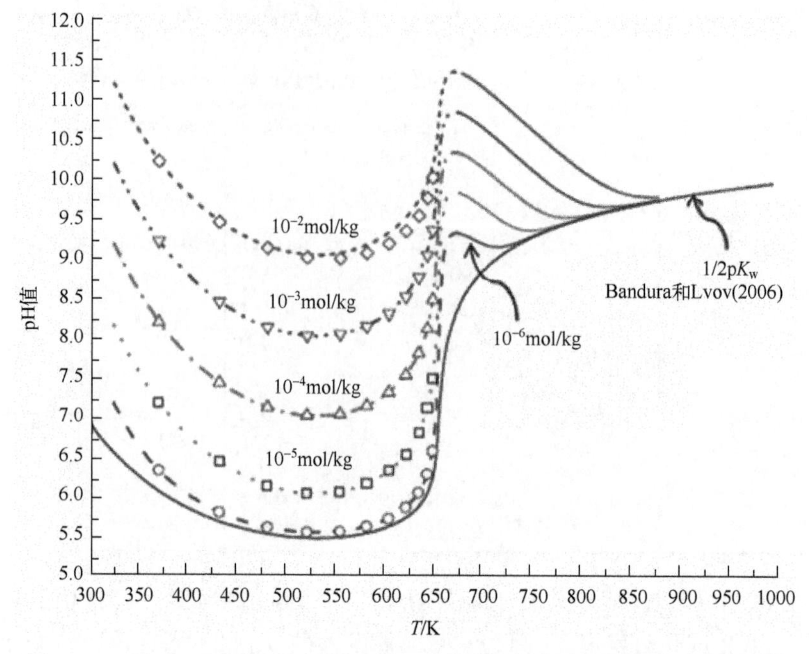


图 4.12　$LiOH_{(aq)}$ 稀溶液的 pH 值沿 25MPa 等压线变化

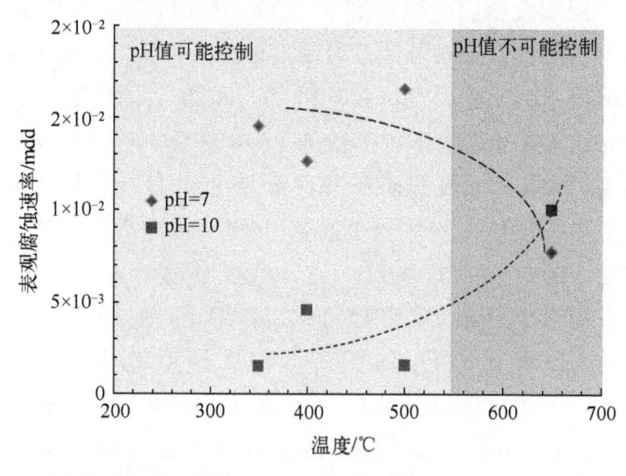


图 4.13　在中性和碱性的超临界水中不同温度下 316 不锈钢的金属释放速率

Beloyarsk 核电站 2 号机组采用氨水控制酸碱度。因为在室温下氨比 LiOH 弱很多,所以需要高浓度的氨(大于 30mmol/kg)使磁铁矿的溶解度梯度为负,以最大限度地减少堆芯内沉积(Cook 和 Olive,2012a;Brosseau,2010)。由于这些物质在高温下的热力学数据稀少或不存在,Brosseau 等(2010)利用$Fe(NH_3)_4^{2+}$络合物的室温数据表明,该络合物在高温中性至碱性条件下占优势,并导致磁铁矿在 SCWR 堆芯条件下的溶解度显著增加。但是随着温度的升高,氨的解离度下降,所以当温度高于 350℃时不能保持磁铁矿正的溶解度梯度。在 SCWR 冷却液中使用高浓度的氨可能会导致下游管道周围的高$^{16}N$ 辐射场,但是 Yurmanov 等(2010)指出,在 Beloyarsk 核电站 2 号机组,不存在活性氮的运输。与氨的使用有关的另一个问题是硝酸盐和亚硝酸盐的辐照分解;在 Beloyarsk 核电站功率为 $130MW_{th}$的反应堆,用氨调节给水系统 pH 值,2 号机组的硝酸盐离子浓度为 1.24mg/kg。使用静态哈氏合金 C 和 625 合金高压釜进行测试,结果显示在含氨的 SCW 中,金属镍的释放量很高(Guzonas 等,2010)。Beloyarsk 核电站的运行经验证实腐蚀产物在氨中的沉积比使用含氧给水进行处理产生的沉积要高(Yurmanov 等,2009)。

联氨在全挥发处理中用作除氧剂,但也被用于原位制氢,以抑制 VVER(俄罗斯"水神"反应堆)中的水辐照分解。Plugatyr 等(2011b)已经研究了温度从室温到 SCWR 运行温度的范围内联氨的热分解。在 SCWR 中,加入到给水系统中的联氨大部分会在到达堆芯前与氧气反应或热分解。但由于联氨的使用受到环境保护的限制,因此用联氨来进行水处理有待考虑。

SCFPP 中最常用的给水化学是氧化处理(OT)(pH 中性和 $O_2$)或者是变化的组合处理方式(注射 $O_2$或 $H_2O_2$,并加氨水,使 $pH_{25}$值为 8.0~8.5),这种方法是在 20 世纪 70 年代前的德国发展起来的,随后被俄罗斯采纳,美国在 20 世纪 90 年代也开始采用。在 Bursik 等(1994)的论文中可以找到关于 OT 及其使用指南的详细说明。该方法的实施细节因国家而异,但一般情况下,氧以空气、气态氧或过氧化氢的形式加入,使氧浓度达到 50~300μg/kg。氧气的存在降低了碳钢给水管系的腐蚀速率,减少了输送到锅炉的腐蚀产物。腐蚀速率的降低是由于形成的氧化物的性质发生了变化(从磁铁到赤铁或氧化铁)。由于在铜合金上形成的氧化物在高温含氧水中溶解,所以 OT 不能用于含有铜合金的工厂。使用 OT 的一个关键先决条件是能够生产高纯度的原料水来防止 SCC(第 6 章),这需要使用全流量冷凝水精处理。

### 4.2.2 其他杂质向堆芯的输运

直接循环的核电站,如 SCWR,有可能将反应堆核心部件暴露在杂质中,这可能会提高腐蚀速率或导致 SCC(Woolsey,1989)。Ru 和 Staehle(2013)强调了氯等杂质对 SCWR 材料降解的重要性,并提到在美国核再热开发计划期间的材料测试发现，即使用最好的方法去除氯，也会产生氯的沉积，而氯的沉积最终导致 SCC 测试的不锈钢失效。

早期在 BONUS 反应堆中对氯离子沉积的研究表明，含氯化物和氧的湿蒸汽导致氯离子诱导的 304 SS 和 347 SS 的 SCC 失效。在测试盒加热器上没有发现可检测到的氯化物沉积，但是在发生冷凝的区域检测到了氯化物的沉积。沉积氯化物的缺乏是由于蒸汽有 1~3℃ 的过热度，在蒸汽通过测试盒式加热器之前，基本上清除了所有含氯的水分(Bevilacqua 和 Brown，1963a)。Bevilacqua 和 Brown(1963b)后续做了 13 个试验，为了提高氯化物的检出限($1\times10^{-6}$g$^{36}$Cl)，他们将$^{36}$Cl 加入到检测回路蒸汽中(含 50μg/kg 氯化物)，并进而量化在不同条件下沉积在过热器包壳表面的氯离子的质量。潮湿蒸汽进行干燥后，所产生的氯化物沉积导致大量的局部附着沉积物。这些大量沉积物以及氧和水的存在促进了氯离子诱导的奥氏体钢 SCC。化学腐蚀产生的表面缺陷内聚集的$^{36}$Cl 的活性比机械加工所产生的零件表面缺陷中的高。但是，应当指出的是，Beloyarsk 核电站 2 号机组使用了许多年，典型的氯化物浓度为 25μg/kg，没有不良影响报告(Yurmanov 等，2010)。

对于不同的 SCWR，在最近的材料测试中，还没有在溶液中加入具有代表性的氯化物浓度(或其他已知的促进构件开裂的杂质)的测试(参见第 6 章的讨论)。目前运行的 BWR 的给水氯化物浓度低至 0. 25μg/kg(典型方法的检测极限)，硫酸盐浓度低至 2μg/kg(Stellwag 等，2011)。测试应在此范围内进行，并在浓度扰动的条件下进行。



NaCl 在高温水中的溶解度随压力的变化已被许多研究小组报道(Galobardes 等，1981；Leusbrock 等，2008)。溶解度($S$)对水密度($\rho$)的依赖关系，通过 $\lg S\approx\lg\rho$ 给出了一个合理的近似(图 4. 14)，而溶解度对温度的依赖性较弱。正如 Palmer 等(2004)所强调的那样，在对此类图表使用的试验数据进行评估时必须谨慎。这些作者给出了一个简洁而全面的可用数据和模型，用于评估 NaCl 在蒸汽中的溶解度。Chialvo 等(2010)讨论了在超临界温度和压力下，特别是在低 SCW 密度下，模拟了 NaCl 缔合和运输过程。图 4. 14 显示，在 25MPa 时的溶解度比 8. 8MPa (Beloyarsk NPP)时的溶解度高一个数量级，比 Bevilacqua 和 Brown (1963a，b) 在蒸汽压力为 6MPa 时的氯化物沉积试验溶解度高一个数量级以上。由于 SCWR 堆芯中没有相变，NaCl 在堆芯条件下的溶解度应该足够高，当给水氯离子浓度为0. 25μg/kg时，即使是在包壳峰值温度，产生的氯离子沉积也较少。主要的危险可能隐藏在燃料包壳表面的氧化层中。然而，随着杂质溶解度的增加，其堆芯沉积的风险也会降低，它还可以导致转移到汽轮机的杂质增加，这是 SCFPP 行业众所周知和研究的现象(Zhou 和 Turnbull，2002)。

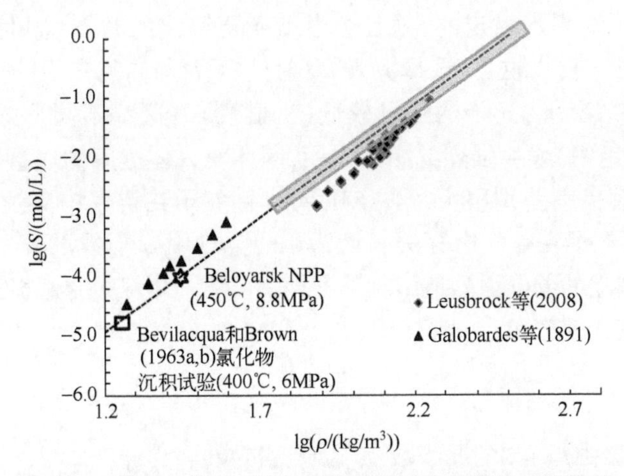


图 4.14　400℃时测定的 NaCl 溶解度与水密度的对数关系

从本质上来说，我们对 SCW 中的隐匿①现象一无所知。在 $T_c$ 附近，溶解度发生的巨大变化表明沉积物中包含多种物质，并在停堆期间重新溶解，导致冷却剂中的杂质浓度出现大的峰值。杂质沉积对局部腐蚀的影响也是一个问题，正如前面提到的氯化物。

Bevilacqua 和 Brown（1963a）指出，回路瞬变，即使是那些短暂的瞬变，只导致回路管道的轻微冷却，也会使管道内 $^{36}Cl$ 的浓度大量增加。在某些情况下，水中的 $^{36}Cl$ 浓度增加到停堆前的 100 倍。这些观测结果突出了反应堆运行暂态期的重要性。当前 WCR 的腐蚀问题通常是这种瞬态（扰动）条件的结果，这些问题需要在未来的 SCWR 研发项目中进行研究。

## 4.3　输运活动

与 SCFPP 相比，SCWR 独有的一个方面是放射性物质从反应堆堆芯运出，这个过程表示放射性物质的运输。自从核反应堆技术出现以来，这一现象已经为人所知和研究，因为它会导致反应堆堆芯外部产生危险的辐射场，并使核电站工作人员吸收辐射剂量，此外，系统管道的放射性冷却剂泄漏可能导致重大的污染风险，并有可能扩散到核电站周围。Cohen（1985）对这个主题进行了很好的介绍。

① 隐匿和隐匿析出是 WCR 中众所周知的现象。隐匿是水污染物浓度和化学控制添加剂浓度在沉积物下和裂缝的局部热点沸腾和沉淀的结果。在这些地区，当系统满负荷运行时，可能会出现非常高的热浓度梯度。当功率降低、温度下降时，污染物会溶解并“返回”冷却剂中。

放射性元素是不稳定的,并且会衰变到更稳定的状态,同时伴随着 α 粒子(He 原子核)或 β 粒子(选择性)的发射,这通常与高能光子(γ 射线)的发射有关,在某些情况下,光子的能量较低(X 射线)。这些不同形式的辐照与物质(包括人体组织)发生强烈的相互作用,因此对人体健康构成重大危害。这 3 种形式的辐射范围差别很大。4~5MeV 的 α 粒子在空气中穿透长度为 3cm,3.0MeV 的 β 粒子在空气中穿透长度为 1m,空气中 γ 射线的穿透长度是无限的。这些范围上的差异强烈地影响了各种辐照对人体健康的危害类型。

放射性衰变速率为

$$\frac{\mathrm{d}n}{\mathrm{d}t}=-\lambda N \tag{4.1}$$

式中:$N$ 为元素在 $t$ 时刻的原子数;$\lambda$ 为衰变常数。

元素放射性衰变常用的特征时间是半衰期,$t_{1/2}$是原子数 $N$ 衰变为 $0.5N$ 所需的时间,有

$$t_{1/2}=\ln 2\cdot\lambda^{-1}=0.691\lambda^{-1} \tag{4.2}$$

不同形式辐照范围的差异导致两种截然不同的危害:体内剂量,即放射性核素通过吃、喝或呼吸进入体内;外部剂量,即放射性核素留在体外,辐照穿透皮肤。除意外事故外,核电厂工人在操作及维修期间所受到的辐照大部分是外源性辐照。



从堆芯释放的放射性核素大致可分为两类。

(1) 由中子(或质子)与堆芯中使用的一种材料(燃料包壳、支撑结构)的原子核发生反应而形成的活化产物,或任何增长或沉积的氧化物的原子核发生反应而形成的活化产物。一个例子是自然产生的非放射性$^{59}$Co 变为$^{60}$Co,即$^{59}$Co(n,γ)$^{60}$Co。

(2) 由可裂变燃料同位素(如$^{235}$U)的核反应形成的裂变产物和锕系元素。锕系元素是由中子吸收产生的,而裂变产物是由裂变过程中产生碎片的放射性衰变产生的。

尽管释放机制不同,两组放射性核素都有可能从反应堆堆芯释放出来。图 4.15 描述了与运输活动相关的一些过程。进入堆芯的杂质可能以溶解的物质或颗粒的形式存在。虽然有些活化发生在冷却剂在堆芯中流动的短暂过程中,更大的问题是杂质通过沉淀,在燃料元件表面沉积,溶液中的离子与燃料元件表面的氧化物发生离子交换。一旦杂质沉积下来,活化就会发生,直到杂质通过溶解、离子交换或粒子释放从表面释放出来。

如果像前面讨论的建模工作所预测的那样,相对于相关金属氧化物,SCWR 冷却剂中的溶解金属保持过饱和度,则没有氧化物溶解,那么在正常运行过程中,很少释放金属离子(冷却剂中的同位素与氧化物中的同位素的动态交换率

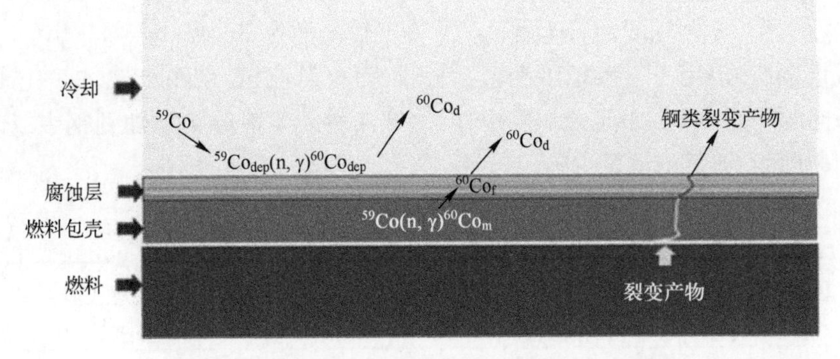


图 4.15　超临界水冷反应堆活性输运的 3 个源项

预计会很低,但这个过程可能会释放出少量的活性)①。在这种情况下,反应堆关闭期间,这些沉积物释放活性的可能性最大,因为当温度低于临界温度时,溶解度增加了几个数量级。其他可能的释放机制是机械磨损(如由于燃料棒的振动)、流动冷却剂的侵蚀或表面氧化物的剥落。

## 4.3.1 堆芯材料的活化

虽然堆芯的外来腐蚀产物通常在较短的时间内沉积和释放②,但有些堆芯腐蚀产物可以在堆芯长期驻留。在此期间,它们受到几乎连续的辐照,导致合金成分发生变化,这是由于嬗变和第 3 章讨论的各种形式的辐照损伤造成的。一些嬗变产物是潜在危险的放射性核素(表 4.2),可以通过腐蚀或机械磨损过程释放到冷却剂中。由于长时间的辐照,合金中这些放射性核素的浓度可能很高。虽然从冶金学的角度来看,这些浓度通常不显著,但其浓度可能足够高,即使是低腐蚀速率也可能释放出有害物质③。此外,将合金从堆芯移除时需要采取重要的预防措施,这增加了电厂退役的成本。

在表 4.2 所列的放射性核素中,$^{60}Co$(来自活化 Co)和$^{58}Co$(来自活化 Ni)一直是现有 WCR 中最棘手的问题。Co 通常作为杂质以 μg/kg 为浓度单位,存在于钢中,但也可能是某些特殊合金的主要成分。$^{60}Co$ 尤其危险,因为它会释放出两种高能量(1MeV)γ 射线,而且它的半衰期相对较长,使它能够在堆芯外管道上大量积累。短寿命的放射性核素不会在堆外外表面积聚,但在断电期间会导

① 冷却剂中同位素与氧化物中同位素的动态变换率预计会很低,但这一过程可能会释放(可能是少量)活性。

② 例如,在一个典型的分批燃料 SCWR 中,每 3 个燃料循环将更换整个燃料装载量(3~4 年)。

③ 举例来说,一个 1mg 的 Co 中子粒子被激活到$^{60}Co$ 的平衡活性($^{60}Co$ 的生成速率和放射性衰变速率达到稳定状态),其活度均为 1Ci(37GBq)。

致高剂量的放射性核素,特别是当它们在反应堆关闭期间被化学瞬态激活时。$^{58}Co$ 在 PWR 中断期间的行为就是一个众所周知的例子。Sb 在许多钢中也是一种微量杂质,也用于一些轴承和密封材料,而放射性 Sb 在许多 WCR 中已成为一个问题。Zr 和 Nb 被记录在表中,因为这两种元素有时被添加到钢中以改善它们的性能,通常浓度在 1%左右。

表 4.2　主要活化产物的母同位素、生产反应及半衰期

| 同位素 | 反应 | 半衰期 |
| --- | --- | --- |
| $^{51}Cr$ | $^{50}Cr(n,\gamma)^{51}Cr$ | 27.8 天 |
| $^{54}Mn$ | $^{54}Fe(n,p)^{54}Mn$ | 312.2 天 |
| $^{55}Fe$ | $^{54}Fe(n,\gamma)^{55}Fe$ | 2.7 年 |
| $^{56}Mn$ | $^{55}Mn(n,\gamma)^{56}Mn$ | 2.58h |
| $^{59}Fe$ | $^{58}Fe(n,\gamma)^{59}Fe$ | 44.5 天 |
| $^{58}Co$ | $^{58}Ni(n,p)^{58}Co$ | 71 天 |
| $^{60}Co$ | $^{59}Co(n,\gamma)^{60}Co$ | 5.27 年 |
| $^{64}Co$ | $^{63}Cu(n,\gamma)^{64}Cu$ | 12.7h |
| $^{65}Zn$ | $^{64}Zn(n,\gamma)^{65}Zn$ | 244 天 |
| $^{95}Zr$ | $^{94}Zn(n,\gamma)^{95}Zr$ | 64 天 |
| $^{95}Nb$ | $^{95}Zr$ 发生 $\beta$ 衰变产生的 $^{95}Nb$ | 35 天 |
| $^{94}Nb$ | $^{93}Nb(n,\gamma)^{94}Nb$ | 20000 年 |
| $^{122}Sb$ | $^{121}Sb(n,\gamma)^{122}Sb$ | 2.7 天 |
| $^{124}Sb$ | $^{123}Sb(n,\gamma)^{124}Sb$ | 60.2 天 |



目前对超冷堆中金属释放的研究很少。Guzonas 和 Cook(2012)报道了 403 SS 在静态高压釜中释放 Fe、Mn、Ni 和 Cr 的数据,以及 Hastelloy C 和合金 625 在高压釜中将各种元素释放到 SCW 的数据。Han 和 Muroya (2009)对 304 SS 样品进行辐照,收集在离子交换树脂上释放的物质,并通过 $\gamma$ 能谱测量收集的量,将来自 304 SS 样品的 $^{60}Co$ 释放的放射性作为时间和温度的函数。他们注意到,304 SS 样品的腐蚀速率随着温度的升高而增大(550℃时的腐蚀速率是 300℃的 10 倍),而 550℃时的 $^{60}Co$ 释放率比 300℃时的低 1%。他们认为这是由于温度高于 $T_c$ 时,氧化物溶解度在降低。Choudhry 等(2016)报道了压力为 25MPa、温度分别为 550℃、650℃、700℃条件下,800H 合金在体积流量为 0.1mL/min 的含氧给水(20μg/kg)中 Fe、Cr、Ni、Ti、Mn、Cu 和 Al 的释放量随时间变化的数据(第 5 章)。在超过几百小时的时间内,每一个元素的释放量都是时间的复杂函数,在某

些情况下，仅仅几个小时释放量就发生了巨大的变化（图 5.36）。

Saito 等（2006）类比 BWR，对 SCWR（1000$MW_e$、$T_{in}=290℃$，$T_{out}=550℃$、$P=25MPa$）中$^{60}Co$输运至汽轮机的速率进行了建模。在缺乏腐蚀速率和冷却剂中溶解的$^{59}Co$浓度数据的情况下，他们检查了一系列相关的输入参数。该模型既包括由燃料包壳腐蚀释放的$^{60}Co$，也包括从包壳表面沉积物释放的$^{60}Co$，沉积物中的$^{59}Co$由给水带入堆芯并进行合金腐蚀。腐蚀释放量由下式计算，即

$$AR = C_R \cdot C_{RR} \cdot C_{Co} \cdot S_A \tag{4.3}$$

式中：$C_R$为包壳合金的腐蚀速率；$C_{RR}$为释放率（被腐蚀的元素释放到冷却剂中的量）；$C_{Co}$为合金中 Co 的含量占比；$S_A$为包壳面积①。

同一合金在沸水堆中的腐蚀速率的差别从 0.0001 到 10 倍不等。根据所选择的输入参数，发现$^{60}Co$的传输速率（相对于 BWR）从 0.1 到大于 100 不等。这项研究强调了需要更多关于腐蚀速率的数据，更重要的是，需要更多关于腐蚀合金进入冷却剂而不是停留在表面氧化层的数据。

得益于近年来 SCW 中腐蚀和金属释放速率的数据，Guzonas 和 Qiu（2013）使用 Carvajal-Ortiz 等（2012b）的金属释放数据，估计两种极限情况下 SCWR 中可能来自作为燃料包壳的不锈钢释放的$^{60}Co$活性。因为释放速率可能需要数百小时才能达到稳定状态，所以在使用试验数据时必须谨慎。Cohen（1985）讨论了在与 Yankee PWR 相关的条件下 304 SS 燃料包壳的活化；经过 3 年的辐照（大致相当于 SCWR 燃料包壳的使用寿命），在一个典型的压水堆的通量下，被活化的 304 SS 燃料包壳合金中$^{60}Co$活性约为 2MBq/mg。一个对于$^{60}Co$的活性释放率 AR（MBq/d）的合理估计为

$$AR = 2S_{ac} \cdot R \tag{4.4}$$

式中：$S_{ac}$为包壳的表面积；$R$为表面金属释放速率（等于式（4.3）中的$C_R \cdot C_{RR}$）。

释放速率将随堆芯位置的不同而变化，因为它取决于腐蚀速率和包壳表面生成的氧化膜的溶解度。该模型假设一种元素从合金中按其浓度的比例释放出来，我们知道这是不正确的（第 5 章），但是考虑到所有其他的不确定性，这样的估计是可以接受的。有如下两种极限情况。

（1）蒸发器部分。冷却剂温度低、密度高，因此氧化物溶解度较高。$^{60}Co$释放率的峰值大约为 50000MBq/d（燃料包壳表面积为 $2000m^2$）。在一个运行的 SCWR 中，冷却剂中金属物质的浓度可能不为零，并且如果冷却剂中相关的溶解金属物质已经达到饱和（虽然不活跃的溶解物和活跃的氧化物之间的离子交换仍然是可能

---

① 这里的$S_A$与 Saito 等（2006）文献中的原始符号有所差异。

的),那么氧化层的溶解度可能很低。因此,有一种观点认为,从活化运输的角度来看,堆芯入口的冷却剂中溶解的金属物质浓度低是十分有益的。

(2) 过热器部分。冷却剂温度高、密度低,氧化物溶解度很低,腐蚀释放出的大部分金属物质仍停留在表面。估计平均释放率为 27500MBq/d;冷却剂流量为 1000kg/s,冷却剂中$^{60}Co$的浓度大约为 320Bq/kg,与目前的 PWR 和 PHWR 的冷却剂中$^{60}Co$浓度的数量级相同。因为目前无法获得有关氧化溶解度的数据,而且堆芯内的化学条件(特别是水的辐射分解)也无法充分描述,因此无法进行更详细的预测。

基本上,关于活化产物在超临界温度下输运的唯一数据来自 Beloyarsk NPP。$^{51}Cr$、$^{54}Mn$、$^{58}Co$、$^{60}Co$、$^{65}Zn$ 和$^{124}Sb$(与当前 WCR 中发现的相同活化产物)在沉积物和冷却剂样品中被报道(Dollezhal 等,1969;Aleksandrova 等,1968;Veselkin等,1968)。在 Beloyarsk AMB-1 NPP 不同汽轮机轮级叶片上测量的表面活动如图 4.16 所示,$^{60}Co$ 略低于$^{124}Sb$,但与$^{54}Mn$ 的表现相似,而$^{54}Mn$ 是铁的活化产物,因此是铁运输的标志(Veselkin 等,1968)。在汽轮机的 11 级或 12 级附近观察到活性$^{60}Co$、$^{124}Sb$、$^{54}Mn$ 的最小沉积。然而,$^{60}Co$ 的沉积情况表现出对时间的依赖性,这使得对这些数据的详细解释变得不可能(Veselkin 等,1971)。汽轮机的设计是复杂的,蒸汽通过汽轮机后温度和压力会下降;利用现实的假设,在汽轮机出口,估计蒸汽密度从 $24kg/m^3$ 下降至 $3.5kg/m^3$ 不等。因此,溶解的金属物会在整个汽轮机中沉积。此外,冷却剂样品表明大部分活动与粒子有关,颗粒会发生迁移和沉积。



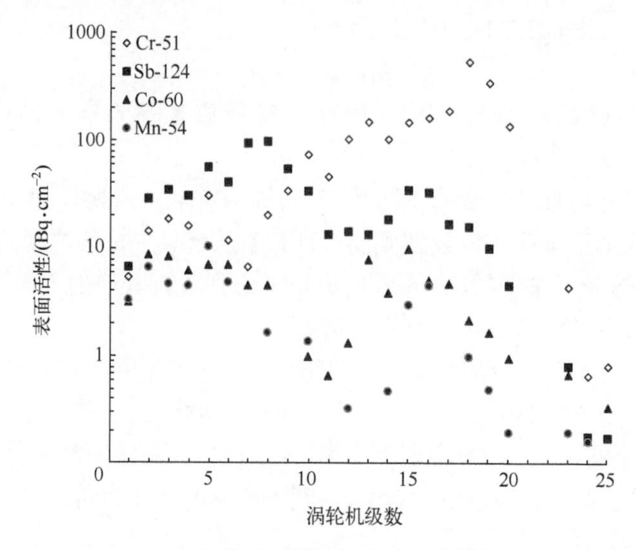


图 4.16 Beloyarsk NPP 汽轮机不同轮级叶片上放射性沉积物的表面活性

有趣的是，$^{51}Cr$ 的表现似乎与图 4.16 所示的其他 3 种放射性核素不同，$^{51}Cr$ 在汽轮机的 11 级和 12 级附近没有明显的最小值。对于这种现象没有给出任何解释，但这一观察表明释放或运输机制的差异。

总体而言，Beloyarsk 的放射性腐蚀产物沉积数据与 SCFPP 中报道的腐蚀产物沉积数据吻合很好。一开始溶解度的降低导致沉积的增加，再热器中溶解度的增加导致沉积量的减少，最终溶解度的进一步降低导致沉积的增加。

据报道，Beloyarsk 靠近二回路设备的辐射剂量没有超过 3mR/h（Veselkin 等，1968）。汽轮机周围的辐射场比其他直接循环核电站的辐射场要低，而且（与其他 WCR 一样）工人受到的大部分辐射都发生在维护期间。

收集反应堆（冷凝水）和汽轮机后的蒸汽冷凝水后，在 SHS 样品中分别检测到 $^{51}Cr$、$^{54}Mn$、$^{58}Co$、$^{60}Co$ 和 $^{124}Sb$（Aleksandrova 等，1968）。据报道，SHS 中 $^{60}Co$ 的活性为 50～150pCi/kg；假设图 4.14 中 SCWR 和 Beloyarsk 的 NaCl 溶解度相差一个数量级的情况也适用于钴氧化物，那么预计 SCWR 中 $^{60}Co$ 的活性为 0.5～1.5nCi/kg。结果表明，SHS 冷凝液中相当一部分的放射性物质是以悬浮物的形式存在（大多数粒子的大小在 3～10μm 内）。然而，在冷却样品时，部分或全部检测到的微粒可能在冷却过程中沉淀在样品线上。据报道，$^{54}Mn$ 以溶解物的形式存在（50%～70%的阳离子，阴离子小于 3%）（Yurmanov 等，2010）。汽轮机后蒸汽冷凝水中杂质的活度明显低于 SHS（但不是零），这证明了大量的活性腐蚀产物沉积在反应堆堆芯和汽轮机之间的管道上以及汽轮机中（图 4.16）。冷凝水中存在放射性表明在所有的 SCWR 设计中，凝结水管线必须返回除盐装置上游的系统，以达到降低给水系统受到污染的目的。

其他仅有的关于活化腐蚀产物的测量报告是由 Mravca 和 Simpson（1961）完成的，他发现 SADE loop15 中的燃料组件 SH-4B 失效 4 个月后①（此时裂变产物活性已基本衰减），回路管道活性的主要来源是腐蚀产物 $^{51}Cr$、$^{58}Co$、$^{60}Co$、$^{65}Zn$ 和 $^{110m}Ag$。

## 4.3.2 超临界水冷反应堆中燃料元件破损

燃料元件的破损使主冷却剂进入燃料与燃料包壳间的缝隙，并允许燃料颗粒和裂变产物逃逸。当前的 WCR 中存在的燃料元件破损现象已经得到充分研究，各种检测、定位、诊断和管理技术也得到发展（如国际原子能机构，2002）。

WCR 中有缺陷的 $UO_2$ 燃料释放是由腐蚀引起的，并因燃料的氧化程度而增

① 最高包层温度为 621℃，过热条件下辐照 617h。包层材料为 28mil（1mil = 25.4μm，28mil = 711.2μm）厚的 304 SS。包层失效归因于氧化物诱发的应力腐蚀开裂（SCC）。

强(Lewis 等,1993)。腐蚀速率受几何形状缺陷的影响(如孔、大的轴向破裂、氢化破裂等),而燃料氧化速率主要受燃料温度的影响。随着 SCWR 的发展,必须研究 SCW 与燃料的相互作用,以及燃料材料和裂变产物在 SCW 中的溶解度和输运。还必须研究 SCW 与燃料包壳内表面之间的相互作用,以确定燃料包壳破口附近的腐蚀行为(如内部包壳氧化)。

一旦裂变产物被释放到冷却剂中,它们就能在系统中被探测到,进而可以对燃料破损进行识别和诊断。根据测量得到的冷却剂的释放速率和冷却剂总量,可用于推断破口数量、破口大小、破口位置、功率、燃料燃耗等各种技术(Locke,1972;Likhanskii 等,2009)。其中一些技术可能适用于 SCWR,当前对 WCR 开发的 FP 释放和传输的理解也将促进新技术的开发。

到目前为止,还很少有对裂变产物和锕系元素在超临界水中的溶解度进行系统的测量。Qiu(2014)回顾了在超临界水中关于 U、Th 的溶解度和一些可能的裂变产物的有限数据。Zimmerman 等(2014)采用高压交流电导技术,测量了 350℃下锶与氢氧化物和氯离子的缔合常数。选择 $Sr(OH)_2+H_2O$ 和 $SrCl_2+H_2O$ 系统是因为它们溶于近临界水,使得它们能作为其他 $M^{2+}$ 物质模型系统。SCWR 条件下的形成 $SrOH^+$ 和 $Sr(OH)_2^0$ 离子对大于 $SrCl^+$ 和 $SrCl_2^0$ 离子对。因此,锶的氢氧化物将在溶液中占主导地位。在温度为 350℃时,中性物质的浓度很大,高于 $10^{-3}$ mol/kg。



#### 4.3.2.1 电厂经验和试验数据

1) 稀有气体和碘

在反应堆运行过程中,核裂变将产生放射性惰性气体(如 $^{133}Xe$、$^{88}Kr$)和碘(如 $^{131}I$),并通过燃料基体扩散到燃料与燃料包壳之间的缝隙。如果包壳被破坏,这些气体会进入冷却剂;在运行瞬态过程中,放射性气体释放量通常会增加。裂变气体在 WCR 中的释放和迁移一直是人们研究的热点,并将裂变气体的在线检测应用于燃料破损的检测与监测。20 世纪 60 年代,美国开展了核能过热蒸汽计划,收集了大量关于 SHS 中惰性气体和碘离子行为的数据。由于气体在超临界水中是无限可溶的,因此在低压下获得的惰性气体输运数据与超临界水堆中惰性气体的输运直接相关。

美国的核能过热蒸汽计划中,ESADE-VBWR 测试项目的化学程序的主要目标是检测燃料包壳的失效并测量裂变产物的释放率。在 ESH-1 辐照期间,尽管通过虹吸检测①(通过对燃料周围的水进行采样,来检测存储在充满水的燃料

① 虹吸是通过对燃料束周围局部的水取样来检测装满水的燃料舱中乏燃料的缺陷(Deng 和 Deng,2012)。

箱内的乏燃料中的缺陷)和金相切片发现了一个小缺陷,但放射化学测量表明,燃料并不存在缺陷。然而,VBWR 本身存在的一个已知的燃料缺陷严重限制了测量的灵敏度(Hazel 等,1965)。在 SADE 回路中 SH-4B 燃料元件失效后(Spalaris 等,1961),裂变产物(主要是放射性碘)使该回路中的放射性达 20mR/h(Mravca 和 Simpson,1961)。在 EVESR 的试验过程中出现了一些燃料元件的破损,这是经过几个月的故意操作,使用严重破损的燃料元件(KB-41)来量化裂变产物的释放(Murray,1965)。在反应堆热功率由 15MW 增加至 17MW 后约 3h,发现 KB-41 束有缺陷出现,并且惰性气体检测器的数值在 15min 内由 1.5mR/h 增加至 15mR/h(Busboom 等,1966)。结果发现,裂变产物的释放高度依赖于反应堆功率和反应堆出口蒸汽温度。在燃料额定温度下运行,导致碲和其他裂变产物沉积在过热蒸汽管道上,这使外部管道的表面剂量率高达 300mR/h 以及反应堆管路上一个阀门的表面剂量率高达 5R/h。辐射场最高的地方是靠近反应堆的地方,以及靠近反应堆的管道连接处或阀门处。半衰期为 78h 的辐射水平随着反应堆的关闭而降低。

一项试验(KB-41-2)是用故意有缺陷的燃料棒进行的(在一个宽为 0.005~0.013 英尺、厚 0.016 英尺的 800 合金包壳中有一个 1 英寸长狭缝)。测试的目的是将已知的缺陷燃料放置在堆芯的已知位置,以研究裂变产物的释放以及这种操作对蒸汽过热系统的后果。Roof(1966)报告了 13 种惰性气体和 5 种碘同位素的释放数据。

在 SCWR 发展的背景下,对这些数据的彻底审查还有待进行,但是随着反应堆内监测设施的投入使用,这样的评估可以帮助设计试验程序来研究 SCWR 中的燃料破损。

2) 燃料浸出测试

Guzonas 等(2016)报道了 SCW 中模拟燃料(SIMFUEL)浸出试验的结果,该试验旨在识别那些能够从燃料基质中溶解的裂变产物种类,并确定燃料本身是否可能在 SCW 中溶解。目前加拿大 SCWR 的标准燃料由含有 13%(质量分数)$PuO_2$的 $ThO_2$组成,而钚的放射性和毒性使新燃料和辐照燃料的处理变得更为复杂,同时也带来了其他挑战。SIMFUEL 中使用天然 $UO_2$代替 $PuO_2$,并且使用了非放射性添加剂来模拟辐照燃料中存在的裂变产物。之所以选择燃耗为 60GW·d/t 的 SIMFUEL 进行测试,是因为它含有浓度最高的裂变产物替代物,增加了裂变产物替代物释放的可能性,并增加了它们在溶液中被检测到的可能性。试验是在压力为 25MPa、温度分别为 400℃和 500℃的 Hastelloy C-276 型静态高压釜中进行的。选择 400℃的温度条件是因为在该温度下,SIMFUEL 组分的溶解度相对较高,而 500℃是高压釜的最高操作温度。

试验检测到高压釜中大量释放的腐蚀产物(Fe、Ni、Cr、Mo)。在模拟的裂变产物中,只有Sr和Ba在显著浓度下被检测到(图4.17)。在测试条件下,这两种物质的溶解度都相对较高,但可能还没有达到平衡溶解度,这可能是因为在烧结的SIMFUEL芯块中的扩散限制了溶解速率。因为Mo也是Hastelloy C-276的主要成分,并且高压釜的浸水面积比试样大得多,所以SIMFUEL释放的Mo无法量化。

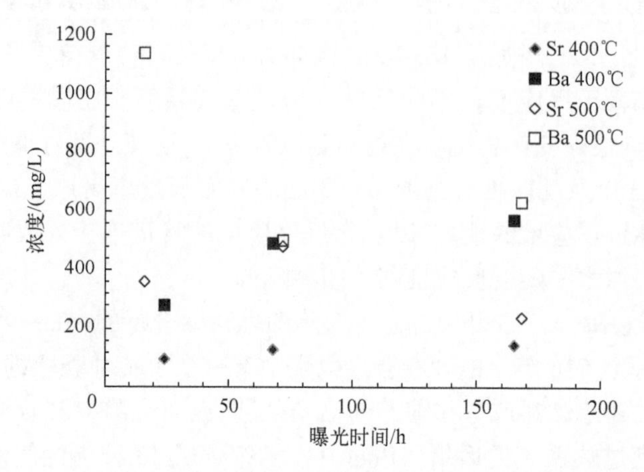


图4.17　在压力为25MPa、温度分别为400℃和500℃时Sr和Ba的浓度

在400℃时,一个或多个接近ICP-MS的MDL的测试中,检测到了Zr、Ru、Pd、Ag、Sn,这表明这些元素在试验条件下的溶解度较低。在500℃时获得了相同的结果(除了Ba的第一个数据点外,其余浓度没有明显高于400℃时的浓度)。500℃时,一个或多个接近ICP-MS的MDL的测试中,检测到了Rh、Ce、La和Nd。在400℃和500℃,除Sr、Ba和高压釜腐蚀产物外,Rh是唯一被检测到的元素,它在500℃的浓度比在400℃低一个数量级。

在两个测试中报告了溶液中可测量的U浓度:温度为400℃时浓度为0.031mg/kg(lg(U/(mol/kg))=-9.9),500℃时浓度为0.064mg/kg(lg(U/(mol/kg))=-9.5)。这些数值与文献值基本一致。例如,Red'kin等(1989,1990)报告了在压力为100MPa、温度为300~600℃,lg(U/(mol/kg))=-9.5条件下的平均值;Guzonas等(2016)也报道了在压力为25MPa、温度为400℃条件下,氧化钍粉末溶解度的简单测量结果。经过冷却和过滤后,溶液中Th的浓度小于0.2μg/L。

## 4.4 水的辐照分解

所有 SCWR 设计中最重要的水化学挑战之一就是预测水的辐照分解对材料性能和腐蚀产物运输的影响，并在可能的情况下制定缓解策略。当 SCWR 冷却剂通过反应堆堆芯时，将受到强烈的辐照。核反应堆堆芯水的辐射化学是由 γ 辐照和快中子辐照导致水的电离。辐照化学可能也受靠近燃料包壳表面的 α 辐照分解和不稳定铀的裂变碎片辐照分解的影响。典型反应堆堆芯的辐照剂量约为 1000kGy/h。不同类型的辐照与水的相互作用不同，因此建模工作最终必须考虑所有相关类型的辐照。

当一个移动的带电粒子在物质中减速时，它就会失去能量，在其运动路径上产生一串被激发的原子和分子，这一过程如图 4.18 所示。电磁能（γ 射线和 X 射线）通过产生电子和正电子也有类似的效果。由于中子是不带电的，它们不会直接导致物质电离，但是许多中微子与物质之间的相互作用确实会产生带电粒子，这些带电粒子随后会充当“基本粒子”。这些不同形式的辐照以不同的速率在介质中失去能量，通常以线性能量传输（LET）的形式表示，LET 定义为“电离粒子通过介质时能量损耗（局部吸收）的线性速率”（Spinks 和 Woods，1990），即 $dE/dx$，$E$ 为能量，$x$ 为距离。光子是低 LET 粒子，而快中子和 α 粒子是高 LET 粒子。已知 LET 会影响辐照分解产物的产量。



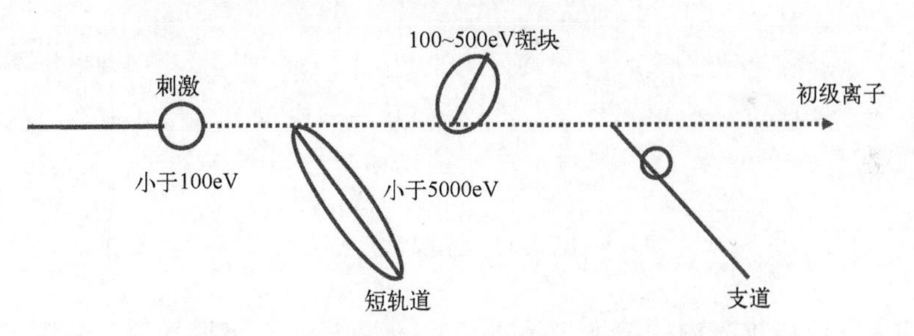


图 4.18　初级电离粒子射入液体后的轨迹结构

如图 4.19 所示，在电离事件后的第一皮秒内，会发生解离、非辐照衰变、质子转移、电子热化和水合、离解电子俘获和复合反应，从而形成水合电子（$e_{aq}^-$）、氢氧自由基（·OH）、氢原子（·H）、氢分子（$H_2$）、过氧化氢（$H_2O_2$）、超氧化物自由基（$HO_2/O_2\cdot^-$）以及质子（$H^+$）（Spinks 和 Woods 1990；Elliot 和 TTBartels，2009）。当这些物质从最初的电离中心扩散开来，形成稳定的氢、氧和过氧化氢等最终产物时，它们会发生其他反应。而且这些稳定的最终产物的浓度主要决定了堆芯下游的放射性水的腐蚀性，以及放射性水与短寿命活性中间体表面的相互作用，因此如羟基(HO·)、水合电子($e_{aq}^-$)、氧自由基($O_2^-$)等都不可忽略。Ishigure 等(1980)报道称，γ 辐照增加了 304 SS 溶解铁的释放速率，但不增加颗粒铁氧化物的释放速率。Christensen(1981)提出 $O_2^-$是在堆芯辐照时，氧化锆-2 腐蚀增加的原因。Moreau 等(2014)指出 HO·和 $e_{aq}^-$能通过反应被金属(316L SS 和哈氏铬)表面清除。中间体与表面氧化物或合金的这种反应可能在有限区域最为显著，如孔或缝隙。

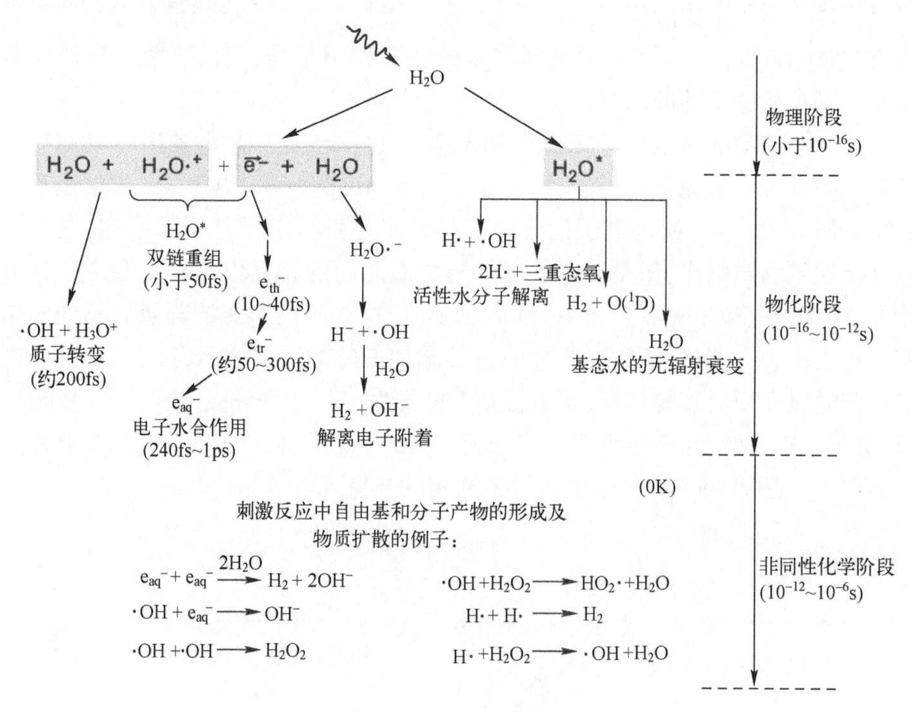


图 4.19　水辐解途径



测量反应堆堆芯内的浓度是不可能的，但是通过对下游的采样和测量，可以获得稳定物质的浓度，同时需要考虑在高温水系统中测量氧化性物质(如氧和过氧化氢)的难度。利用电化学腐蚀电位探头在 BWR 中进行间接测量，这些最终产物在堆芯和下游的浓度是确定材料测试所需的氧化剂浓度的关键。最近大多数用于 SCWR 开发的材料测试都是在 DO 浓度为 8mg/kg 或更低浓度的情况下进行。

迄今为止，对 SCW 中的水辐照分解的研究一直是基于实验室条件进行的(有关设施的描述可参阅第 2 章)，而不是基于反应堆的。因此，目前我们是基于模型来探究水在 SCWR 中可能产生的影响。这种模拟需要了解氧化产物①( $\cdot OH$、$H_2O_2$、$HO_2/O_2$ ) 的温度和密度的相关性,减少辐解产物,并了解在支线管路中发生反应的速率常数和产生这些主要物质的过程。这些信息可以用于当前 WCR 以亚临界水运行时的温度范围,而对 SCW 中这些参数的了解是不完整的,此时必须考虑建模工作的结果。Lin 等(2010)对近期的试验工作进行了总结。

主要的辐照分解产物包括离子和自由基两种,因此,有理由认为在临界点附近,水的性质(如介电常数)的迅速变化会影响这些物质的各种反应。此外,水密度的变化还会改变各种活性物质相互远离的程度。最后,SCW 的非均匀性意味着物质可以根据它们与水的相互作用是否容易,划分成局部的高密度区域或低密度区域。在近临界区,水辐照分解的模拟是最困难的。

在从室温到 $T_c$ 以上的温度范围内,通过使用低 LET ($\gamma$ 或快速电子) 来辐照主要物质所得到的 $g$ 值已经被多个机构研究。这些速率常数对温度的依赖关系并不遵循 Arrhenius 关系,因此,它们在 SCW 中的值不能通过现有亚临界水数据的简单外推来预测。与本章前面讨论的热力学量一样,高温速率常数的测定,特别是在 SCWR 条件下,极具挑战性。除了试验上的挑战,使用化学"清除剂②(离子或自由基清除剂是添加到测试溶液中的化学物质,与特定的初级辐照分解产物发生定量反应,从而可以测量其浓度)"来跟踪中间辐照分解产物反应的技术可能存在问题,因为我们对 SCW 中清除剂分子的行为了解很少。更复杂的是,不同的试验技术需要测量在不同的时间尺度上(即在水分子与辐照的初始相互作用和最终活化状态结束并返回到均匀动力学状态之间)的产量(Sanguanmith 等,2013)。例如,Haygarth 和 Bartels(2010) 通过试验得出水合电子($e_{aq}^-$)产率③的结果与其消失速度很接近,而 Lin 等(2005)的试验测量结果更接近初始活化事件。

许多相关反应对温度的依赖关系在临界点附近显示出局部的最小值(对于轻水和重水,图 4.20)。据报道,这个最小值发生在流体可压缩性达到最大的热力学条件下(Ghandi 等,2002、2003;Percival 等,2007;Marin 等,2005;Bonin 等,2007)。正如 Alcorn 等(2014)讨论的,这种最小值可能来自密度不均匀性(Tucker,

---

① 产率(每吸收单位能量产生或消耗的物种数量) 用 $g$ 或 $G$ 值表示,常规单位为"分子/100eV"(Spinks 和 Woods,1990)。初级产率(通常在加速膨胀的末端测量(图 4.19 中的非均匀化学阶段结束))用小写字母 $g$ 表示,如 $g(e_{aq}^-)$,稳定产物的产率用大写字母 $G$ 表示(如 $G(H_2)$,两者进行区分。

② 离子或自由基清除剂是添加到测试溶液中的化学物质,特定的初级放辐照溶解产物定量反应,从而可以测量其浓度。

③ 刺激寿命结束,回归均匀动力学。

1999;Ghandi 等,2012),热力学临界点上反映的“临界慢化”(Procaccia 和 Gitterman,1981)和冷却剂笼效应(指溶剂团簇内分子间的接触倾向于持续更长的时间,在每次碰撞中为反应物提供更多的碰撞机会。气相碰撞由于密度低,通常只发生一次碰撞。当流体结构随温度变化时,笼效应对速率常数的温度依赖性有重要影响)①(Ghandi 等,2002、2003;Percival 等,2007)等因素的影响。Liu 等(2016)提出了笼效应模型,并将其用于预测 SCW 中水辐照分解中涉及的一系列平衡反应速率常数。

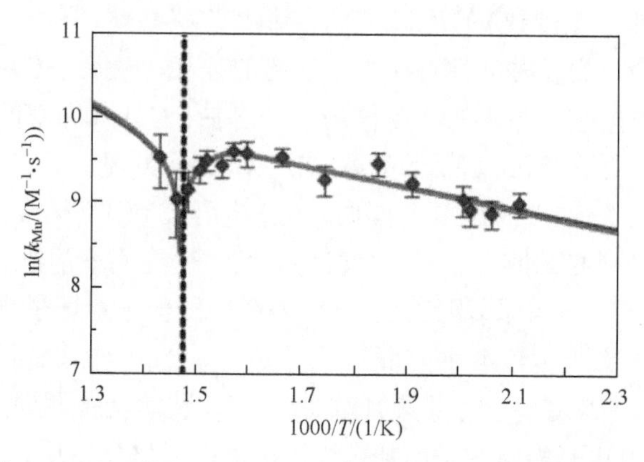


图 4.20　压力为 30MPa、温度为 35~425℃,μ 介子素(氢的同位素)与水的反应速率常数



我们知道,密度波动和冷却剂不均匀性发生在冷却剂可压缩性很大的临界点附近(Tucker,1999;Ghandi 等,2012)。在临界点处,水分子的局部聚集导致其密度分布不均(图 4.2)。疏水物质,如 H·、Mu·,在能量上有利于形成水分子簇之间的空隙,而亲水物质,如离子,将位于水分子簇内。像 H·+$H_2O$ 这样的反应,通过减少 H·与水的接触次数,以降低空隙中水分子的浓度,使该反应速率常数为最小值。

## 4.4.1 模型建立的方法

从 SCWR 设计者的角度来看,需要的不是一次和二次瞬态辐照分解产物的浓度(图 4.18),而是最终产物(氢、氧和过氧化氢)的稳态浓度。SCWR 的发展采用了两种方法:一种是利用试验数据和计算机模拟来开发预测模型,这些模型基于详细微观已知的辐照分解产量,以及它们对产生影响的温度、密度、结构及

① 笼效应是指溶剂团簇中分子之间的碰撞持续时间更长,在每次碰撞中为反应物提供更多的碰撞机会。由于密度低,气相碰撞通常由一次碰撞组成。当流体结构随温度变化时,笼效应对速率常数的温度依赖性有重要影响(Leblanc 等,2014)。

其他参数;另一种是半经验方法,类似于当前 WCR 使用的方法。

#### 4.4.1.1 微观模型

蒙特卡罗模拟为定量描述 SCW 辐照溶解过程中温度和压力(密度)效应、识别关键参数以及阐明辐射与 SCW 相互作用的机制提供了一个很好的途径。分子动力学(MD)模拟提供了对 SCW 的微观理解,以及准确描述在大的密度范围内发生辐照分解时其分子结构受到的影响(Metatla 等,2016)。与腐蚀产物输运模型一样,这种类型的辐照分解模型需要在 SCW 条件下检测试验数据的可用性。

这种反应速率变化的重要性可以用氢和羟基自由基反应生成氢自由基和水来说明:

$$H_2+HO\cdot \longleftrightarrow H\cdot+H_2O \tag{4.5}$$

根据该反应,可以通过在 WCR 的冷却剂中加入氢来抑制水的净辐解①。这是唯一发生在非均相化学阶段的反应,其反应速率足够快(通过加入高浓度的氢),在·OH 自由基遇到并氧化另一种物质之前,将氧化的·OH 自由基转化为还原的·H 自由基。式(4.5)中反应物和产物的相对浓度可以通过调节 $H_2$的浓度来改变。因为这是所有反应中唯一一个除去 $H_2$的反应,所以通常将氢分子注入压水堆和水冷堆冷却系统中,以抑制氧化辐解产物的形成。在高温水的脉冲辐照分解试验中发现该反应的逆反应是 $H_2$ 浓度增加的原因(Swiatla-Wojcik 和 Buxton,2005;Bartels,2009;Swiatia-Wojcik 和 Buxton,2010)。

式(4.5)的速率常数显示了在 250℃ 的温度范围内的 Arrhenius 行为;当高于 250℃时,反应速率减慢,并且逆向反应可能变得很重要。由于目前对超临界流体中瞬态物质溶解能的研究资料较少,水热条件下的反应速率常数一直是人们讨论的热点。在近临界区,离子和自由基的相对稳定性随温度的变化会改变各种反应途径的相对重要性(如扩散和重组)。使用蒙特卡罗方法模拟从室温到 350℃的温度范围内式(4.5)的速率常数对 $H_2$产率的影响,结果表明,当温度为 300℃时,模拟结果与试验结果十分吻合,速率常数 $k=10^4M^{-1}\cdot s^{-1}$。

为了解释式(4.5)对所测温度的依赖性,Alcorn 等(2014)提出了两个相互竞争的反应,一个是低温下活化能较低的反应(式(4.6)),另一个是高温下具有高得多的活化能(式(4.5)中的逆向反应)。

$$H\cdot+H_2O \longleftrightarrow H_3O^++e_{aq}^- \tag{4.6}$$

这种反应方式随温度的变化与水的性质随温度的变化相一致。根据

① 净辐照分解的抑制意味着核心出口处的氧化剂浓度无法检测。

式(4.6),在低温下,较高的水密度使所产生的离子稳定下来;但是随着温度的升高,有利于中性物质的增加(式(4.5));在更高的温度下,该反应的活化能接近抽氢反应的气相活化能,因此可能会变成气态。由于在近临界区以外的冷却剂性质没有明显的波动,因此在这一区域将再次出现 Arrhenius 现象。

为支持SCWR的发展,辐照分解模型的一个主要关注点是SCW密度对辐照产率的影响。Guzonas 等(2012)认为在低密度 SCW 中,主要是类气态组分的辐射分解,其基础是辐照水蒸气和辐照液态水的物理轨迹结构过程、化学反应和产物产率的巨大差异。例如(Meesungnoen 等,2013),在电子热化之前,次激励电子与母体阳离子在接近相反的电荷后会重新结合(图4.19左上角的阴影框)。之前 Goulet 等(1990)的模拟表明,重组发生在次激励电子开始随机移动时。重新结合的电子往往是那些在没有正离子的情况下,在离起始位置相对较近的地方被加热的电子,但不适合 SCW 密度较低、水分子相对较少的情况。低密度 SCW 中阴、阳离子对复合的减少将提高 $e_{aq}^-$ 的初始产率(时间尺度为 $10^{-12}$s)。Meesungnoen 等(2013)假设电化法的阴、阳离子对复合概率与密度成简单的线性依赖关系,结果计算得到的低密度下 $e_{aq}^-$ 产率的值有显著的提高(图4.21)。在1ns时,模拟结果与试验值吻合较好,但在60ps和低密度时①,模拟结果仍略低于试验值。这项工作说明,要详细了解在受辐照的超临界水(特别是低密度时)中产率与密度的关系,需要详细考虑密度对水结构的影响。



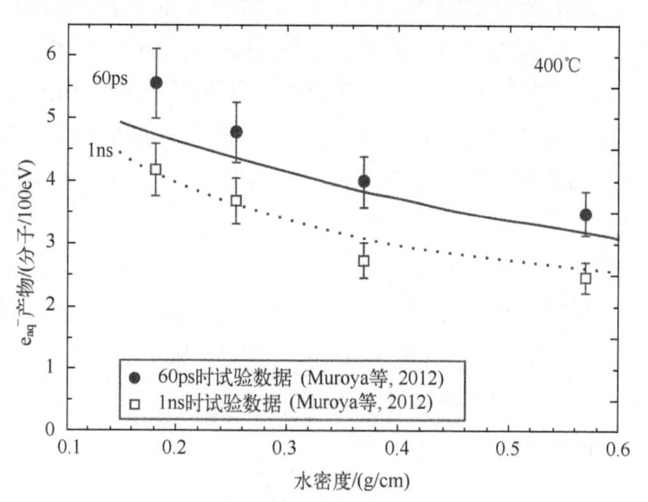


图4.21　在400℃的超临界水中 $e_{aq}^-$ 产率与水密度的关系

① 作者注意到,他们正在将 $e_{aq}^-$ 对 $H_2O$ 的计算产率与 $D_2O$ 中试验得到的值进行比较。Bartels 等(2001)预测 $D_2O$ 中的"初始"产率 $e_{aq}^-$ 比 $H_2O$ 的大7%;进行这种修正,可以使在60ps下测量和计算的 $e_{aq}^-$ 值非常一致。

虽然 SCW 辐照分解的微观模型提供了有价值的参考，如水密度等参数对辐照分解产率的影响，但是微观模型还没有发展到能够提供定量预测氧化物质浓度的地步，而这是定义 SCWR 堆芯化学所必需的。

#### 4.4.1.2 半经验模型

有几个研究小组已经采用了一种更宏观、更少的第一性原理的方法来建立 SCWR 中氢、氧和过氧化氢稳态浓度的模型，类似于对当前 WCR 冷却剂中氧化剂浓度进行预测的方法。这些模型要求在瞬态辐照分解产物的各种反应中选择一组基本反应以及这些反应的速率常数。通过用适当的参数值求解耦合的时变速率方程来计算稳态浓度。

第一次这样的模拟数据（Yeh 等，2012；Yeh 和 Wang，2013）来自一个美国 SCWR 型压力容器，并使用 400℃ 的 $g$ 值，在某些情况下的 $g$ 值是由 300℃ 的值使用 Arrhenius 关系外推得到的（是无效的）。尽管如此，最终的结果还是很有价值的，因为它们表明在 SCWR 堆芯中可能存在高浓度的氧化剂。他们还模拟了在堆芯入口添加浓度为 1mg/kg 的 $H_2$，发现 $H_2O_2$ 和 $O_2$ 的浓度降低了几个数量级。

随后 Subramanian 等（2016）试图更好地解释近临界区域中水的性质变化产生的一些影响，这使得近临界区域水辐照化学模型得到发展。现在建立了两种化学动力学模型，分别从类液态和类气态接近临界区密度。类液态辐射分解模型（LLRM）使用的反应装置与现有的适用于室温条件下液态水的辐照分解模型中使用的反应装置类似（Wren 和 Ball，2001；Wren 和 Glowa，2000；Joseph 等，2008；Yakabuskie 等，2010、2011）。类蒸汽辐照分解模型（VLRM）使用了为水蒸气辐照分解而开发的反应集（Arkhipov 等，2007），并应用于高温（900K）条件下的水蒸气化学，但与 SCWR 中使用的反应集相比，压力（0.01～1MPa）和蒸汽密度（0.25～1kg/$m^3$）要低得多。

给定一组物质初始浓度，开始进行计算，持续 $10^4$s；这个时间范围限定了冷却剂在 SCWR 堆芯中的停留时间，大约为 1s。计算范围为 25～400℃，比较 VLRM 和 LLRM 模型的预测结果。在连续辐照通量（堆芯内 $\gamma$ 辐照场估计约为 1000kGy/h（Janik 等，2007））的阶跃函数启动之前，用纯水（pH 中性，无添加剂，除氧）开始计算。对于液态水和水蒸气，系统压力为目标温度下的饱和压力（Alexandrov和 Grigoreiv，1999）。在 LLRM 和 VLRM 的计算中，水密度的变化会影响单位体积吸收的辐照能量。在 VLRM 模型计算中，水密度也会影响涉及水分子的三体反应的伪二阶速率常数。

不同辐照分解产物的浓度随时间的变化速率不同，在堆芯停留时间的长短对不同辐照分解产物的浓度影响较大。LLRM 预测，在堆芯停留 1s 时间内，随着温度的变化，对辐照分解产物浓度的影响很小，因为水相中更多依赖于温度的化学反应只有在较长时间内才变得重要。VLRM 预测，当 $T>50℃$ 时，温度变化对辐照分解产物浓度的影响更大（图 4.22）。这些初步的计算是在沿饱和压力曲线的压力和温度下进行的，结果水蒸气密度发生了很大的变化。水蒸气密度的升高增加了单位体积水蒸气的能量吸收速率，增加了初级辐照分解产物的生成速率，增加了涉及水分子的三体反应速率。随着时间的增加，LLRM 和 VLRM 预测结果之间的差异越来越大，这与两种模型处理辐照分解和化学所采用不同的方式相一致。在短时间内，氧化物质的浓度应该主要取决于初级辐照分解产率，而不是次级辐照分解产物的反应。

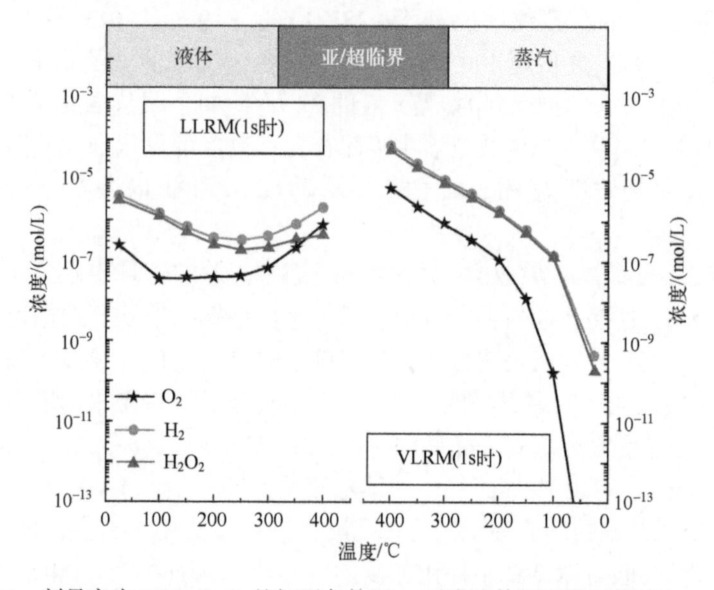


图 4.22 剂量率为 1000kGy/h 的辐照条件下 1s 时类液体辐照分解模型（LLRM）和类蒸汽辐照分解模型（VLRM）对辐照产物浓度随温度变化的预测



在接近临界的区域，两个模型对在堆芯短时间停留条件下的预测结果相差 2~3 个数量级。同时，这些结果表明，与腐蚀产物沉积一样，近临界区将是建模和化学控制的最大挑战。

### 4.4.1.3 大规模回路或反应堆内研究

用于乌克兰的加拿大-乌克兰电子辐照测试设施（CUEITF）的 SCW 对流回路和在 Centrum Výzkumu Řež 的用于材料和化学测试的 SCW 回路（两者在第 2 章都有描述）都已经开始建造，并将用于研究水的辐照分解对材料性能的影响。乌克兰设施已经投入使用，并提供了一些关于水辐照效应的宝贵数据。几乎所有被考虑用于 SCWR 堆芯的合金上形成的氧化膜都是 Cr 的氧化物,因此在选择 SCWR 堆芯材料时应特别关注氧化条件下 Cr 的性能。氧化铬在氧化条件下是可溶的(Cook 和 Olive,2012b),形成 Cr (Ⅵ)物质。在 BWR 中发生正常的水化学反应中(没有加入氢),观察到氧化铬溶解为 $H_2CrO_4$(转位溶解),这是因为水中含有因辐照分解产生的高浓度氧化物。如果水的辐照分解无法控制,SCWR 中也会出现类似的现象(Guzonas 等,2012;Karasawa 等,2004;Fujiwara 等,2007)。

在 25MPa 的氧化条件下,观察到可溶性 Cr(Ⅵ)的形成(第 5 章)。在温度高于 500℃的氧气和水的混合物中,对色度蒸发的等效气相过程进行了详细研究(Halversson 等,2006;Holcomb,2008、2009)。色度蒸发是由下列反应引起的,即

$$\frac{1}{2}Cr_2O_3(s)+\frac{3}{4}O_2(g)+H_2O \longrightarrow CrO_2(OH)_2(g) \tag{4.7}$$

$CrO_2(OH)_2$的化学性质与 $H_2CrO_4$相同。发现这个过程对流动有很强的依赖性,这表明它依赖于挥发性(溶解)物从物体表面的质量转移。

值得注意的是,在美国核能蒸汽再热的发展计划中,绝大多数材料测试是在 DO 浓度为 20~50mg/kg 的条件下进行的。然而,最近对 SCWR 的大多数测试是在浓度为 8mg/kg 或更低的条件下进行的。使用 CUEITF 设备在临界点附近的温度条件下,进行电子辐照测试(Bakai 等,2013),测试 500h 的结果为氧化铬在水辐照分解氧化条件下的溶解提供了明确的证据。出口水样本含 3~5μg/kg 的氧①(图 4.23)。试验段出口水的电导率以(0.03±0.1)μS/(cm·h)的速度累积增加至 23μS/cm,这表明金属因腐蚀而释放到冷却剂中。ICP-OES 元素分析表明,水中 Cr 的浓度高达 54μg/kg;但在暴露于电子束之前,水中没有检测到任何 Cr。

除了可能增强堆芯合金的腐蚀外,水的辐照分解也可能影响腐蚀产物的运输。Yakabuskie 等(2011b)曾报道,受 γ 辐照的亚临界脱氧 $FeSO_4$溶液导致形成大小均匀的 γ-FeOOH 胶体颗粒。Mayanovic 等(2012)发现放射性溶解作用可能导致 SCW 中形成钨离子的胶体沉淀,并且根据压力、温度和辐照条件,铁可能被氧化、还原或形成沉淀。同时还应考虑水在沉积或腐蚀氧化物内的孔隙或缝隙中辐照分解的可能性。

如前所述,式(4.5)是控制 WCR 水辐照分解的关键。在冷却剂中加入足够高浓度的氢(或氨,它会分解成氢)(根据不同的反应堆设计不同)可以减少由放

---

① 由于将化学监测仪与辐照区隔离需要很长的样品线长度,试验段的实际氧浓度可能更高。

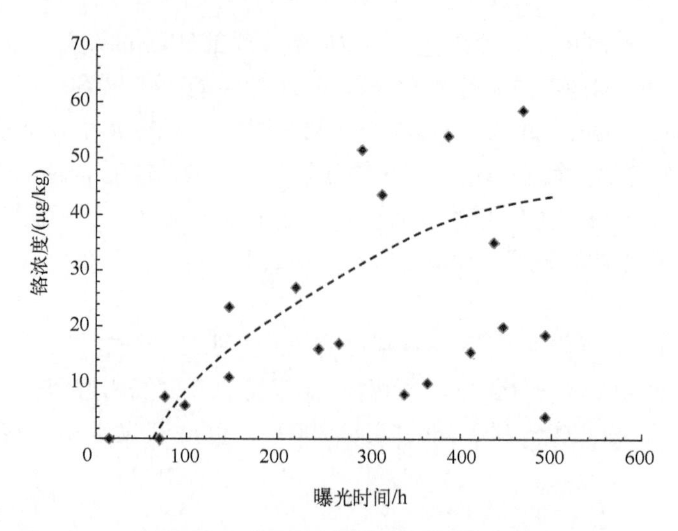


图 4.23　辐照期间循环水中铬的含量

射性导致的氧化物。在 Beloyarsk 核电站中研究了高温蒸汽条件下抑制净辐照分解(Yurmanov 等,2010;Gruzdev 等,1970)。起初,1 号机组单独使用氨来控制 pH 值,通过氨的辐照分解制氢来间接抑制净辐照分解。尽管在蒸汽进入过热通道时,氢浓度为 45~88mL/kg(在 NTP),但 SHS 中氧气浓度仍为 2.3mg/kg。1 号机组测试了通过注入氢来抑制过热通道中的净辐照分解的方法,但在研究期间停止了对给水中氨的处理。注氢成功地将蒸汽中的氢浓度维持在 1.2~6.2mL/kg,导致 SHS 中的氧浓度小于 0.15mg/kg。进一步的测试表明,当蒸汽氢浓度大于 45mL/kg 时,SHS 中的氧浓度可以维持在 0.03mg/kg 以下。Bartels 等(2010)假设,向 SCWR 中加入浓度约为 PWR 中当前值 10 倍的氢可能足以抑制氧化物的形成,这与 Yurmanov 等(2010)的结果近乎一致。因此,在 SCWR 堆芯出口的低密度 SCW 中控制水的辐照分解可能与在 Beloyarsk 低密度蒸汽中的控制方式相同。



Yurmanov 等(2010)也注意到,在 Beloyarsk 核电站 1 号机组的过热器通道中添加了氢,汽轮机喷射器产生了由 62%的氢和 8%的氧组成的爆炸性蒸汽混合物,这就需要用空气进行稀释,而类似的问题也可能出现在 SCWR 中。通过腐蚀产生额外的氢(Choudhry 等,2013)也必须考虑在内。

作为联合英国原子能机构(UKAEA)/加拿大原子能有限公司(AECL)WCR 系统开发计划的一部分,在 AECL 的国家研究实验核反应堆(NRX)中的一个压水回路(X-4 回路) 使用蒸汽进行冷却(LeSurf 等,1963),用以考察使用蒸汽作为核动力反应堆冷却剂的可行性。由于在这样的反应堆中,通过添加碱来控制

pH 值是不合适的(因为添加碱会使燃料元件表面发生沉积),因此用氨来进行 pH 值控制,研究了氨在两种沸水中的辐照分解,并研究了蒸汽循环中各种气体成分对 SHS 的影响。采用蒸汽冷却的反应堆,其沸腾区蒸汽中含有高浓度的氧(10~30mg/kg)和氢(14mg/kg)。我们希望能使该反应堆沸腾区产生的蒸汽中的氧和氢在过热区发生催化复合,以降低进口蒸汽的氧含量。以过氧化氢作为氧的来源,研究了添加到 X-4 回路中的氧和氢的行为。在试验段前后连续取样,并进行氧分析。操作条件因燃料成本的不同而不同,以下是一组典型的条件:

(1) 工作压力(锅炉)$57kgf/cm^2$(约 5.7MPa);

(2) 试验段蒸汽流量 320kg/h;

(3) 通过泵的总的水流量为 2100kg/h;

(4) 试验段入口蒸汽温度 300℃;

(5) 试验段出口蒸汽温度 430℃;

(6) 燃料包壳温度 550~650℃。

试验中加入了高浓度的氢(1110mL/kg),回路中没有燃料泄漏。预计不会在整个试验段建立反应平衡,因为达到平衡需要燃料元件热表面的催化协助。结果显示,在反应堆的过热区域内,如果加入的氢浓度远远超过化学计量比,那么在蒸汽中的氧浓度为 20~30mg/kg 时,可以实现氧与氢的辐照驱动复合反应。

Subramanain 等(2016)建立的模型用于预测在 400℃时,添加 $H_2$后 1s,SCW 辐照分解的变化(蒸汽 24MPa、$148kg/m^3$)(图 4.24)。LLRM 预测,$H_2$的加入可以减少(而不是抑制)辐照分解生成的 $O_2$;在加入 $H_2$的初始浓度为大于 $10^{-4}mol/dm^3$(约 10mL/kg)的条件下,1s 时 DO 浓度低于 $10^{11}$ $mol/dm^3$。然而,加入 $H_2$ 对减少 $H_2O_2$产率的效果较差,即便在加入 $H_2$ 的初始浓度为大于 $10^{-2}$ $mol/dm^3$(约 1000mL/kg)的条件下,1s 时 $H_2O_2$的浓度保持在约 $10^{-7}$ $mol/dm^3$;这种氢浓度远远高于现有的 WCR。这些结果与 Yeh 和 Wang(2013)的结果相似。VLRM 预测,添加任何适量的 $H_2$都不会抑制 $O_2$或 $H_2O_2$的产生(图 4.24,在更长的时间,$O_2$和 $H_2O_2$的稳定浓度会更高)。添加 $H_2$(浓度大于 $10^{-4}mol/L$)后,唯一能观测到的影响就是·O 浓度的降低,这是因为发生了如下反应,即

$$\cdot O + H_2 \longleftrightarrow OH + H\cdot \quad (4.8)$$

这两个模型都预测,加入 $H_2$可能不会有效地减少 $H_2O_2$的辐照分解产物。图 4.24 还显示了 Beloyarsk NPP 添加氢和添加氨后测量的氧浓度数据。测量的氧浓度与 VLRM 预测的氧浓度之间的良好一致性表明了该方法基本可靠。

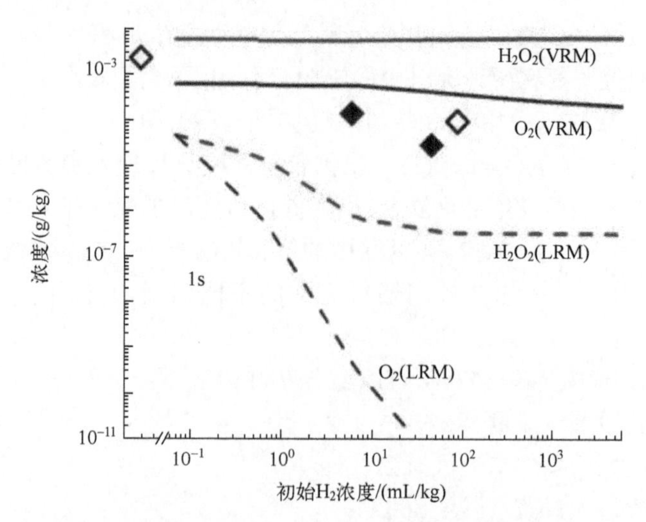


图 4.24　400℃、1000kGy/h 的条件下照射 1s,用类液体辐照分解模型(LLRM)和类蒸汽辐照分解模型(VLRM)预测辐照分解产物浓度与初始 $H_2$加入量的关系

## 4.5 超临界水冷反应堆的化学控制



SCWR 的部署预计要在 2020—2030 年才能完成,由于 SCWR 内环境对候选堆芯材料的腐蚀和 EAC 测试有很强的影响,因此需要一套具有坚实技术基础的水化学规范。回到图 4.6,本章对腐蚀产物运输到堆芯、活化物质从堆芯输出以及水的辐照分解进行了讨论,得出以下 3 个结论,分别对应图中确定的 3 个过程中的一个。

(1) 通过采用 SCFPP 和 BWR 中给水系统的材料选择和化学控制的最佳方法,可以最大限度地减少 SCWR 中杂质和腐蚀产物的堆芯内沉积。这些数据表明,与氨、联氨或氢氧化钠相比,含氧给水化学反应产生的问题更少。需要使用预期腐蚀产物和杂质的实际浓度进行长期试验,首先在反应堆外回路进行试验,最后使用反应堆内的试验设施进行测试,以评估这些物质输运和堆内沉积的影响。

(2) 水的辐照分解仍然是阻碍定义 SCWR 对内环境的关键问题。数据清楚地表明,在水的辐照分解未被抑制的情况下,所产生的高浓度氧化剂会导致 Cr 从合金表面释放出来。Beloyarsk NPP 采用核蒸汽再热技术的运行经验表明,加氢可以抑制氧化物的净辐解生成。模型表明,在较低的 SCW 密度下,在堆芯入口添加氢不能“抑制”水辐射分解产生的氧和过氧化氢,但它们的浓度可以得到降低。因此,建议加入氢气,但不能确定加氢的具体浓度。添加氢意味着“双重”水化学,其中含氧的给水化学在堆芯入口被氢的给水化学取代(Kysela 等,2009)。因此,材料测试应该在一定的 DO 和氢浓度范围内进行,以限制这些物质的预期浓度。在 SCWR 设计的任何阶段,都需要考虑堆芯出口的 SCW 中存在氢和氧爆炸混合物的可能性。

(3) 简单的计算和合理的假设表明,在 SCWR 堆芯出口的冷却剂中,$^{60}Co$ 的浓度可能与当前的 WCR 没有明显不同。虽然金属氧化物在低密度 SCW 中的溶解度较低,但仍远高于 BWR 中相同氧化物在低压蒸汽中的溶解度。有关物质在 SCWR 条件下的溶解度需要更多的数据,微粒物质(如前面提到的胶体)的释放和运输也必须加以考虑。必须强调的是,与 BWR 不同,BWR 的相变导致蒸汽中很少有活性物质从堆芯向外转移,而 SCW 中大多数金属的溶解度虽然低,但非零;单相冷却剂还可以方便地输送微粒,因此需要屏蔽主蒸汽管道和高压汽轮机。

Kysela(2009)、Yurmanov(2009)、Guzonas(2010),以及 Guzonas 和 Cook 等(2015)提出了 SCWR 中可能的水化学控制方法。表 4.3 列出了由 Guzonas 和 Cook(2015)以及 Yurmanov 等(2010)提出的初步 SCWR 化学规范。考虑到上面第(2)项关于水辐照分解的观点,这两套规范相似,为材料测试奠定了良好的基础。

表 4.3 为超临界水冷反应堆提出的化学规范

| 参数 | 标准 | | 单位 |
|---|---|---|---|
| | Guzonas 和 Cook(2015) | Yurmanov 等(2010) | |
| 氯化物 | 0.25 | <2 | μg/kg |
| 硫酸盐 | 2.0 | — | μg/kg |
| 铁 | 0.1 | <5 | μg/kg |
| 硅 | 300 | <2 | μg/kg |
| 钠 | — | <2 | μg/kg |
| 溶氧的初始给水 | 200 | | μg/kg |
| 溶氧的最终给水 | 50 | <20 | μg/kg |
| 氢化物 | 待定 | 待定 | — |
| pH 值 | 7.0 | 6.8~7.1 | — |
| 电导率 | 8 | <0.1 | μS/m |
| 油 | — | <100 | μg/kg |

## 4.6 分子动力学模拟

最后一节将考虑物质表面上的 SCW 结构，作为本书最后两章的过渡。现在已经证实，在亚临界温度下，由于带电表面对水分子和离子进行排序，造成了阳离子和阴离子密度的局部不平衡，并随着距离表面距离的增加而逐渐衰减，从而在金属表面形成了双电层。腐蚀反应包括两个半反应。氧化反应（$M = M^{2+} + 2e^-$）产生的金属离子可以在溶液中发生水解反应；氧化反应产生的电子在金属中运输；在阴极和阳极位置产生或消耗离子将会引起溶液中局部电荷的不平衡，但通过溶液中的离子运输可以平衡电荷。

当 $T \gg T_c$ 时，SCW 密度低，不存在双电层。实验和模拟研究都表明，离子周围 SCW 的局部结构不同于非离子主流 SCW（Tucker，1999）。考虑到这种局部结构和 SCW 的非均质、团簇性（图 4.2），Guzonas 和 Cook（2012）提出了一种在温度高于 500℃时的可能表面结构，SCW 密度较低，氧化反应为 $M+O_2 \longrightarrow MO_2$（直接化学氧化，见第 5 章）。据预测，近表面区域由被类气态水簇隔开的、小的类液态水簇组成。Guzonas 和 Cook（2012）提出了一个接近临界温度的过渡区，在该过渡区中仍然存在不连续的阳极和阴极位点，但它们之间的距离更近，随着溶液中导电离子的稳定性下降，离子对成为溶液中的主要物质。因此，对溶液中的电荷平衡（或不平衡）的研究必须更加局部化。非离子主流的水在近临界区出现的大密度波动也可能存在于物质表面，因为连续的、稳定的双电层可能不存在于整个表面。双电层结构的转变作为一种表面相变，其转变过程可能是突然的，也可能是渐进的①。可以预测，金属或氧化物表面上流体团簇的随机性质将会不同，与物质表面直接接触的 SCW 的结构将与远离表面的主流流体的结构有很大的不同，它更聚集、密度更高。有人提出，在温度远高于 $T_c$ 的情况下，表面可能存在连续的类液体膜。



MD 模拟是获得 SCW 中水与溶质反应动力学详细信息的重要工具（Kallikragas 等，2015b）。水模型可以精确地再现物理和热力学性质，包括在大范围的温度和压力下的扩散系数，这使模拟 SCWR 环境成为可能，并利用所获得的结果来指导腐蚀和 EAC 的试验方案。通过 MD 模拟，研究在温度分别为 715K、814K 和 913K 时，被限制在呈静电中性的氢氧化铁（II）亲水表面的 SCW 行为。在模拟的前 250ps 内，表层水迅速形成，达到最终密度的约 90%。

氢氧化铁体系的原子密度分布表明，大多数被吸附的水分子都位于氢氧化

① 正如 I. Svishchev 教授建议的那样。

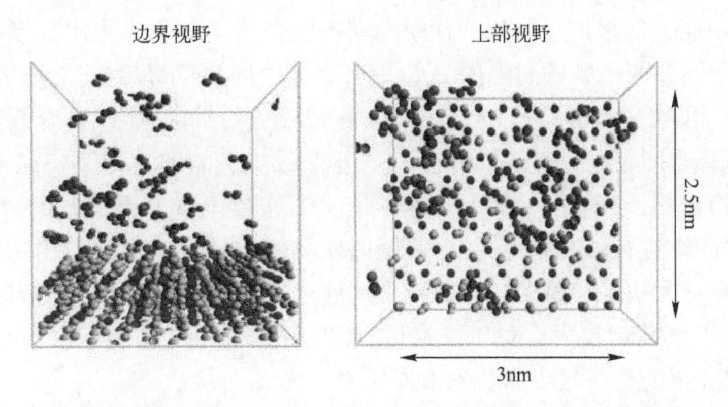


图 4.25　715K 时超临界水在氢氧化铁(Ⅱ)界面的原子密度分布(表面铁原子用深蓝色表示,表面氧原子用黄色表示,水氧原子用红色表示,水氢原子用紫色表示)(见彩插)

铁表面边缘约 1.5Å 内,其偶极子远离表面。正如 Guzonas 和 Cook(2012)预测的那样,水结构的图像(图 4.25)显示了超临界条件下表面离子的局部聚集,形成了一个有较高密度的非均匀吸附层。由亚临界水向超临界水转变时,氢氧化铁表面水的密度与非氢氧化铁表面(以下简称非表面)水的密度之比增大,当温度大于 $T_c$,该比率为 2~3.5,视温度和水的体积密度而定(图 4.26)。氢氧化铁表面水浓度的提高可能影响近临界区腐蚀机理(第 5 章),特别是对压力(密度)依赖性的影响。由于大部分 SCW 的密度会低于氢氧化铁表面水密度,因此不能准确反映氢氧化铁表面的化学条件。由于水的辐照分解也受到密度的影响,应该考虑近表面区域(以及在孔隙和裂缝中)各种辐照分解产物的产量增加(或减少)。暴露表面 OH 基团的水分子数随温度的升高而减少,由 567K 时的 5.71 降至 913K 时的 0.4(水密度为 79kg/m$^3$)。

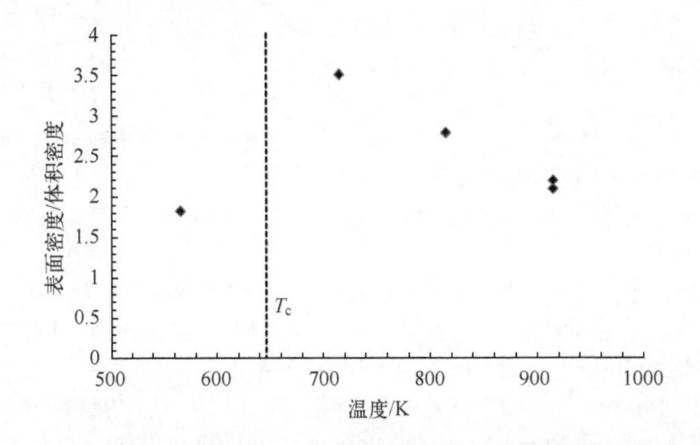


图 4.26　在 567K、715K、814K 和 913K 处氢氧化铁(Ⅱ)表层水密度与水体积密度之比

Kalligragas 等报道了使用 MD 模拟来确定在高温亚临界水中以及在超临界水中，被限制在两个紧密间隔的 $Fe(OH)_2$表面之间的稀释后氯离子(Kalligragas 等,2015a)和稀释后氧分子(Kalligragas 等,2016)的扩散系数、水合系数和密度分布。选择小间距(1nm、2nm 和 8nm)模拟裂纹尖端的环境;正如 Staehle(2006,2010)所指出的,需要用一个分子而不是一个连续体来分析这样的环境。1nm 间距对水的结构有限制作用,8nm 间距时水的结构消失。水化作用随间距增大而增大,随密度减小和温度升高而减小。缝隙的大小对氯离子的扩散影响不大,但对水的扩散影响较大。对于氧来说,当温度 $T>T_c$ 时,$O_2$位于水簇的外面,因此找到水分子的概率最小而找到氧原子 $O_2$的概率最大。

对于水,最大的扩散系数出现在 8nm 体系中,对于氯离子,最大的扩散系数出现在 1nm 体系中。在这两种情况下,在狭小系统中的扩散系数都小于在大系统中的扩散系数。水在 1nm 间距处发生聚集,并在间隙中间处增加,这种现象在更大的间隙中消失。在更大的 8nm 系统中,水接近一种体积行为。分子构型显示氯离子扩散到靠近表面的位置,并携带少量发生合水作用的水。$O_2$的扩散系数一般随着表面(缝隙)间距的增大而增大(与氯离子相反)。这项工作的一个关键结果是在超临界温度下,$O_2$在纳米尺寸的裂纹中比在大体积溶液中更容易扩散。



利用 MD 研究了 NaCl 纳米粒子在超临界水中的成核动力学(Svishchev 和 Guzonas,2011)。有 15~30 种离子(粒子)对检测 SCW 的状态至关重要。成核率大约在 $10^{28}cm^{-3}\cdot s^{-1}$的数量级,这表明成核速度非常快。在 SCW 中发现临界盐核具有非晶体结构。模拟结果表明,HCl 和 NaOH 可以在共存的超临界流体和 NaCl 团簇之间进行配分,这与水热条件下存在酸性凝聚物和碱性沉积物的现象是一致的。

## 参考文献

Adschiri,T.,Sue,K.,Arai,K.,Watanabe,Y.,2001. Estimation of metal oxide solubility and understanding corrosion in supercritical water. In:*Corrosion 2001*,Paper 01354.

Akiya,N.,Savage,P. E.,2002. Roles of water in chemical reactions in high-temperature water. Chem. Rev. 102,2725-2750.

Alcorn,C.,Brodovitch,J.-C.,Ghandi,K.,Kennedy,A.,Percival,P. W.,Smith,M.,2011. Kinetics of the reaction between H· and superheated water probed with muonium. In:5th Int. Sym. SCWR(ISSCWR-5),Vancouver,British Columbia,Canada,March 13-16,2011. Paper P130.

Alcorn,C. D.,Brodovitch,J.-C.,Percival,P. W.,Smith,M.,Ghandi,K.,2014. Kinetics of the reaction between H· and superheated water probed with muonium. Chem. Phys. 435,29-39. https://doi.org/10.1016/j.chemphys.2014.02.016.

Aleksandrova, V. N. , Veselkin, A. P. , Levich, A. A. , Lyutov, M. A. , Sklyarov, V. P. , Khandamirov, Yu. E. , Shchapov, G. A. , 1968. Investigation of the radioactivity of long-lived isotopes in the coolant of the I. V. Kurchatov nuclear power station at Beloyarsk. At. Energiya 24, 222.

Alexandrov, A. A. , Grigoriev, B. A. , 1999. The Tables of Thermophysical Properties of Water and Steam. MEI Press(in Russian).

Anisimov, M. A. , Sengers, J. V. , Levelt Sengers, J. M. H. , 2004. Near-critical behavior of aqueous systems. In: Palmer, D. A. , Fernàndez-Prini, R. , Harvey, A. H. (Eds.), Aqueous Systems at Elevated Temperatures and Pressures. Elsevier Academic Press(Chapter 2).

Arkhipov, O. P. , Verkhovskaya, A. O. , Kabakchi, S. A. , Ermakov, A. N. , 2007. Development and verification of a mathematical model of the radiolysis of water vapor. At. Energy 103(5), 870-874. https://doi.org/10.1007/s10512-007-0138-4.

Bakai, A. S. , Guzonas, D. A. , Boriskin, V. N. , Dovbnya, A. N. , Dyuldya, S. V. , 2013. Supercritical water convection loop for SCWR materials corrosion tests under electron irradiation: first results and lessons learned. In: 6th International Symposium on Supercritical Water-cooled Reactors(ISSCWR-6), March 03-07, 2013, Shenzhen, Guangdong, China.

Bartels, D. M. , Gosztola, D. , Jonah, C. D. , 2001. Spur decay kinetics of the solvated electron in heavy water radiolysis. J. Phys. Chem. A 105(34), 8069-8072. https://doi.org/10.1021/jp012153u.

Bartels, D. M. , Jonah, C. , Edwards, E. , Janik, D. , Haygarth, K. , Sims, H. , Marin, T. , Takahashi, K. , Janik, I. , Kanjara, K. , 2010. Hydrogen water chemistry: can it work in a supercritical water reactor? . In: Proceedings of the 8th Int' l Radiolysis, Electrochemistry & Materials Performance Workshop, Quebec City, Canada, October 8, 2010.

Bartels, D. M. , 2009. Comment on the possible role of the reaction H · + $H_2O$ - $H_2$ + · OH in the radiolysis of water at high temperatures. Radiat. Phys. Chem. 78(3), 191-194.

Bergeron, A. , Hamilton, H. , 2013. Fabrication and characterization of Canadian SCWR SIMFUEL. In: 6th International Symposium on Supercritical Water-cooled Reactors(ISSCWR6), March 03-07, 2013, Shenzhen, Guangdong, China. Bevilacqua, F. , Brown, G. M. , 1963a. Chloride Deposition from Steam onto Superheater Fuel Clad Materials. General Nuclear Engineering Corporation Report GNEC 295.

Bevilacqua, F. , Brown, G. M. , 1963b. Chloride corrosion effects on BONUS superheater materials. In: Terminal Report Under Job No. 66-14, Chloride Corrosion Program. USAEC Report GNEC 257.

Bohnsack, G. , 1987. The Solubility of Magnetite in Water and in Aqueous Solutions of Acid and Alkali. Hemisphere Publishing Corporation, Washington.

Bondarenko, G. V. , Gorbaty, Yu. E. , 1973. v3 infrared spectra of HDO at high pressures and temperatures. Dokl. Phys. Chem. 210, 369.

Bonin, J. , Janik, I. , Janik, D. , Bartels, D. M. , 2007. Reaction of the hydroxyl radical with phenol in water up to supercritical conditions. J. Phys. Chem. A 111(10), 1869-1878.

Brazhkin, V. V. , Fomin, Yu. D. , Lyapin, A. G. , Ryzhov, V. N. , Tsiok, E. N. , 2011. Widom line for the liquid-gas transition in Lennard-Jones system. J. Phys. Chem. B 115, 14112-14115.

Brosseau, F. , Guzonas, D. A. , Tremaine, P. , 2010. The solubility of magnetite and nickel ferrite under supercritical water reactor coolant conditions. In: 2nd Canada-China Joint Workshop on Supercritical Water-cooled Reactors(CCSC-2010), Toronto, Ontario, Canada, April 25-28, 2010.

Brosseau, F. J. R., 2010. Complexation and Hydrolysis of Aqueous Iron and Nickel With Chloride and Ammonia Under Hydrothermal Conditions (MSc. thesis). Faculty of Graduate Studies, University of Guelph.

Burns, W. G., Sims, H. E., 1981. Effect of radiation type in water radiolysis. J. Chem. Soc. Faraday Trans. 77, 2803-2813.

Burrill, K. A., 2000. Water chemistries and corrosion product transport in supercritical water in reactor heat transport systems. In: Proceedings of the 8th BNES Conference on Water Chemistry of Nuclear Reactor Systems, Bournemouth, UK, 2000, p. 357.

Bursik, A., Dooley, R. B., Gunn, D., Larkin, B. A., Oliker, I., Pocock, F. J., Ryan, D., Webb, L. C., 1994. Cycle Chemistry Guide for Fossil Plants: Oxygenated Treatment. Electric Power Research Institute Report TR-102285.

Burton, M., 1969. Chem. Eng. News 47, 86.

Busboom, H. J., Boyle, R. F., Harling, G., Hazel, V. E., 1966. Post-irradiation Examination of EVESR Mark III Superheat Fuel (0.008 Inch Cladding Failure). General Electric Report GEAP5135.

Carvajal-Ortiz, R. A., Choudhry, K. I., Svishchev, I. M., Guzonas, D. A., 2012a. Chemical species in fluences under flow conditions relevant to PWR and SCWR. In: Nuclear Plant Chemistry Conference (NPC 2012), Paris, France, September 23-28, 2012.

Carvajal-Ortiz, R. A., Plugatyr, A., Svishchev, I. M., 2012b. On the pH control at supercritical water-cooled reactor operating conditions. Nucl. Eng. Des. 248, 340-342.

Chialvo, A. A., Gruszkiewicz, M. S., Cole, D. R., 2010. Ion-pair association in ultrasupercritical aqueous environments: successful interplay among conductance experiments, theory, and molecular simulations. J. Chem. Eng. Data 55, 828-1836.



Choudhry, K. I., Carvajal-Ortiz, R. A., Svishchev, I. M., 2013. Thermochemical hydrogen production under SCWR conditions. In: 6th International Symposium on Supercritical Water-cooled Reactors (ISSCWR-6), March 03-07, 2013, Shenzhen, Guangdong, China. Paper ISSCWR6-13079.

Choudhry, K. I., Guzonas, D. A., Kallikragas, D. T., Svishchev, I. M., 2016. On-line monitoring of oxide formation and dissolution on Alloy 800H in supercritical water. Corros. Sci. 111, 574-582.

Christensen, H., 1981. Effects of water radiolysis on corrosion in nuclear reactors. Radiat. Phys. -Chem. 18, 147-158.

Chudnovskaya, I. I., Myakas, V. I., Buchis, S. V., Shtern, Z. Yu, 1988. The state of internal tube deposits when hydrazine water chemistry treatment is practised. Therm. Eng. 35, 96-99.

Cialone, H. J., Wright, I. G., Wood, R. A., Jackson, C. M., 1986. Circumferential Cracking of Supercritical Boiler Water-Wall Tubes. Electric Power Research Institute Report CS-4969.

Cohen, P., 1985. Water Coolant Technology of Power Reactors. American Nuclear Society.

Cook, W. G., Olive, R. P., 2012a. Corrosion product deposition on two possible fuel geometries in the Canadian-SCWR concept. In: 3rd China-Canada Joint Workshop on Supercritical-water-cooled Reactors (CCSC-2012), Paper 12064, Xi'an, China, April 18-20, 2012.

Cook, W. G., Olive, R. P., 2012b. Pourbaix diagrams for chromium, aluminum and titanium extended to high-subcritical and low-supercritical conditions. Corros. Sci. 58, 291.

Cook, W. G., Olive, R. P., 2013. Corrosion product transport and deposition in the Canadian supercritical water-cooled reactor. In: Proceedings of the 16th International Conference on the Properties of Water and Steam,

September 1-5,2013. University of Greenwich,London,UK.

Cowan,R.,Ruhle,W.,Hettiarachchi,S.,2011. Introduction to Boiling Water Reactor Chemistry,vol. 1. Advanced Nuclear Technology International,Sweden.

Cowan,R.,Ruhle,W.,Hettiarachchi,S.,2012. Introduction to Boiling Water Reactor Chemistry,vol. 2. Advanced Nuclear Technology International,Sweden.

Daigo,Y.,Watanabe,Y.,Sue,K.,2007a. Effect of chromium ion from autoclave material on corrosion behavior of nickel-based alloys in supercritical water. Corrosion 63,277.

Daigo,Y.,Watanabe,Y.,Sue,K.,2007b. Corrosion mitigation in supercritical water with chromium ion. Corrosion 63,1085.

Deeva,Z. V.,Saichuk,L. E.,Panova,G. N.,Moreva,T. V.,Belyaeva,M. V.,1986. The results of operation and investigation of neutral/hydrazine water treatment in supercritical power-generating units. Therm. Eng. 33(5), 265-267.

Deng,J.,Deng,F.,2012. The setting of sipping test devices for irradiated fuel in nuclear power plant. J. Energy Power Eng. 6,1935.

Dickinson, N. L., Keilbaugh, W. A., Pocock, F. J., 1958. Some physico - chemical phenomena in supercritical water. In:ASME Paper No. 58-A-267 Presented at the ASME Annual Meeting of Committee on Boiler Feedwater Studies and Power Division,New York.

Dollezhal, N. A., Krasin, A. K., Aleshchenkov, P. I., Galanin, A. N., Grigoryants, A. N., Emelyanov, I. Ya.,Kugushev, N. M., Minashin, M. E., Mityaev, U. I., Florinsky, B. V., Sharpov, N. N., 1958. Uranium-graphite reactor with superheated high pressure steam. In:Proceeding 2nd Int. Conf. on the Peaceful Uses of Atomic Energy,8,p. 398. UN,Geneva.

Dollezhal,N. A.,Aleshchenkov,P. I.,Evdokimov,Yu. V.,Emel' yanov,I. Ya.,Ivanov,B. G.,Kochetkov, L. A.,Minashin,M. E.,Mityaev,Yu. I.,Nevskii,V. P.,Shasharin,G. A.,Sharapov,V. N.,Orlov,K. K.,1969. Operating experience with the Beloyarsk nuclear power station. At. Energiya 27,379.

Dooley,R. B.,Ball,M.,Bursik,A.,Rziha,M.,Svoboda,R.,2004. Water chemistry in com-mercial water-steam cycles. In:Palmer,D. A. Fernàndez-Prini,R.,Harvey,A. H. (Eds.),Aqueous Systems at Elevated Temperatures and Pressures. Elsevier Academic Press(Chapter 17).

Edwards,M.,Rousseau,S.,Guzonas,D.,2014. Corrosion wall loss and oxide fi lm thickness on SCWR fuel cladding. In:2014 Canada-China Conference on Advanced Reactor Development(CCCARD-2014) Niagara Falls,Ontario Canada,April 27-30,2014. Paper CCCARD2014-49.

Elliott,A. J.,Bartels,D. M.,2009. The Reaction Set,Rate Constants and G-Values for the Simulation of the Radiolysis of Light Water over the Range 20 to 350℃ Based on Information Available in 2008. Atomic Energy of Canada Ltd. Report AECL-153-127160-001. EPRI,2004. Radiation Field Control Manual. Electric Power Research Institute Report 1003390,Palo Alto CA.

Franck,E. U.,Roth,K.,1967. Infrared absorption of deuterium hydroxide in water at high pressures and temperatures. Discuss. Faraday Soc. 43,108-114.

Fujiwara,K.,Watanabe,K.,Domae,M.,Katsumura,Y.,2007. Stability of chromium oxide film formed by metal organic chemical vapor deposition in high temperature water up to supercritical region. In:Corrosion 2007, Nashville,Tennessee,March 11-15,2007.

Gabrielli,F.,Schwevers,H.,2008. Design factors and water chemistry practices-supercritical power cycles.

In:Proceedings of the 15th International Conference on the Properties of Water and Steam, September 7-11, 2008, Berlin, Germany.

Galkin, A. A., Lunin, V. V., 2005. Subcritical and supercritical water: a universal medium for chemical reactions. Russ. Chem. Rev. 74, 21-35.

Galobardes, J. F., Van Hare, D. R., Rogers, L. B., 1981. Solubility of sodium chloride in dry steam. J. Chem. Eng. Data 26, 363.

Ghandi, K., Addison-Jones, B., Brodovitch, J.-C., McKenzie, I., Percival, P. W., 2002. Near-diffusion-controlled reactions of muonium in sub- and supercritical water. Phys. Chem. Chem. Phys. 4, 586-595.

Ghandi, K., Addison-Jones, B., Brodovitch, J.-C., Kecman, S., McKenzie, I., Percival, P. W., 2003. Muonium kinetics in sub- and supercritical water. Phys. B 326(1-4), 55-60.

Ghandi, K., McFadden, R. M. L., Cormier, P. J., Satija, P., Smith, M., 2012. Radical kinetics insub-and supercritical carbon dioxide: thermodynamic rate tuning. Phys. Chem. Chem. Phys. 14, 8502-8505.

Gorbaty, Yu. E., Kalinichev, A. G., 1995. Hydrogen bonding in supercritical water. 1. Experimental results. J. Phys. Chem. 99, 5336-5340.

Goulet, T., Patau, J. P., Jay-Gerin, J. P., 1990. In fluence of the parent cation on the thermalization of sub-excitation electrons in solid water. J. Phys. Chem. 94(18), 7312-7316.

Gruzdev, N. I., Shchapov, G. A., Tipikin, S. A., Boguslavskii, V. B., 1970. Investigating the water conditions in the second unit at Beloyarsk. Therm. Eng. 17(12), 22-25(Translated from Russian).

Guzonas, D. A., Cook, W. G., 2012. Cycle chemistry and its effect on materials in a supercritical water-cooled reactor: a synthesis of current understanding. Corros. Sci. 65, 48.

Guzonas, S., Cook, W., 2015. Water chemistry specifications for the Canadian supercritical water-cooled reactor concept. In: 7th International Symposium on Supercritical Water-cooled Reactors(ISSCWR-7), March 15-18, 2015, Helsinki, Finland. Paper ISSCWR7-2089.



Guzonas, D. A., Qiu, L., 2004. A predictive model for radionuclide deposition around the CANDU heat transport system. In: Proceedings of International Conference Water Chemistry of Nuclear Reactor Systems, San Francisco, October 11-14, 2004.

Guzonas, D. A., Qiu, L., 2013. Activity transport in a supercritical water-cooled reactor. In: 6th International Symposium on Supercritical Water-cooled Reactors, March 03-07, 2013, Shenzhen, Guangdong, China.

Guzonas, D. A., Wills, J., Do, T., Michel, J., 2007. Corrosion of candidate materials for use in a supercritical water CANDU ® reactor. In: Canadian Nuclear Society-13th International Conference on Environmental Degradation of Materials in Nuclear Power Systems 2007, pp. 1250-1261.

Guzonas, D. A., Wills, J., Dole, H., Michel, J., Jang, S., Haycock, M., Chutumstid, M., 2010. Steel corrosion in supercritical water: an assessment of the key parameters. In: 2nd Canada-China Joint Workshop on Supercritical Water-cooled Reactors(CCSC-2010), Toronto, Ontario, April 25-28, 2010.

Guzonas, D., Tremaine, P., Brosseau, F., Meesungnoen, J., Jay-Gerin, J.-P., 2012. Key water chemistry issues in a supercritical-water-cooled pressure-tube reactor. Nucl. Technol. 179, 205.

Guzonas, D., Qiu, L., Livingstone, S., Rousseau, S., 2016. Fission product release under supercritical water-cooled reactor conditions. J. Nucl. Eng. Rad. Sci 021010-021011, 021010-021016.

Guzonas, D. A., 2010. Chemistry control strategies for a supercritical water-cooled reactor. In: Nuclear Plant Chemistry Conference 2010, Quebec City, Canada, October 3-7, 2010.

Guzonas, D., 2014. Extreme water chemistry-how Gen IV water chemistry research improves Gen III water-cooled reactors. In: 19th Pacific Basin Nuclear Conference (PBNC 2014), Vancouver, B. C., Canada, August 24-28, 2014.

Halvarsson, M., Tang, J. E., Asteman, H., Svensson, J. - E., Johansson, L. - G., 2006. Microstructural investigation of the breakdown of the protective oxide scale on a 304 steel in the presence of oxygen and water vapour at 600℃. Corros. Sci. 48, 2014.

Han, Z., Muroya, Y., 2009. Development of a new method to study elution properties of stainless materials in subcritical and supercritical water. In: 4th International Symposium on Supercritical Water-cooled Reactors, Heidelberg, Germany, March 8-11, 2009.

Haygarth, K., Bartels, D. M., 2010. Neutron and β/γ radiolysis of water up to supercritical conditions. 2. $SF_6$ as a scavenger for hydrated electron. J. Phys. Chem. A 114, 7479-7484.

Hazel, V. E., Boyle, R. F., Busboom, H. J., Murdock, T. B., Skarpelos, J. M., Spalaris, C. N., 1965 Fuel Irradiations in the ESADE-VBWR Nuclear Superheat Loop. General Electric Report GEAP-4775.

Hazelton, R. F., 1987. Characteristics of Fuel Crud and its Impact on Storage, Handling and Shipment of Spent Fuel. Pacific Northwest Laboratory. PNL-6273.

Henshaw, J., McGurk, J. C., Sims, H., Tuson, A., Dickinson, S., Deshon, J., 2006. A model of chemistry and thermal hydraulics in PWR fuel crud deposits. J. Nucl. Mater. 353, 1-11.

Ho, P. C., Palmer, D. A., Wood, R. H., 2000. Conductivity measurements of dilute aqueous LiOH, NaOH and KOH solutions to high temperatures and pressures using a fiow-through cell. J. Phys. Chem. B 104, 12084-12089.

Hodes, M., Marrone, P. A., Hong, G. T., Smith, K. A., Tester, J. W., 2004. Salt precipitation and scale control in supercritical water oxidation-Part A: Fundamentals and research. J. Supercrit. Fluids 29, 265-288.

Hoffmann, M. M., Conradi, M. S., 1997. Are there hydrogen bonds in supercritical water? J. Am. Chem. Soc. 119, 3811-3817.

Holcomb, G. R., 2008. Calculation of reactive-evaporation rates of chromia. Oxid. Met. 69, 163.

Holcomb, G. R., 2009. Steam oxidation and chromia evaporation in ultrasupercritical steam boilers and turbines. J. Electrochem. Soc. 156, 292.

Holster, W. T., Schneer, C. J., 1961. Hydrothermal magnetite. Geol. Soc. Am. Bull. 72, 369-386.

IAEA, 1994. Decontamination of Water Cooled Reactors. International Atomic Energy Agency Technical Report 365, Vienna.

IAEA, 2002. Fuel failure in water reactors: causes and mitigation. In: Proceedings of a Technical Meeting Held in Bratislava, Slovakia, IAEA-TECDOC-1345, June 2002.

Imre, A. R., Deiters, U. K., Kraska, T., Tiselj, I., 2012. The pseudocritical regions for supercritical water. Nucl. Eng. Des. 252, 179-183.

Ishigure, K., Kawaguchi, M., Oshoma, K., 1980. The effect of radiation on the corrosion product release from metals in high temperature water. In: Proceedings of the 4th BNES Conference on Water Chemistry of Nuclear Reactor Systems, Bournemouth, UK. Paper 43.

Janik, D., Janik, I., Bartels, D. M., 2007. Neutron and beta/gamma radiolysis of water up to supercritical conditions. 1. Beta/gamma yields for H(2), H(.) atom, and hydrated electron. J. Phys. Chem. A 111, 7777-7786.

Joseph, J. M., Seon Choi, B., Yakabuskie, P., Wren, J. C., 2008. A combined experimental and model analysis on the effect of pH and $O_2(aq)$ on γ-radiolytically produced $H_2$ and $H_2O_2$. Radiat. Phys. Chem. 77, 1009-1020.

Kalinichev, A. G., Bass, J. D., 1997. Hydrogen bonding in supercritical water. 2. Computer simulations. J. Phys. Chem. A 101, 9720-9727.

Kallikragas, D. T., Plugatyr, A. Yu., Guzonas, D. A., Svishchev, I. M., 2015a. Effect of confinement on the hydration and diffusion of chloride at high temperatures. J. Supercrit. Fluids 97, 22-30.

Kallikragas, D. T., Guzonas, D. A., Svishchev, I. M., 2015b. Properties of aqueous systems relevant to the SCWR via molecular dynamics simulations. AECL Nucl. Rev. 4, 9-21.

Kallikragas, D. T., Choudhry, K. I., Svishchev, I. M., 2016. Diffusion and hydration of oxygen confined in $Fe(OH)_2$ nano-cracks at high temperatures. J. Supercrit. Fluids 120, 345-354.

Karakama, K., Rogak, S. N., Alfantazi, A., 2012. Characterization of the deposition and transport of magnetite particles in supercritical water. J. Supercrit. Fluids 71, 11-18.

Karasawa, H., Fuse, M., Kiuchi, K., Katsumura, Y., 2004. Radiolysis of supercritical water. In: 5th Int. Workshop on LWR Coolant Water Radiolysis and Electrochemistry, San Francisco, USA, October 15, 2004.

Kontorovich, L. Kh., Rogal' skaya, I. A., 1987. Effects of water-chemical conditions on growth of deposits and metal corrosion in supercritical pressure boilers. Energomashinostoenie 6(1987), 24-26 in Russian (English translation in 1987. Sov. Energy Technol. 6, 48-52).

Kontorovich, L. K., Vasilenko, G. V., Sutotskii, G. P., Rogal' skaya, I. A., Evtushenko, V. M., 1982. The effect of water chemistry conditions on the resistance to corrosion of steel in the condensate-feed circuit of supercritical power generating units. Therm. Eng. 29(6), 337-339.



Kritsky, V. G., 1999. Water Chemistry and Corrosion of Nuclear Power Plant Structural Materials. American Nuclear Society.

Kysela, J., Růžičková M., Petr, J., 2009. Water chemistry specification for supercritical water cooled reactors-possible options. In: 4th International Symposium on Supercritical Water-cooled Reactors, Heidelberg, Germany, March 8-11, 2009. Paper No. 68.

Larsen, O. H., Blum, R., Daucik, K., 1996. Chemical and mechanical control of corrosion product transport. In: Proceedings from Power Plant Chemical Technology, Kolding, Denmark, September 4-6, 1996, pp. 11.1-11.17.

Leblanc, R., Ghandi, K., Hackman, B., Liu, G., 2014. Extrapolation of rate constants of reactions producing $H_2$ and $O_2$ in radiolysis of water at high temperatures. In: 19th Pacific Basin Nuclear Conference (PBNC 2014), Vancouver, British Columbia, Canada, August 24-28, 2014.

LeSurf, J. E., Johnson, R. E., Allison, G. M., 1963. Radiolysis in a Steam Loop. Atomic Energy of Canada Limited Report AECL-1800.

Leusbrock, I., Metz, S. J., Rexwinkel, G., Versteeg, G. F., 2008. Quantitative approaches for the description of solubilites of inorganic compounds in near-critical and supercritical water. J. Supercrit. Fluids 47, 117.

Lewis, B. J., MacDonald, R. D., Ivanoff, N. V., Iglesias, F. C., 1993. Fuel performance and fission product release studies for defected fuel elements. Nucl. Technol. 103, 220.

Likhanskii, V. V., Evdokimov, I. A., Sorokin, A. A., Kanukova, V. D., 2009. Applications of the RTOP-CA code for failed fuel diagnosis and predictions of activity level in WWER primary circuit. In: Proceedings of Top

Fuel 2009, Paris, France, 2009, p. 2054.

Lin, M., Katsumura, T., He, H., Muroya, Y., Han, Z., Miyazaki, T., Kudo, H., 2005. Pulse radiolysis of 4,4′-bipyridyl aqueous solutions at elevated temperatures: spectral changes and reaction kinetics up to 400℃. J. Phys. Chem. A 109, 2847-2854.

Lin, M., Muroya, Y., Baldacchino, G., Katsumura, Y., 2010. Radiolysis of supercritical water. In: Wishart, J. F., Rao, B. S. M. (Eds.), Recent Trends in Radiation Chemistry. World Scientific Publishing Co. PTE. Ltd.

Liu, G., Du, T., Toth, L., Beniger, J., Ghandi, K., 2016. Prediction of rate constants of important reactions in water radiation chemistry in sub- and supercritical water: equilibrium reactions. CNL Nucl. Rev. 5, 345-361.

Locke, D. H., 1972. The behaviour of defective reactor fuel,. Nucl. Eng. Des. 21, 318.

Majer, V., Sedlbauer, J., Wood, R. H., 2004. Calculation of standard thermodynamic properties of aqueous electrolytes and non-electrolytes. In: Palmer, D. A., Fernandez-Prini, R., Harvey, A. H. (Eds.), Chapter 12 in "Aqueous Systems at Elevated Temperatures and Pressures". Elsevier Academic Press (Chapter 4).

Marchaterre, J. F., Petrick, M., 1960. Review of the Status of Supercritical Water Reactor Technology. Argonne National Laboratory Report ANL-6202.

Margulova, T. K., 1978. An investigation of 'neutral' water treatment in supercritical power generating units. Therm. Eng. 25(10), 39-45.

Marin, T. W., Jonah, C. D., Bartels, D. M., 2005. Reaction of hydrogen atoms with hydroxide ions in high-temperature and high-pressure water. J. Phys. Chem. A 109(9), 1843-1848.

Martynova, O. I., 1973. Solubility of inorganic compounds in subcritical and supercritical water. In: High Temperature, High Pressure Electrochemistry in Aqueous Solutions": International Corrosion Conference Series, NACE-4, University of Surry, England, January 7-12, 1973, pp. 131-138.

Mayanovic, R. A., Anderson, A. J., Dharmagunawardhane, H. A. N., Pascarelli, S., Aquilanti, G., 2012. Monitoring synchrotron x-ray-induced radiolysis effects on metal (Fe, W) ions in high-temperature aqueous fluids. J. Synchrotron Rad. 19, 797.

Meesungnoen, J., Sanguanmith, S., Jay-Gerin, J. -P., 2013. Density dependence of the yield of hydrated electrons in the low-LET radiolysis of supercritical water at 400℃: influence of the geminate recombination of subexcitation-energy electrons prior to thermalization. Phys. Chem. Chem. Phys. 15, 16450.

Metatla, N., Jay-Gerin, J. -P., Soldera, A., 2011. Molecular dynamics simulation of sub-critical and supercritical water at different densities. In: 5th International Symposium on Supercritical Water-cooled Reactors (ISSCWR-5), Vancouver, British Columbia, Canada, March 13-16, 2011.

Metatla, N., Lafond, F., Jay-Gerin, J. -P., Soldera, A., 2016. Heterogeneous character of supercritical water at 400℃ and different densities unveiled by simulation. RSC Adv 6, 30484-30487.

Moreau, S., Fenart, M., Renault, J. P., 2014. Radiolysis of water in the vicinity of passive surfaces. Corros. Sci. 83, 255-260.

Mravca, A. E., Simpson, D. E., 1961. Nuclear Superheat Meeting Number 5. September 13-15, 1961, San Jose, California. United States Atomic Energy Commission Div. of Technical Information Report TID-7630.

Muroya, Y., Lin, M., de Waele, V., Hatano, Y., Katsumura, Y., Mostafavi, M., 2010. First observation of picosecond kinetics of hydrated electrons in supercritical water. J. Phys. Chem. Lett. 1(1), 331-335.

Muroya, Y., Sanguanmith, S., Meesungnoen, J., Lin, M., Yan, Y., Katsumura, Y., Jay-Gerin, J. -P., 2012. Time-dependent yield of the hydrated electron in subcritical and supercritical water studied by ultrafast pulse radiolysis and Monte-Carlo simulation. Phys. Chem. Chem. Phys. 14(41),14325-14333.

Murray,J. L.,1965. EVESR-nuclear Superheat Fuel Development Project Thirteenth Quarterly Report. General Electric Report GEAP-4941.

Oliker,I.,Armor,A. F.,1992. Supercritical Power Plants in the USSR. Electric Power Research Institute Report TR-100364.

Olive,R. P.,2012. Pourbaix Diagrams,Solubility Predictions and Corrosion-Product Deposition Modelling for the Supercritical Water-cooled Reactor. PhD Thesis. University of New Brunswick,Department of Chemical Engineering.

Palmer,D. A.,Simonson,J. M.,Jensen,J. P.,2004. Partitioning of electrolytes to steam and their solubilities in steam. In:Palmer,D. A.,Fern andez-Prini,R.,Harvey,A. H. (Eds.),Aqueous Systems at Elevated Temperatures and Pressures. Elsevier Academic Press(Chapter 12).

Pencer,J.,Edwards,M. K.,Guzonas,D.,Edwards,G. W. R.,Hyland,B.,2011. Impact of corrosion product deposition on CANDU-SCWR lattice physics. In:The 5th Int. Sym. SCWR(ISSCWR-5) Vancouver, British Columbia,Canada,March 13-16,2011.

Percival,P. W.,Brodovitch,J.-C.,Ghandi,K.,McCollum,B. M.,McKenzie,I.,2007. H atom kinetics in superheated water studied by muon spin spectroscopy. Rad. Phys. Chem. 76,1231-1235.

Plugatyr,A.,Carvajal-Ortiz,R. A.,Svishchev,I. M.,2011a. Ion-pair association constant for LiOH in supercritical water. J. Chem. Eng. Data 56,3637-3642.

Plugatyr,A.,Hayward,T. M.,Svishchev,I. M.,2011b. Thermal decomposition of hydrazine in sub and supercritical water at 25MPa,. J. Supercrit. Fluids 55,1014-1018.

Procaccia,I.,Gitterman,M.,1981. Slowing down of chemical reactions near thermodynamic critical points. Phys. Rev. Lett. 46(17),1163.



Qiu,L.,Guzonas,D. A.,2010. An overview of corrosion products solubilities in subcritical and supercritical water. In:2nd Canada-China Joint Workshop on Supercritical Water-cooled Reactors(CCSC-2010),Toronto, Ontario,Canada,April 25-28,2010.

Qiu,L.,2014. Solubilities of fission products under supercritical water-cooled reactor conditions. In:Canada-China Conference on Advanced Reactor Development(CCCARD-2014),Niagara Falls,Ontario Canada,April 27-30,2014.

Red' kin,A. F.,Savelyeva,N. I.,Sergeyeva,E. I.,Omel' yanenko,B. I.,Ivanov,I. P.,Khodakovsky, I. L.,1989. Investigation of uraninite $UO_2$(c) solubility under hydrothermal conditions. Sci. Géologiques Bull. 42,329.

Red' kin,A. F.,Savel' yeva,N. I.,Sergeyeva,E. I.,Omel' yaneko,B. I.,Invanov,I. P.,Khodakovsky, I. L.,1990. Experiment 89,Informative Volume,p. 79. Moscow,Nauka.

Roduner,E.,2005. Hydrophobic solvation,quantum nature,and diffusion of atomic hydrogen in liquid water. Rad. Phys. Chem. 72(2-3),201-206.

Roof,R. R.,1966. EVESR-nuclear Superheat Fuel Development Project Seventeenth Quarterly Report. General Electric Report GEAP-5269.

Ru,X.,Staehle,R. W.,2013. PART I-review:historical experience providing bases for predicting corrosion and stress corrosion in emerging supercritical water nuclear technology. Corrosion 69(3),211-229.

Saito,N.,Tsuchiya,Y.,Yamamoto,S.,Akai,Y.,Yotsuyanaji,T.,Domae,M.,Katsumura,Y.,2006.

Chemical thermodynamic considerations on corrosion products in supercritical water-cooled reactor coolant. Nucl. Technol. 155,105-112.

Sanguanmith,S. ,Muroya, Y. ,Meesungnoen,J. ,Lin,M. ,Katsumura,Y. ,Mirsaleh Kohan,L. ,Guzonas, D. A. ,Stuart,C. R. ,Jay-Gerin,J. -P. ,2011. Low-linear energy transfer radiolysis of liquid water at elevated temperatures up to 350℃ Monte-Carlo simulations. Chem. Phys. Lett. 508(4-6),224-230.

Sanguanmith,S. ,Meesungnoen,J. ,Guzonas,D. A. ,Stuart,C. R. ,Jay-Gerin,J. -P. ,2013. Effects of water density on the spur lifetime and the "escape" $e_{aq}^-$ yield in $\gamma$ -irradiated supercritical water at 400℃ :a re-examination. In:6th International Symposium on Supercritical Water Cooled Reactors(ISSCWR-6),Shenzhen,Guangdong,China,March 03-07,2013. Paper ISSCWR6-13093.

Sanguanmith,S. ,2012. Low-linear Energy Transfer Radiolysis of Liquid Water at Elevated Temperatures up to 350℃ :Monte-Carlo Simulations(M. Sc. thesis). Faculty of Medicine and Health Sciences,Université de Sherbrooke,Sherbrooke,Québec,Canada.

Shields,K. ,Cutler,F. M. ,Sadler,M. A. ,2006. Condensate Polishing State of Knowledge Assessment:Technology Needs for Fossil Plants. Electric Power research Institute Report 1012208.

Simeoni,G. G. ,Bryk,T. ,Gorelli,F. A. ,Krisch,M. ,Ruocco,G. ,Santoro,M. ,Scopigno,T. ,2010. The widom line as the crossover between liquid-like and gas-like behavior in supercritical fluids. Nat. Phys. 6,503-507.

Solberg,H. L. ,Brown,T. R. ,1968. The effect of high-temperature steam corrosion upon the heat transfer through superheater tube alloys. In:Lien,G. E. (Ed.),Behavior of Superheater Alloys in High Temperature, High Pressure Steam. The American Society of Mechanical Engineers,New York.

Soper,A. K. ,Bruno,F. ,Ricci,M. A. ,1997. Site-site pair correlation functions of water from 25 to 400℃ : revised analysis of new and old diffraction data. J. Chem. Phys. 106,247-254.

Spalaris,C. N. ,Boyle,R. F. ,Evans,T. F. ,Esch,E. L. ,1961. Design,Fabrication and Irradiation of Superheat Fuel Element SH-4B in VBWR. General Electric Report GEAP-3796.

Spinks,J. W. T. ,Woods,R. J. ,1990. An Introduction to Radiation Chemistry,third ed. Wiley,New York.

Staehle,R. ,2006. The small dimensions of stress corrosion cracking-tight cracks. In:Proc. Int. Conf. on Water Chemistry of Nuclear Plant Systems,October 23-26,2006,Jeju Island,Korea.

Staehle,R. ,2010. Fundamental approaches to predicting stress corrosion:quantitative micro-nano approach to predicting stress corrosion cracking in water cooled nuclear plants. In:Proceedings of the Nuclear Plant Chemistry Conference(NPC2010),Quebec City,Canada.

Stellwag,B. ,Landner,A. ,Weiss,S. ,Huttner,F. ,2011. Water chemistry control practices and data of the European BWR fleet. Powerpl. Chem. 13,167.

Styrikovich,M. A. ,1969. Steam solutions. Vestn. Akad. Nauk. SSSR 39,1969.

Subramanian,V. ,Joseph,J. M. ,Subramanian,H. ,Noël,J. J. ,Guzonas,D. A. ,Wren,J. C. ,2016. Steady-state radiolysis of supercritical water:model predictions and validation. J. Nucl. Eng. Rad. Sci. 2,021021, 021021-021026.

Svishchev,I. M. ,Guzonas,D. A. ,2011. Supercritical water and particle nucleation:implications for water chemistry control in a GEN IV supercritical water cooled nuclear reactor. J. Supercrit. Fluids 60,121-126.

Svishchev,I. M. ,Carvajal-Ortiz,R. A. ,Choudhry,K. I. ,Guzonas,D. A. ,2013a. Corrosion behavior of stainless steel 316 in sub- and supercritical aqueous environments:effect of LiOH additions. Corros. Sci. 72,

20-25.

Svishchev, I. M., Kallikragas, D. T., Plugatyr, A. Y., 2013b. Molecular dynamics simulations of supercritical water at the iron hydroxide surface. J. Supercrit. Fluids 78, 7-11.

Sweeton, F. H., Baes Jr., C. F., 1970. Solubility of magnetite and hydrolysis of ferrous ion in aqueous solutions at elevated temperatures. J. Chem. Thermodyn. 2, 479-500.

Swiatla-Wojcik, D., Buxton, G. V., 2005. On the possible role of the reaction $H\cdot + H_2O \rightarrow H_2 + \cdot OH$ in the radiolysis of water at high temperatures. Radiat. Phys. Chem. 74(3-4), 210-219.

Swiatla-Wojcik, D., Buxton, G. V., 2010. Reply to comment on the possible role of the reaction $H\cdot + H_2O \rightarrow H_2 + \cdot OH$ in the radiolysis of water at high temperatures. Radiat. Phys. Chem. 79, 52-56.

Tabor, D., 1969. Gases, Liquids and Solids: And Other States of Matter. Cambridge University Press.

Tremaine, P., Arcis, H., 2014. Solution calorimetry under hydrothermal conditions. Rev. Mineral. Geochem. 76, 219-263.

Tremaine, P. R., LeBlanc, J. C., 1980. The solubility of magnetite and the hydrolysis and oxidation of $Fe^{2+}$ in water to 300℃. J. Solut. Chem. 9, 415.

Tremaine, P., Zhang, K., Bénézeth, P., Xiao, C., 2004. Ionization equilibria of acids and bases under hydrothermal conditions. In: Palmer, D. A., Fern andez-Prini, R., Harvey, A. H. (Eds.), Aqueous Systems at Elevated Temperatures and Pressures. Elsevier Academic Press (Chapter 13).

Tse, J. S., Klein, M. L., 1983. Are hydrogen atoms solvated by water molecules? J. Phys. Chem. 87(25), 5055-5057.

Tucker, S. C., 1999. Solvent density inhomogeneities in supercritical fl uids. Chem. Rev. 99, 391-418.

Vasilenko, G. V., 1976. The effect of the pH of the feedwater on corrosion of the low pressure heaters of supercritical power generating units. Therm. Eng. 23(8), 51-52.



Vasilenko, G. V., 1978. The pattern of precipitation of iron compounds in supercritical steam generators under different water conditions. Therm. Eng. 25(3), 26-30.

Veselkin, A. P., Lyutov, M. A., Khandamirov, Yu. E., 1968. Radioactive deposits on the surfaces of the technological equipment of the I. V. Kurchatov nuclear power station at Beloyarsk. At. Energiya 24, 219.

Veselkin, A. P., Beskrestnov, N. V., Sklyarov, V. P., Khandamirov, Yu. E., Yashnikov, A. I., 1971.

Radiation safety in the design and operation of channel power reactors. At. Energiya 30, 144.

Walther, J. V., 1986. Mineral solubilities in supercritical $H_2O$ solutions. Pure Appl. Chem. 58, 1585-1598.

Wesolowski, D. J., Palmer, D. A., Machesky, M. L., Anovitz, L. M., Hyde, K. E., Hayashi, K. -I., 2000. Solubility and surface charge of magnetite in hydrothermal solutions. In: Tremaine, P. R., Hill, P. G., Irish, D. E., Balakrishnan, P. V. (Eds.), Steam, Water and Hydrothermal Systems: Physics and Chemistry Meeting the Needs of Industry, Proc. of the 13th Int. Conf. on the Properties of Water and Steam. NRC Research Press.

Wesolowski, D. J., Ziemniak, S. E., Anovitz, L. M., Machesky, M. L., Bénézeth, P., Palmer, D. A., 2004. Solubility and surface adsorption characteristics of metal oxides. In: Palmer, D. A., Fernàndez-Prini, R., Harvey, A. H. (Eds.), Aqueous Systems at Elevated Temperatures and Pressures. Elsevier Academic Press (Chapter 14).

Woolsey, I. S., 1989. Review of Water Chemistry and Corrosion Issues in the Steam-Water Circuits of Fossil-Fired Supercritical Plants. Central Electricity Generating Board Report RD/L/3413/R88, PubID 280118.

Wren, J. C., Ball, J. M., 2001. LIRIC 3.2 an updated model for iodine behaviour in the presence of organic impurities. Radiat. Phys. Chem. 60,577-596.

Wren, J. C., Glowa, G. A., 2000. A simplified kinetic model for the degradation of 2-butanone in aerated aqueous solutions under steady-state gamma-radiolysis. Radiat. Phys. Chem. 58,341-356.

Yakabuskie, P. A., Joseph, J. M., Wren, J. C., 2010. The effect of interfacial mass transfer on steady-state water radiolysis. Radiat. Phys. Chem. 79,777-785.

Yakabuskie, P. A., Joseph, J. M., Stuart, C. R., Wren, J. C., 2011a. Long-term γ-radiolysis kinetics of NO3(-) and NO2(-) solutions. J. Phys. Chem. A 115,4270-4278.

Yakabuskie, P. A., Joseph, J. M., Keech, P., Botton, G. A., Guzonas, D. A., Wren, J. C., 2011b. Iron oxyhydroxide colloid formation by gamma-radiolysis. Phys. Chem. Chem. Phys. 13,7198-7206.

Yeh, T. K., Wang, M. Y., 2013. Evaluation of the effectiveness of hydrogen water chemistry in a supercritical water reactor. In: Proceedings of the 16th International Conference on the Properties of Water and Steam, September 1-5, 2013. University of Greenwich, London, UK.

Yeh, T. K., Wang, M. Y., Liu, H. M., Lee, M., 2012. Modeling coolant chemistry in a supercritical water reactor. In: Topsafe 2012, Helsinki, Finland, April 22-26, 2012.

Yetisir, M., Gaudet, M., Rhodes, D., 2013. Development and integration of Canadian SCWR concept with counter-flow fuel assembly. In: ISSCWR-6, March 3-7, 2013, Shenzhen, Guangdong, China.

Yurmanov, V. A., Belous, V. N., Vasina, V. N., Yurmanov, E. V., 2009. Chemistry and corrosion, issues in supercritical water reactors. In: IAEA Int. Conf. on Opportunities and Challenges for Water-cooled Reactors in the 21st Century, October 27-30, 2009, Vienna, Austria, IAEA-CN-164-5S12.

Yurmanov, V. A., Belous, V. N., Vasina, V. N., Yurmanov, E. V., 2010. Chemistry and corrosion, issues in supercritical water reactors. In: Proceedings of the Nuclear Plant Chemistry Conference 2010 (NPC 2010), October 3-8, 2010, Quebec City, Canada. Paper 11.02.

Zenkevitch, Yu. V., Sekretar, V. E., 1976. The formation of iron oxide deposits on the tubes of supercritical boilers. Therm. Eng. 23,60.

Zhou, S., Turnbull, A., 2002. Steam Turbine Operating Conditions, Chemistry of Condensates, and Environment Assisted Cracking-A Critical Review. National Physical Laboratory Report MATC(A) 95.

Ziemniak, S. E., Jones, M. E., Combs, K. E. S., 1995. Magnetite solubility and phase stability in alkaline media at elevated temperatures. J. Solut. Chem. 24,837-877.

Zimmerman, G. H., Arcis, H., Tremaine, P., 2014. Ion-pair formation in aqueous strontium chloride and strontium hydroxide solutions under hydrothermal conditions by AC conductivity measurements. Phys. Chem. Chem. Phys. 16.

# 第5章
# 腐　蚀

## 5.1 引　言

本书介绍的超临界水冷反应堆(SCWR)材料的各种降解模式中,似乎可以肯定地说,超临界水的一般腐蚀问题是迄今为止被调研最多的(Longton,1966;Montgomery 和 Karlsson,1995;Fry 等,2002;Kritzer,2004;Was 和 Teysseyre,2005;Was 等,2007;Sun 等,2009c;Wright 和 Dooley,2010;Allen 等,2012;Ru 和 Staehle,2013a、b、c;Sarrade 等,2017;Guzonas 等,2017)。近年来,对超临界水冷反应堆材料腐蚀情况的综述,跨越了 10 多年,概述了与超临界水冷反应堆有关现象理解方面所取得的重大进展。2005 年,Was 和 Teysseyre (2005)简明扼要地总结了这一现状:“在为超临界水冷反应堆提出的候选材料中,公开文献中很少报告超临界水腐蚀问题,并且现有数据库非常有限”。时至 2017 年,采用“基于大量耐腐蚀性测量的结果得到……结论”这种表述更为合适。

在过去 10 年中,尽管与超临界水冷反应堆腐蚀相关的研发工作量有所增加,但在此之前很多工作就已经得到了开展。Boyd 和 Pray(1957)报道了在温度为 427℃、538℃和 738℃,压力为 34.5MPa 的超临界水中暴露 130 天(3168h)后 12 种合金重量的增加情况。Spalaris(1963)基于在 Vallecitos 沸水反应堆的“过热高级示范试验”回路中(1959—1962 年)进行的测试发现,即使是 304 不锈钢,其一般性腐蚀速率也足够低,可以在表面温度为 677℃、峰值熔覆温度为 732℃、溶解氧(DO)浓度为 15~20mg/kg 的条件下,用作核过热器的燃料包壳材料,使用寿命可达 3~4 年。Longton(1966)对这项早期工作(直到 1966 年)进行了全面总结,涵盖了 6 项研究结果,涉及温度 427~765℃、压力 6.9~34.5MPa,测试持续时间 1000~10100h,其中一些测试还验证了添加氧气和传热的影响,整个报告涵盖了 65 种合金的数据。Longton 的结论很简洁,并且其中许多结论经受了时间的考验。报告中指出,最初的高腐蚀速率最终基本上是随时间呈线性的速率下降,压力只有很小的影响,等温和传热测试之间只有2倍的差异。他详细讨论了增重和减重测量之间的区别,并强调前者“只有在确定没有氧化物从表面脱落并消失到系统中时”,才能产生可靠的信息。他指出,氧化物损失高达50%的情况是存在的,而当合金存在时,在600℃条件下3年后的熔深仅为13mm,在700℃条件下仅为41mm,这一结论具有重要意义。

支持超临界水冷反应堆开发的腐蚀研究已经测试了多种不同类型的材料,包括铁素体-马氏体(F/M)钢、奥氏体钢、镍基合金、锆基合金和钛基合金。如果将超临界化石燃料电厂(SCFPP)和研究超临界水氧化(SCWO)系统①同时考虑,获得的腐蚀数据大约涵盖100种合金材料(如Gu等,2010)。尽管几乎没有足够完整的数据集能够可靠地预测寿命终期腐蚀渗透的合金,但确实存在一些顶级候选合金的大数据集。例如,Guzonas等(2016a)对800H合金在500~800℃的温度下暴露时间为500~14000h进行的14项研究进行了仔细评估,以建立随时间和温度变化的腐蚀经验模型。值得注意的是,研究之间的测试条件经常存在显著差异。在前述的800H合金数据中,压力范围为0.1~34.5MPa,表面粗糙度范围包含从样品来料时的状态到抛光后的状态。如第2.3.2.1节中所述,由于流速、样品安装等方面的差异,即使在名义上相同的测试条件下,不同实验室间的结果也可能不同。因此,了解各种测试参数对测得的腐蚀速率的影响很重要。

尽管在某些超临界水冷反应堆的概念中使用了陶瓷组件,但本书不考虑超临界水中陶瓷的降解,除非是陶瓷涂层。Sun等(2009c)在此方面提供了有用的总结。

在下文中,所用合金名称与原始参考文献②中所用的合金名称相同。如有可能,使用mg/dm或mg/(dm·d)(mdd)作为重量变化单位以便比较,并且压力单位已转换为MPa。

### 5.1.1 性能标准

一般性腐蚀工程研究的最终目标是在寿命期末定义总金属损失(金属渗透),使设计人员可以根据其他材料特性(通常是机械),在设计阶段对于零件的厚度增加腐蚀余量。例如,超临界水冷反应堆燃料包壳的内部加压对屈服强度和蠕变强度提出了要求,这些要求定义了最小壁厚;从这些考虑出发,高性能轻

① Kritzer(2004)列出了约80篇与SCWO系统中的腐蚀有关的文献。

② 一些特殊的合金有多个名称,如具有相同成分的HCM12A和T122。

水堆燃料包壳的最大允许腐蚀渗透定义为工作时间 20000h 后 140mm(Schulenberg,2013)。张等(2012)讨论了加压燃料包壳的强度要求,并建议对于 3~4 个工作循环后,对于 0.5~0.6mm 的包壳厚度,包壳渗透率(包括一般腐蚀和应力腐蚀开裂)不应超过 25~30mm。在加拿大的超临界水冷反应堆概念中,燃料包壳是可折叠的,并且必须具有约 0.4mm 的最小厚度以防止纵向起伏(Xu 等,2014)。但是,为了中子经济性,包层厚度必须不超过约 0.6mm。结果,在使用寿命(约 3 年)结束时,最大允许金属损失(包括沿晶界的平均氧化物渗透以及微动和其他形式的磨损)为 0.2mm(Guzonas 等,2016a)。

## 5.2 合金组成成分

合金组成成分可能是抗腐蚀性的关键材料参数。如图 5.1 所示,已发现超临界水中合金的耐腐蚀性与合金等级有很大的关系,图 5.1 显示了 5 种合金(钛基、锆基、F/M 钢、奥氏体钢和镍基合金)的腐蚀测试数据。由不同的研究小组在 500℃、25MPa 的超临界水中进行了 1000h 的测试;在大多数测试中,测试段入口处的 DO 浓度小于 25μg/kg。在图中的 5 种材料中,F/M 钢的氧化速率最快,而镍基合金的氧化速率最低。在每种合金类别中,某些合金表现出更好的抗氧化性,例如,对于图 5.1 中的 F/M 钢,腐蚀速率按 T91> T92> T122 的顺序依次降低。



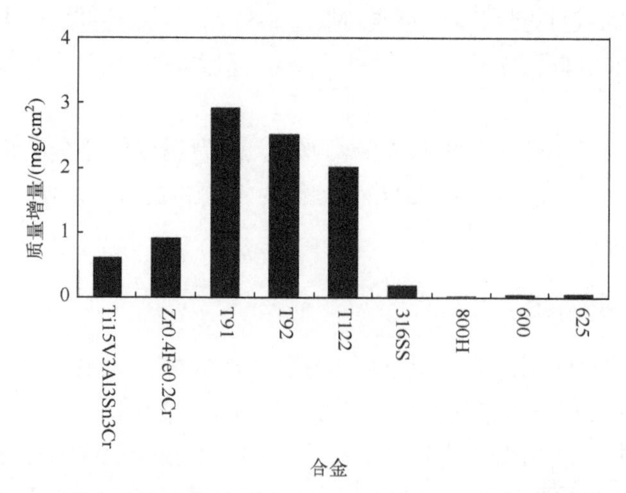


图 5.1 选定的钛基合金、锆基合金、F/M 钢、奥氏体钢和镍基合金在 500℃、25MPa 的超临界水中 1000h 后的质量增量(DO 浓度小于 25μg/kg,钛基合金(8mg/kg)和 316 SS(8mg/kg)除外)

长期以来,人们已经熟知较高的 Cr 含量可以改善钢的耐腐蚀性。如图 5.1 所示,奥氏体钢的腐蚀速率远低于 F/M 钢。图 5.2 显示了在 650℃、700℃和 750℃时各种钢的腐蚀速率常数 $K_p$ 与 Cr 质量分数的抛物线关系。F/M 钢(在 Cr 质量分数为 9%时速率下降较大)和奥氏体钢(在 Cr 质量分数为 19%时速率下降较大)与 Cr 质量分数的相关性是很明确的。与铁素体钢相比,由于在 fcc 奥氏体晶格中 Cr 的扩散率较低,因此在奥氏体钢上形成保护性铬氧化物需要较高的 Cr 质量分数。这种保护膜可减少 Fe 向外扩散,并防止形成保护性较低的 $(Fe,Cr)_3O_4$外层。

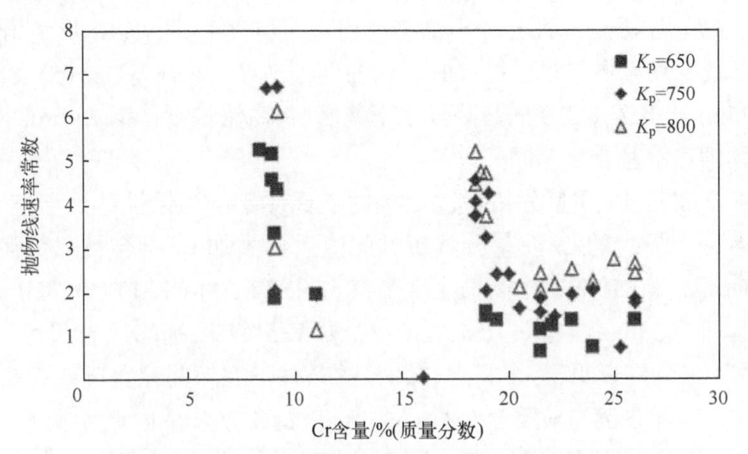


图 5.2 抛物线腐蚀速率常数 $K_p$ 与各种 F/M 和奥氏体不锈钢在 650℃、700℃和 750℃下 Cr 质量分数的相关性

(引自文献 Sarver, J. , 2009. The oxidation behavior of candidate materials for advanced energy systems in steam at temperatures between 650℃ and 800℃. In: 14th Int. Conf. on Environmental Degradation of Materials in Nuclear Power Systems, Virginia Beach, VA, August 23-27, 2009).

Wright 和 Dooley(2010)很好地总结了 Cr 对铁素体钢和 F/M 钢腐蚀的影响,并基于这些合金在 SCFPP 锅炉中的广泛应用建立了数据库。数据清楚地表明,要使超临界水中的氧化速率降低,就需要至少 9%的 Cr 质量分数。Jang 等(2005a)在饱和空气压力为 25MPa 条件下的超临界水中,分别在 500℃和 627℃下测量了 4 种 Cr 质量分数为 9%~20%(打磨至 600 号砂粒尺寸)的铁素体钢,发现在 Cr 质量分数为 9%~12%时质量降低至 1/5,而将 Cr 质量分数提高到 20%时质量增加了 14 倍。Zurek 等(2008 年)建议,应将 Cr 质量分数在 10%~12%以内的 F/M 钢与 Cr 质量分数大于 12%的铁素体钢区分开,前者表现出的氧化行为随测试时间、温度、表面处理以及微量合金元素质量分数等因素的变化而变化。这种差异归因于这些合金在氧化物形成初期所对应的氧化物性质,其中仅包含形成连续保护性氧化物所需的 Cr 质量分数。初始条件的微小变化会损害或改善连续氧化物层的形成,这将对耐蚀性产生长期影响。高 Cr 质量分数(大于 12%)F/M 钢作为支持超临界水冷反应堆发展的重要材料,相关研究很少。Artymowicz 等(2010)和 Cook 等(2010)报道了在铸态、轧制和轧制以及均质化条件下,Cr 质量分数为 14%和 25%的铁素体钢在超临界水(500℃、25MPa)中暴露 500h 的氧化初步数据。Cr 质量分数为 14%的材料均显示出质量增加,遵循幂律动力学规律,指数约为 0.9。Cr 质量分数为 25%的材料显示出轻微的质量损失,氧化限于表面上的小结节。Cho 等(2004)表明,Cr 质量分数大于 13%可以减少氧化物弥散强化(ODS)钢的腐蚀速率。Brar 等(2015)发现,当在 550℃、25MPa、低浓度 DO 的超临界水下暴露时,高铬铁素体钢 E-brite① 的性能与奥氏体钢和镍基合金相同。

图 5.2 表明,Cr 质量分数超过 20%的奥氏体钢的耐蚀性达到平稳状态,这意味着有一个最大的 Cr 质量分数超过此值后再添加 Cr 不会带来任何额外的好处。Otsuka 等(1989)测量了过热蒸汽(SHS)(500~900℃、0.1MPa)中多种奥氏体钢(13%~25%Cr、15%Ni、0.5%Si、1.5%Mn,平均晶粒尺寸 110μm)的氧化特性,结果表明,为了在 $T<750$℃的条件下在粗晶粒奥氏体钢表面形成均匀的 $Cr_2O_3$外表面,Cr 质量分数必须大于 21%。当晶粒尺寸减小至 20μm 时,在高达 900℃下形成保护膜。与 F/M 钢一样,当 Cr 质量分数刚好超过形成连续的富 Cr 保护层所需的最低量时,可能对初始条件极为敏感,这些变化可能会影响保护性氧化物的形成,从而损害长期的耐蚀性。Mahboubi(2014)发现 33 合金(33.4% Cr)比 800HT 合金(20.6%Cr)能够更好地形成 $Cr_2O_3$氧化物。



尽管已发现奥氏体钢在超临界水中具有比 F/M 钢更好的耐蚀性,但几乎没有关于 Ni 对耐蚀性影响的系统研究。Brar 等(2015)在超临界水(550℃、25MPa)中测量了一组市售的 Fe 和 Ni 基合金在 2500h 内质量随时间的变化,其中 Cr 质量分数相对相似,Ni 质量分数为 0.5%~72%。除 304 SS 以外,所有合金的质量增量几乎没有差异,而 304 SS 的质量增量更大(图 5.3)。Peraldi 和 Pint(2004)研究了多种铁素体和奥氏体合金在空气和 10%水蒸气的混合物中的氧化,发现 Ni 质量分数较高时可延迟(但不能阻止)加速腐蚀和/或减少剥落过程。

---

① 72%(质量分数)Fe,26%(质量分数)Cr,0.5%(质量分数)Ni,0.4%(质量分数)Si,0.7%~15%(质量分数)Mo,0.4%(质量分数)Mn。

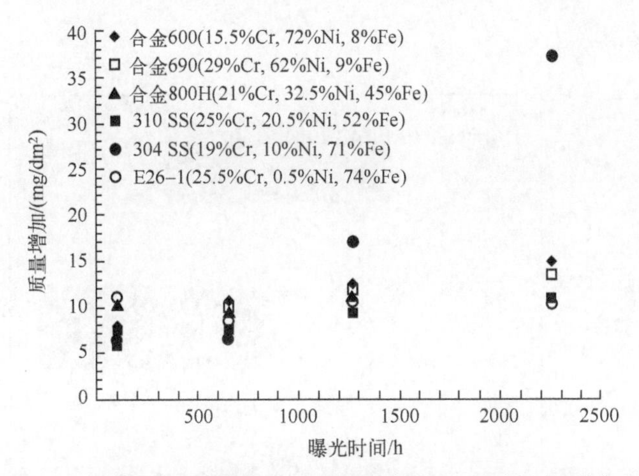


图 5.3　在静态高压釜中 550℃、25MPa 条件下,6 种具有不同 Ni 含量合金的质量增加与暴露时间的函数关系

(引自文献 Brar, C., Davis, W., Mann, E., Ridley, C., Wirachowsky, B., Rangacharyulu, C., Guzonas, D., Leung, L., 2015. GEN IV SCWR cladding analysis project: nickel content in SCWR cladding material. In: Proceedings 5th Int. Youth Conference on Energy, Pisa, Italy, May27-30, 2015, pp. 103-110)

众所周知,Si 可以提高钢的耐腐蚀性。Wright 和 Dooley(2010)简要介绍了 F/M 钢的数据。Li(2012)研究了在 500℃、25MPa 和 5μg/L 或 200μg/L DO 条件下,添加 1.5%的 Si、Al、Mn、Ti 或 V 对 Cr 质量分数为 9%和 12%的合金氧化的影响。研究发现,添加 Si 具有的益处最大,这归因于在氧化前形成了富 Si 颗粒。这些颗粒作为成核位点,促进了 Fe-Cr 混合氧化物的快速形成。Mn、Ti、Al 或 V 的添加对质量增量的影响很小,尽管这些三元合金的氧化物比二元氧化物薄。

Liu(2013)提出了有关 F-M 模型钢(12%Cr)的氧化数据,其中 Si 质量分数为 0.6%~2.2%,Mn 质量分数为 0.6%~1.8%。将样品研磨至 1200 目,并在 500℃、25MPa 下暴露于超临界水中。结果显示增加 Si 质量分数可提高 F/M 钢的抗氧化性,并促进形成更均匀的氧化物层(图 5.4)。然而,增加 Mn 质量分数会降低耐腐蚀性,并导致形成不均匀的腐蚀膜。Mn 是一种强烈的尖晶石形成元素,可以促进含 Cr 合金中 $MnCr_2O_4$ 尖晶石①的形成。Gilewicz-Wolter 等(2006)通过测量 $^{51}Cr$、$^{54}Mn$ 和 $^{59}Fe$ 在 $MnCr_2O_4$ 和 $FeCr_2O_4$ 中的扩散速率,发现 $MnCr_2O_4$ 中

① 尖晶石是指在立方(等距)晶体体系中结晶分子式为 $A^{2+}B_2^{3+}O_4^{2-}$ 的化合物统称。氧阴离子排列在立方密排晶格中,阳离子 A 和 B 占据晶格中八面体和四面体的部分或全部位置。铁氧化物具有(反)尖晶石结构,如磁铁矿和镍铁氧体。

的金属扩散较高。

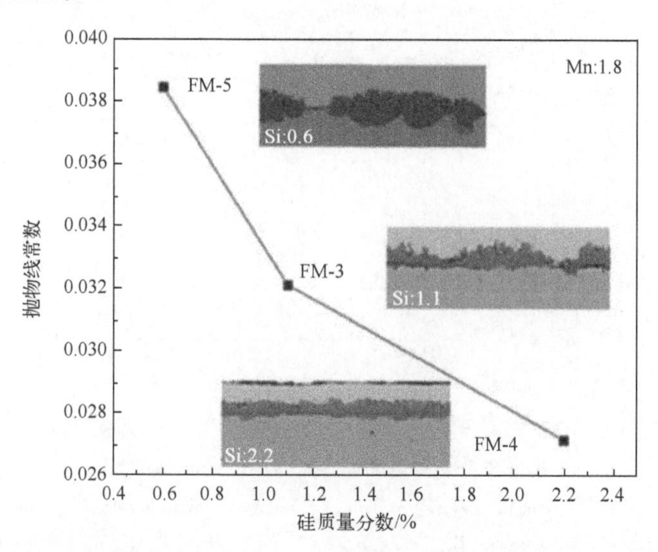


图 5.4 F/M 型钢(约 12%Cr 和 1.8%Mn)氧化时抛物线常数随 Si 质量分数变化的曲线图(Dong 等,2014)(测试的 3 个样品的 SEM 横截面显示为插图)

富锰氧化物的挥发速率小于纯的 $Cr_2O_3$ 氧化物的挥发速率(Holcomb 和 Alman,2006)。Mahboubi 等(2014)报道,在 550℃、25MPa 于 8mg/kg 的氧化超临界水中暴露 100h 后,在 310S SS 上形成的氧化物在氧化物/超临界水界面处含有更多的 Fe 和 Mn,而 Cr 较少。他们提出,含有 Mn 的保护性氧化物层,将富铬氧化物从超临界水中分离出来,并降低了其溶解速率,暴露 250h 后这种益处消失了。Mn 在 $Cr_2O_3$ 中的扩散要快于 Fe、Cr 和 Ni。Choudhry 等(2016)报道了 800H 合金的金属释放数据支持这一假设;在短的暴露时间形成的氧化物,似乎含有足够的 Mn 以防止 Cr 释放(图 5.4),但是随着暴露时间的增加,Mn 溶解在冷却剂中被铁替代。块状合金中的 Mn 质量分数不足以维持氧化物中所需的 Mn 质量分数,并且假定 $x>y>w$ 时组成变为 $Cr_xFe_yMn_wO_z$。数据表明,在较高温度下,Mn 提供的保护作用可能更大。Abe 和 Yoshida(1985)发现在暴露于过热蒸汽(800℃、4MPa)条件下,800 合金上 $Cr_2O_3$ 层上方较厚的 $MnCr_2O_4$ 层的形成,可防止形成球状氧化物,而 Mn 质量分数较低的 Ni 基合金在相同条件下则呈球状腐蚀。

Kish 和 Jiao(未发表的数据)描述了 UNS S21400 的行为,这是一种研磨至 800 粒度,并在静态高压釜中于 550℃、25MPa 的超临界水中暴露 542h 形成的奥氏体铁-Cr-Mn SS(16.7%Cr、12.4%Mn、0.7%Ni、0.5%Si、0.3%N、0.06%C)。溶解氧初始质量分数为 8mg/L。用 Mn(也是奥氏体形成剂)代替 Ni 对于超临界水冷反应堆是非常具有吸引力的，因为 Ni 基合金会受到辐照损伤的强烈影响(第 3 章)。作者认为，在 Mn 贫化区中，奥氏体相转变为铁素体相可降低覆层发生 SCC 的风险，因为已知铁素体钢不易受到环境辅助开裂(EAC)的影响(第 6 章)。但是，该合金显示出过度的氧化物剥落，并且总体腐蚀速率高得令人无法接受。建议可以优化合金的 Mn 质量分数，以平衡一般腐蚀与表面铁素体的形成。

ODS 已被用于改善高温下使用合金(特别是 F/M 钢)的力学性能，从而带来高蠕变强度和良好的抗辐照性。与同一种合金的非 ODS 版本相比，这些合金的耐腐蚀性也有所提高(Chen 等，2007b；Motta 等，2007b)。Cho 等(2004)发现 9%～12% Cr ODS 钢表现出与同等非 ODS 合金相似的腐蚀行为。Zhang 等(2012)报道了在 550℃、25MPa 下暴露于超临界水中的 ODS 钢测试结果(Cr 含量为 12%、14% 和 18%)，并且发现这些合金的耐腐蚀性并不比非 ODS 钢更好。Cho 等(2004)、Zhang 等(2012)和 Gong 等(2015)均报道，在 ODS 合金中添加 Al 可提高其耐腐蚀性。

Ti 和 Zr 基合金的有限工作总结如下。本章的其余部分将重点介绍各种 Fe 基和 Ni 基合金，这些合金已成为大多数超临界水冷反应堆腐蚀研究的重点。

### 5.2.1 锆基合金



Zr 基合金具有良好的中子经济性，因此已被认为可用于超临界水冷反应堆堆芯(所有设计中的燃料包壳和压力管式设计方案中的压力管均采用 Zr 基合金)，尽管目前的商用合金缺乏能够承受超临界水温度下的强度。如第 3 章所述，加拿大超临界水冷概念堆通过绝缘燃料通道保留了 Zr 合金压力管的使用。自 20 世纪 70 年代以来，人们已经知道 Zr 合金在超临界水中表现出非常高的腐蚀速率(Cox，1973)。对 Zr-2 和 Zr-4 的初步研究(Cox，1973；Jeong 等，2005；Motta 等，2007a；Peng 等，2007；Li 等，2007；Khatamian，2013)发现这些合金在 $T>450$℃ 时具有极高的氧化速率，因此在超临界水冷反应堆堆芯温度下使用是不可接受的。许多 Zr 合金类型(Zr-Nb，Zr-Fe-Cr 和 Cr-Cu-Mo)的进一步研究结果表明，存在优化的合金成分，其氧化速率低于 F/M 钢，但仍比奥氏体钢高约 3 倍。所研究的大多数合金的腐蚀动力学大致遵循立方速率定律。Khatamian(2013)发现，这些合金还可能表现出分离腐蚀。Zr0.4Fe0.2Cr 的氧化结果如图 5.1 所示。

### 5.2.2 钛基合金

Ti 基合金已被发现是用于超临界水氧化系统的良好候选材料，因此已被研究用于超临界水冷反应堆。关于超临界水中几种 Ti 基合金腐蚀的数据已有报道(Kasahara,2003;Kaneda 等,2005),检验的合金包括 Ti-15Mo-5Zr-3Al、Ti-3Al-2.5V、Ti-6Al-4V 和 Ti-15V-3Al-3Sn-3Cr。这些合金在 290℃、380℃和 550℃,以及 8000μg/kg DO 条件下,在静态高压釜中暴露 500h。尽管在 290℃和 380℃下暴露的样品增重相似,但在 550℃下暴露的样品增重明显更高。在 550℃下,Ti-15V-3Al-3Sn-3Cr 和 Ti-15Mo-5Zr-3Al 的质量增加系数比 Ti-3Al-2.5V和 Ti-6Al-4V 合金减少 3 倍。Kaneda 等(2005)报道,Ti-15V-3Al-3Sn-3Cr 和 Ti-15Mo-5Zr-3Al 的质量增加与在相同条件下暴露的 304 SS 和 316 SS 的质量增加大致相同。用于超临界水冷反应堆开发的 Ti 基合金后续相关工作较少。

## 5.3 关键参数的影响

对于特定的合金,在 350~600℃温度范围内,各种物理和化学环境参数对总体腐蚀速率的相对贡献按以下顺序排列(重要性递减)(Guzonas 等,2010)①:

温度≈表面粗糙度≈颗粒尺寸>水化学>超临界水密度

辐照会通过改变水的化学性质(水的辐解作用),辐照与合金表面区域的相互作用而影响堆芯材料的腐蚀速率,前者的影响预计会更大。时至今日,对超临界水冷反应堆条件下的流量、传热和热老化影响的系统研究还很少,这些因素未包括在排名中。在以下各节中将探讨各个参数的影响。



### 5.3.1 温度

通常,发现在 $T \gg T_c$ 时温度对材料的氧化/腐蚀影响最大。随着暴露温度的升高,大多数腐蚀特征(重量损失或增加,氧化物厚度)显著增加。由于超临界水冷反应堆堆芯的温度变化很大,因此必须在以下 3 种情况下评估堆芯所用材料的耐腐蚀性:高温亚临界水(约 260℃$<T<T_c$)②、近临界和低温超临界水($T_c<T<$450~500℃)和高温超临界水(450~500℃$<T<$峰值包壳温度)。由于腐蚀是一种表面现象,因此关注的是燃料包壳表面冷却剂温度和密度,而不是大量的冷却剂值。堆芯可分为 3 个区域,这 3 个区域由周围流体的温度和密度定义。

(1) 低于 $T_c$ 的温度下,燃料元件入口下游的区域 1。

(2) 相对恒定,但超临界温度、密度迅速下降的区域 2。

---

① 堆芯水化学条件下的超临界水冷反应堆预计为中性至微碱性,具有低浓度侵蚀性杂质(如 $Cl^-$、$SO_4^-$)的氧化性(DO 浓度为 1~20μg/L,一些 $H_2O_2$),参见第 4 章。

② 较低的温度将由每个 SCWR 概念的堆芯入口温度来定义。

(3) 温度迅速升高且密度较低,相对恒定的区域3。

区域1和区域2对应于“蒸发器”区域(图2.4),区域3对应于“过热器”区域。为支持超临界水氧化系统的开发而进行的腐蚀研究表明,在临界点附近的温度(Kritzer,2004),水密度显著增加(第4章),腐蚀速率达到了最大值。但是,这些研究通常是在接近临界温度的条件下进行的,使用化学侵蚀性条件(例如极酸性的pH和高浓度的氧化剂和侵蚀性阴离子,例如氯),超临界水的特性(例如密度、介电常数)在较小的温度范围内会有很大变化。

在超临界水密度高且流体呈“类液体”的区域1和区域2,预计在亚临界高温水中会发生“传统”电化学过程。Lillard(2012)总结了高温亚临界水中腐蚀的基本原理。在亚临界水中,通常观察到重量的减少,特别是在表面氧化物的高溶解度,且表面不再是钝化的条件下。在该区域中,速率控制步骤可以是电荷转移反应(活化极化)或试剂向表面的质量传输(浓度极化)。这种情况变得更加复杂,因为电化学腐蚀反应涉及两个耦合的半电池(阳极和阴极),这些半电池可以在表面上进行空间分离,并且需要使用混合电势理论。在发生这种主动腐蚀的条件下,合金的使用寿命将相对较短。在高温水中利用金属合金的能力取决于钝化膜的形成,该钝化膜将氧化速率降低到可接受的值(Macdonald,1999)。

与辐照分解反应速率常数一样(第4章),在近临界区域,腐蚀反应速率显示出非阿仑尼乌斯行为。Zhou等(2002)使用电化学噪声分析来测量在150~390℃范围内,流量为0.375~1.00mL/min时,含有0.1mol/kg的NaCl和0.01mol/kg的HCl,25MPa的水中304 SS的腐蚀速率(作者报告了另外两种流速的数据,但对温度的相关性表现相似,为清楚起见,这些数据也从图中省略了)。研究发现,在350℃时测得的流速最大值(图5.5中仅展示了最高流速和最低流速)与后来的建模工作一致(如在5.6节中讨论的Kriksonuv和Macdonald(1995a))。Guan和Macdonald(2009)随后报告了在25MPa、温度范围为25~500℃,在0.01m脱气HCl,0.01m脱气NaOH和脱气纯水中测得的电化学噪声数据,发现$H_2O$和NaOH的最大值远弱于HCl。然而这在当时无法给出任何解释。

图5.6和图5.7显示了在更大温度范围内测得的质量增量数据,结果表明,随着近临界区域的逐渐扩大,观察到了预期的Arrhenius行为,且质量增量随温度升高而迅速增加。在图5.6中,从质量减小到质量增加的转变是在高于$T_c$的温度下观察到的。图5.7显示了不同作者编写的316 SS腐蚀数据。在Arrhenius曲线中很容易看到$T_c$附近速率的明显变化。温度高于$T_c$时,不同来源的数据之间非常吻合。在$T \gg T_c$时,超临界水中研究的所有合金均观察到类似的行为(随着温度增加而呈指数增加)。有关合金氧化速率的温度相关性示例,

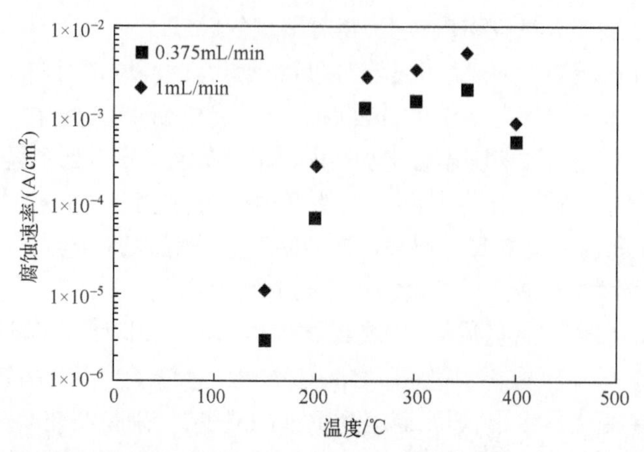


图 5.5　在两个线性流速下，在 25MPa 下含有 0.1mol/kg NaCl 和 0.01mol/kg HCl 的水中，电化学噪声分析得出的 304 SS 腐蚀速率与测试温度的关系

（引自文献 Zhou, Z. Y., Lvov. S. N., Wei, X. J., Benning, L. G., Macdonald, D. D., 2002. Quantitative evaluation of general corrosion of Type 304 stainless steel in subcritical and supercritical aqueous solutions via electrochemical noise analysis. Corros. Sci. 44, 841-860）

可以在 Allen 等（2012）的文章以及本章开头列出的其他综述中找到。



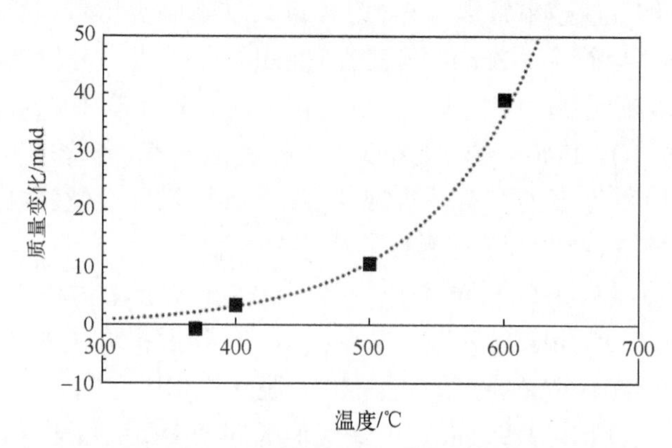


图 5.6　25MPa、pH 值为 6.5 的条件下，T91 在脱气水中（DO<10μg/kg）的腐蚀速率（取样片打磨至 600 目，虚线是温度高于 $T_c$ 时的指数拟合）

（引自文献 Hwang, S. S., Lee, B. H., Kim, J. -G., Jang, J., 2005. Corrosion behavior of F/M steels and high Ni alloys in supercritical water. In: Proceedings of Global 2005, Tsukuba, Japan, October 9-13, 2005, Paper 043）

Potter 和 Mann（1963）提出，在温度为 $T_c$ 时氧化过程性质发生了变化。Glebov 等（1976）提出在超临界水中同时存在电化学氧化（EO）和化学氧化（CO）机理，以解释观察到的 pH 值对 $T_c$ 以上腐蚀的影响（早在 1959 年就将

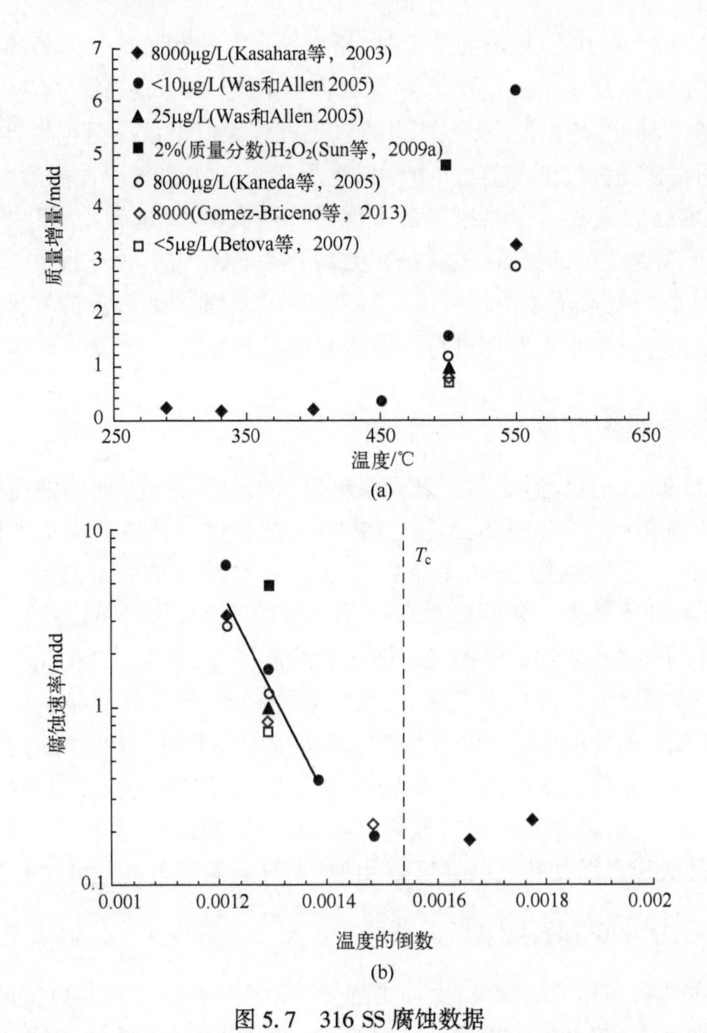


图 5.7 316 SS 腐蚀数据

(a)316 和 316L SS 暴露于含有各种浓度 DO 的超临界水中的质量增量与温度的关系(源引用在图例中列出);(b)腐蚀速率与 $1/T$ 的关系(垂直虚线表示 $T_c$;实线是与 450℃以上的数据最小二乘法拟合的)。

类似的观点,这归功于 N. D. Tomashov)。Robertson(1991)研究了亚临界水和 SHS 中 304 SS 的温度相关性数据,并提出在抛物线速率常数中观测到的局部最大值,可能是由于存在两种具有不同活化能的平行腐蚀机理而导致的。Kriksunov 和 Macdonald(1995a)以及 Guan 和 Macdonald(2009)对近临界腐蚀行为进行了建模,并将其最大值归因于腐蚀机理的变化,从较高水密度(和较高介电常数)的电化学氧化到低于某个阈值水的化学氧化的变化密度。化学氧化机理是基于金属与氧气和/或水的分子相互作用,以及氧化物中阳离子和阴离子缺陷的迁移。Yi 等(2006)得出的结论是,超临界水中的腐蚀与气体条件下的腐蚀相似,是在没有金属溶解的情况下发生氧化物形成。Betova 等(2007)得出的结论是,在高达 500℃的温度下,不锈钢的氧化过程类似于在高温亚临界水中的氧化过程,而在较高温度下,氧化动力学似乎更接近于水蒸气。在接近临界区域,预计在 $T_c$附近的温度下腐蚀速率会出现局部最大值(5.6 节)。超临界水冷反应堆条件下腐蚀速率的测量,通常是在更高温度下进行的,因此很少观察到这样的最大值。由于超临界水冷反应堆冷却剂将通过超临界水冷反应堆堆芯的临界点,因此对临界点附近的腐蚀进行其他研究将很有帮助。

### 5.3.2 表面粗糙度

长期以来,人们已经认识到,表面粗糙度(如研磨、抛光)对腐蚀速率具有很强的影响(如 Ruther 等,1966;Berge,1997)。尽管进行严格的冷加工可减少超临界水中的一般腐蚀,但提供无应变表面会导致较高的一般腐蚀。Maekawa 等(1968)测量了暴露于亚临界水和过热蒸汽中的 304 SS 的质量增量,暴露前对表面进行了研磨、酸洗或电抛光。图 5.8 绘制了 $\lg(WG_{酸洗}/WG_{研磨})$作为温度的函数。可以看到,温度高于 $T_c$时,与酸洗表面相比,研磨表面显示出更高的耐腐蚀性,其效果随着温度的升高而增加。温度为 600℃时,研磨表面的耐蚀性比酸洗表面高约 10 倍。据报道,在 25MPa 的压力下,超临界水中 304 SS 的耐磨性与抛光表面的耐腐蚀性有相似的提高(Guzonas 等,2010)。如图 5.9 所示,磨光和抛光样品的质量增量在近临界区域相似,差异随温度升高而增加。与抛光后的样品相比,在 600℃时,磨光后样品的质量增量约减少了$\frac{1}{4}$。温度低于 $T_c$时,在图 5.8 中观察到相反的效果,冷加工降低了耐腐蚀性。在近临界区域中,该比率处于平稳状态(略大于 1)。Zemniak 等(2008)发现与机械加工的样品相比,电抛光样品在亚临界水中 304 SS 的腐蚀速率以抛物线常数降低。Robertson(1991)、Zemniak 等(2008)以及 Guzonas 和 Cook(2012)提出,在温度 $T_c$上下的不同(相反)行为与腐蚀机理的变化有关。电抛光对降低亚临界水中的腐蚀速率和降低表面吸收放射性的好处是众所周知的,并且已被核工业所利用(Asay,1990;Hudson 和 Ocken,1996;Sawochka 和 Leonard,2006;Hsu 等,1994)。

Leistikow(1972)测量了过热蒸汽(600℃、0.1MPa、DO=(6±2)mg/kg、流速为 2~4cm/s)中多种 Fe 和 Ni 基合金的质量变化,作为酸洗或磨光表面暴露时间的函数,其长达 5000h。酸洗过的表面均显示出质量增加,除了 304 SS 外,其他金属在 2000h 后基本没有质量变化,磨光的表面均显示出质量损失。对于磨碎的样品,由于溶解而损失到冷却剂中的金属比例要高得多,并且显示出对合金中 Ni 质量分数

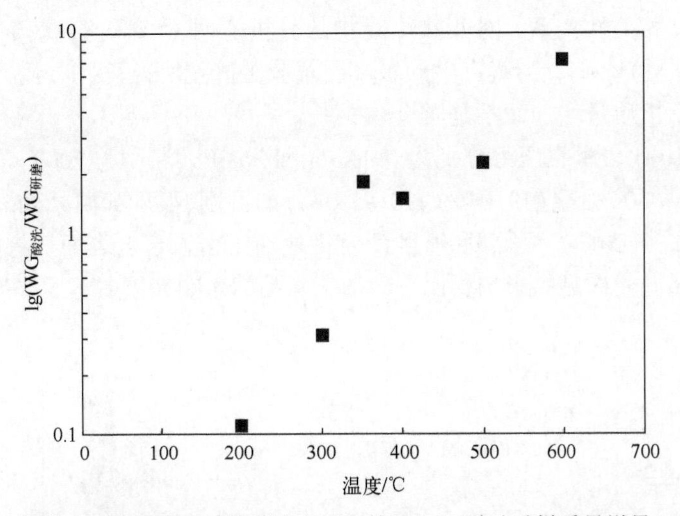


图 5.8　暴露于亚临界水和过热蒸汽(SHS)的 304 SS 酸洗试样质量增量($WG_{酸洗}$)与研磨试样质量增量($WG_{研磨}$)比值的对数与温度的关系

(引自文献 Maekawa, T., Kagawa, M., Nakajima, N., 1968. Corrosion behaviors of stainless steel in high-temperature water and superheated steam. Trans. Jpn. Inst. Metals 9 (2), 130-136)

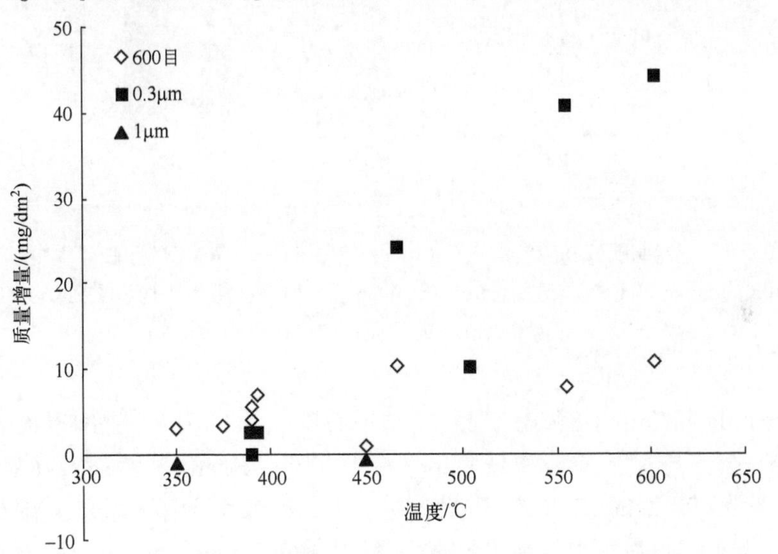


图 5.9　304 SS 在 25MPa 静态高压釜中暴露于超临界水的 3 种表面处理的质量增量与温度的关系

(引自文献 Guzonas, D., Wills, J., Dole, H., Michel, J., Jang, S., Haycock, M., Chutumstid, M., 2010. Steel corrosion in supercritical water: an assessment of the key parameters. In: $2^{nd}$ Canada-China Joint Workshop on Supercritical Water-cooled Reactors (CCSC-2010), Toronto, Ontario, Canada, April 25-28, 2010)

的单调增加相关性关系。对蒸汽冷凝物的分析表明,大部分金属损失是 Cr,可能是含 Cr(Ⅵ)物质。当蒸汽中的氧气质量分数降至 1mg/kg 时,磨碎的 800 合金样品显示出质量增加而不是质量减少。

图 5.10 比较了在压力分别为 0.1MPa 和 26MPa、温度为 700℃条件下暴露 50h 的 Super 304 合金(19.04%Cr,8.93%Ni)的磨削、喷砂和抛光处理对增重的影响(Yuan 等,2013)。研磨和喷砂都会将冷加工引入合金表面。有趣的是,观察到压力对抛光样品的影响要比对冷加工样品的影响大得多。

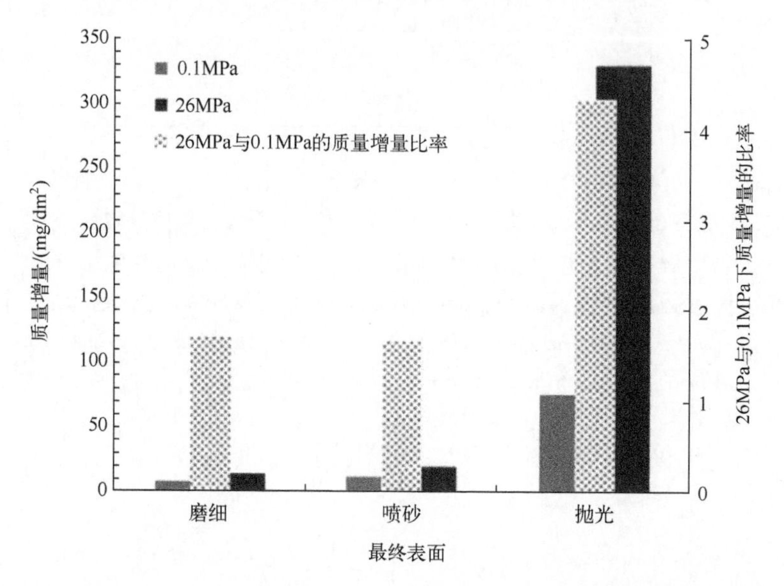


图 5.10 磨削、喷砂和抛光对在 0.1MPa 和 26MPa 下于 700℃的超临界水中暴露 50h 的 Super 304 SS 质量增量的影响(Yuan 等,2013)(阴影线显示了在 0.1MPa 和 26MPa 时的质量增量比率)



Penttila 和 Toivonen(2013)报告,与平面铣削 316L 相比,铣削退火的 316L 和 316Ti 管子样品的质量增量高出一个数量级。Huang 和 Guzonas(2014)发现,研磨至 600 目的 Ni 20%(质量分数)Cr 5%(质量分数)铝合金在 500℃、25MPa 下于超临界水中暴露 6300h,其腐蚀量可忽略不计。用金刚石糊剂抛光的样品在暴露的前 2000h 内显示出非常高的质量增量,随后在更长的暴露时间下质量增量减少,这归因于在高压釜关闭过程中厚氧化物的剥落以去除样品。对于型号 Fe-20Cr-6Al-Y 合金,发现质量增量也有类似的减少(Huang 等,2015a)。

图 5.11 说明了通过晶粒细化可以提高 800H 合金的抗氧化性(Tan 等,

2008a;Allen 等,2012),而喷丸处理引入了 70μm 深的表面变形层,喷丸处理后,800H 合金的表面晶粒尺寸约为 15nm,并且在暴露于超临界水期间,晶粒尺寸增长到约 180nm。喷丸处理和随后在近表面区域的晶粒减小将氧化率降低了约 60%,并且还改善了氧化物附着力,从而显著减少了剥落(注意图 5.11 中所接收材料的质量增量不定)。研究发现,喷丸材料的微观结构比收到的样品要简单,并且主要由富铬氧化物控制。

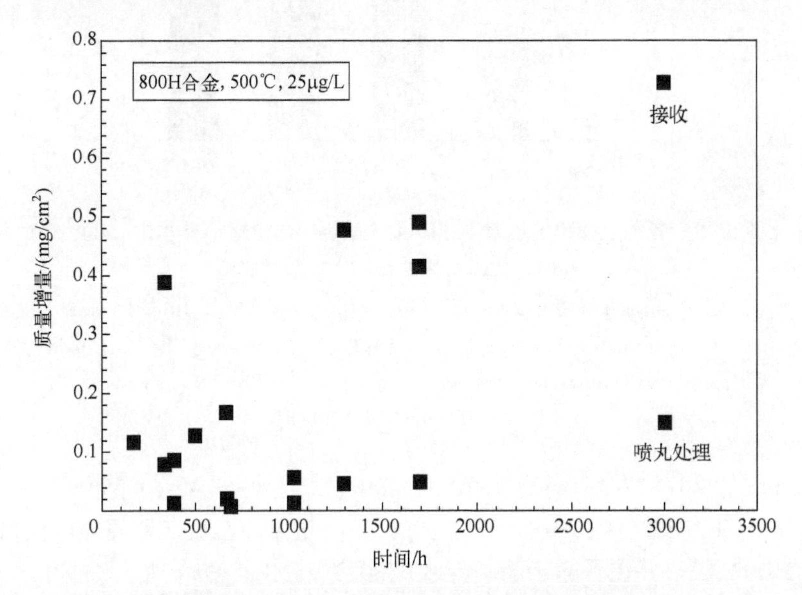


图 5.11　大晶粒和小晶粒两种冶炼条件的 800H 合金暴露于 500℃的超临界水中 3000h 后质量增量情况比较

尽管早期的研究表明冷加工对 F/M 钢的耐腐蚀性没有好处(Eberle 和 Kitterman,1968),但最近的研究表明情况并非如此。Ren 等(2008)报告,T91、HT9、NF616 和两种型号的 Fe-Cr 合金(Fe-15%Cr 和 Fe-18%Cr)的喷丸处理可减少超临界水(500℃、DO<25μg/kg,最多暴露 3000h)中 20%~40%的腐蚀。Artymowisc 等(2014)研究了各种表面粗糙对在超临界水(500℃、25MPa、DO=5μg/kg、250h)中 Fe14Cr1.5Si 合金腐蚀的影响。发现表面覆盖有两种氧化物的混合物,分别是薄而紧凑的氧化物(在与金属氧化物界面相邻的 50nm 厚的富硅层的顶部上 200nm 厚的 $Cr_2O_3$)和由 $Fe_3O_4$层组成的厚氧化物层。该氧化物层由$(Fe、Cr)_3O_4$层上的表面向外生长,而$(Fe、Cr)_3O_4$层是由于 Cr 质量分数太低而无法形成保护性 $Cr_2O_3$层的典型表面向内生长。氧化物覆盖的表面份额随着表面冷加工程度的降低而增加(图 5.12)。

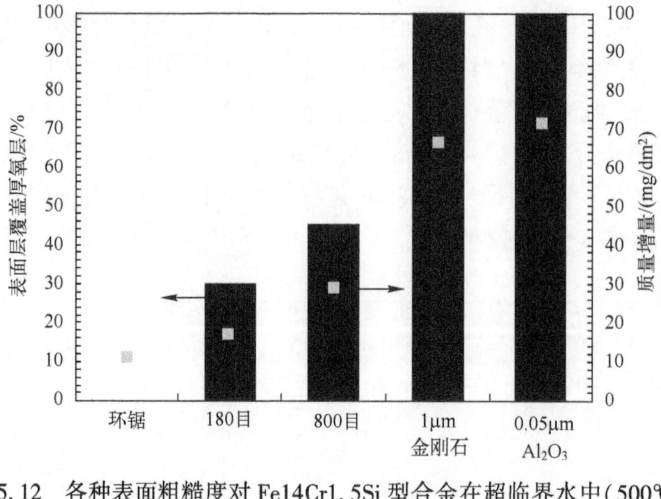


图 5.12　各种表面粗糙度对 Fe14Cr1.5Si 型合金在超临界水中(500℃、25MPa、DO=5μg/L、250h)腐蚀的影响

(引自文献 Artymowicz,D.,Andrei,C.,Cook,W.G.,Newman,R.C.,2014. Influence of the cold work on the supercritical water oxidation of Fe14Cr1.5Si model alloy. In:Canada-China Conference on Advanced Reactor Development(CCCARD-2014),Niagara Falls, Canada,April27-30,2014)



Dong 等(2015)发现,具有 1.7%~1.8%(质量分数)Mn、0.6%或 2%(质量分数)Si、12%(质量分数)Cr 的 F/M 型钢,采用超声冲击喷丸处理可以使氧化物更加均匀和致密,并在暴露于超临界水(500℃、25MPa、初始 DO=8mg/kg)1000h 后质量增加 2 倍。图 5.13 显示了原始样品和喷丸加工样品的横截面显微硬度数据,喷丸样品表面的硬度显著增加并且随深度而降低,直至达到原始材料的值(在150~200mm 的深度)为止。研究发现,原始样品上的氧化物优先在板条形马氏体相上形成。Li(2012)指出,马氏体和铁素体晶粒之间的硬度差异会影响变形层的深度。

图 5.14 显示了超临界水冷反应堆暴露 2000h 后 4 种奥氏体钢的最小和最大氧化物厚度的数据。这些钢具有磨光(1200 目)或机加工的表面粗糙度。较高的 Cr 质量分数和表面冷加工对减少腐蚀的好处是显而易见的,在较高的 Cr 含量下,表面冷作的好处可能会减少。在图中还可以看到减小晶粒尺寸(347H 对 347HFG)的有益效果。

图 5.15 显示了在暴露于超临界水(650℃、25MPa、3000h)下具有加工和抛光表面的 316L 管道样品的照片和 SEM 截面图。机械加工表面具有薄、均匀且致密的表面氧化物层,而抛光的表面氧化物则很厚,并具有明显的内部氧化。机械加工表面上的最大氧化物厚度约为 1μm,而抛光后的样品上的最大氧化物厚

度约为 100μm。

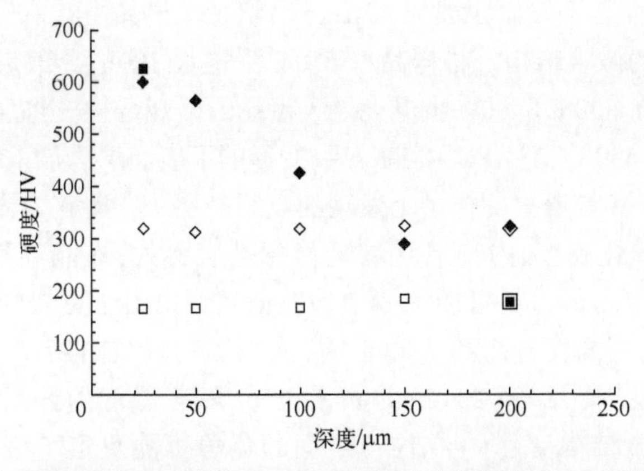


图 5.13 F/M 型合金原始样品(空心)和超声冲击硬化处理后的样品(实心)的横截面显微硬度数据,合金质量分数为 2.2%Si、1.7%Mn、12%Cr

(引自文献 Dong, Z., Liu, Z., Li, M., Lou, J. L., Chen, W., Zheng, W., Guzonas, D., 2015. Effect of ultrasonic impact peening on the corrosion of ferritic-martensitic steels in supercritical water. J. Nucl. Mater. 457, 266-272)

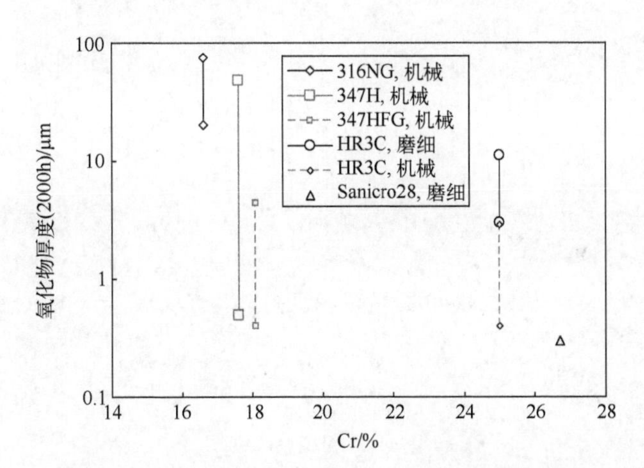


图 5.14 在 650℃、25MPa 和 DO = 150μg/kg 条件下暴露 2000h 后从 4 种奥氏体钢的 SEM 横截面获得的最小和最大氧化物厚度与 Cr 质量分数的关系(对这些钢分别做磨光(1200 目)或机加工的表面粗糙度处理)

温度高于 $T_c$ 的冷加工表面腐蚀速率降低,归因于近表面缺陷密度的增加,这增加了 Cr 向表面的扩散,并有助于形成均匀的保护性氧化膜。Sarver 和 Tanzosh(2013)报道,温度在 750℃和 800℃条件下,在过热蒸汽中未观察到奥氏体不锈钢喷丸强化的好处。显微硬度测试表明,在高于 750℃的温度下,近表面冷加工迅速降低。Li 等(2014)使用 FIB 技术检查机加工或研磨至 600 目的 316L 样品的表面。下表面区域(1~2μm 厚)机加工样品中有一部分具有带剪切带的严重变形的微观结构。将样品在 650℃下退火 100h,导致表面层重结晶为薄而细的层。在 850℃下进一步退火导致粗糙的微观结构。将加工过的样品暴露于超临界水(650℃、25MPa、1000h)后,变形的下表面层已被重结晶层所替代,该层类似于在 650℃退火的样品上发现的重结晶层。在磨碎的样品上形成了一层较厚的富铁氧化物层,内部氧化层延伸到基底深处,在晶界形成了空隙,并严重氧化了晶粒内部。尽管图 5.8 中的数据显示出由于更快的扩散速率和更多的扩散途径,当温度升高到 $T_c$ 以上时,冷加工的好处有所增加,但 Sarver 和 Tanzosh(2013)以及 Li 等(2014)的数据说明,随着温度的升高保护性氧化物的形成速率与微结构变化(再结晶)之间的竞争可能决定了表面冷加工对耐蚀性的影响。



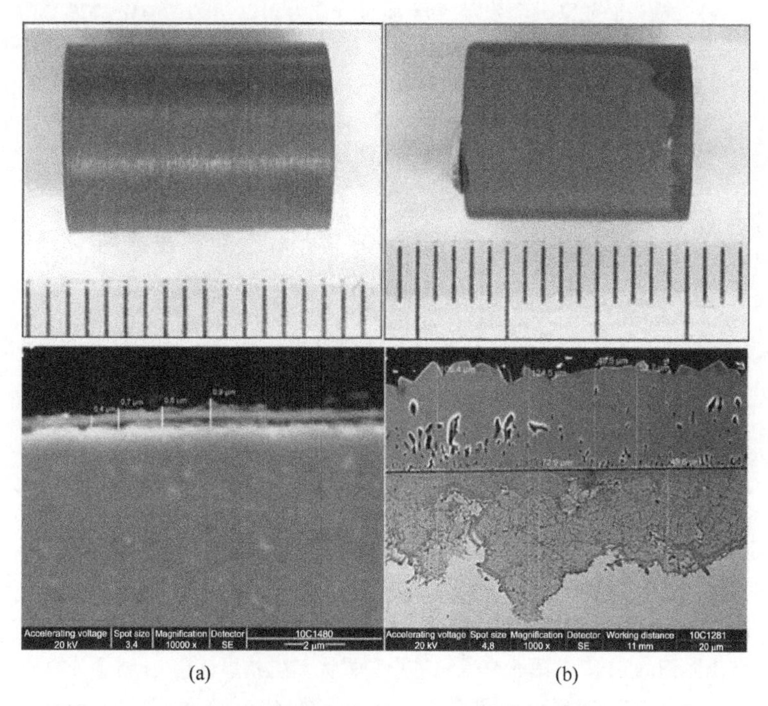


(a) (b)

图 5.15 316L SS 管道样品在 650℃、25MPa 条件下暴露 3000h 后表面的照片和相应的 SEM 横截面图像

(a) 机加工;(b) 1200 号砂纸打磨。

由于晶界是快速扩散的重要途径,减小整体晶粒尺寸还将使形成保护层的元素(如 Cr)迅速扩散到表面。块状金属中保护性氧化形成元素的质量分数通常用于确定较小的晶粒是否有利于减轻氧化。Montgomery 和 Karlsson(1995)总结了347 SS 数据,结果表明,晶粒尺寸的减小使保护层在 650℃和 700℃下的耐蚀性增强。图 5.14 比较了 347H 和 347HFG(细晶粒)上的氧化物厚度,显示出细晶粒合金的最大氧化物厚度降低至 1/10。Tsuchiya 等(2007)比较了大晶粒尺寸(约 30μm)和细晶粒尺寸(约 3μm)304L 和 310L SS 在 550℃、25MPa、DO=8mg/kg 超临界水条件下的腐蚀。细颗粒 304L 样品上的氧化物厚度约为大颗粒材料厚度的 1/10。进入测试溶液的金属损失(从缩小的重量损失数据获得)表明,大部分氧化金属从细颗粒材料释放到溶液中(304L 和 310S 的损失份额均为 50%)。这被认为是导致金属向表面的传输速率提高的原因。Ren 等(2010)发现,等径弯曲通道变形(ECAP)[①]在 T91 上产生较小的晶粒,在 3000h 后,超临界水(500℃、25MPa、DO<25mg/kg)的腐蚀降低了约 30%,与喷丸处理量大致相同。在晶粒细化处理之前,晶粒尺寸为 5μm,在处理之后,ECAP 材料的平均晶粒尺寸为 1.25μm,喷丸材料的晶粒尺寸约为 1.3μm。

通过晶界工程(GBE)能够有效减少氧化物剥落,这一能力在合金 800H 上完成了测试,测试过程中应用一个循环的热机械处理(以约 7%的厚度减薄进行冷轧),然后退火(1050℃,90min),以提高保护氧化物层的抗剥落性(Tan 等,2006a)。氧化物剥落一直是 SCFPP 面临的一个关键技术问题,因为它会导致流道堵塞和涡轮机损坏,这是该研究领域备受关注的主题(Wright 和 Tortorelli,2007)。在 SCWR 中,氧化物剥落也会导致流道堵塞,尤其对燃料组件影响最大,还会损坏涡轮机,并且放射性输运的一种机制。此外,已经裸露的表面可能会受到更大的腐蚀。在 Tan 等(2006a)的工作中,将样品抛光至 1mm 的粗糙度。SCW 腐蚀试验(500℃,25MPa,DO=25μg/kg)后,通过测量样品质量变化和 SEM 评估表明,在收到的样品中发生了拉伸性氧化物剥落,但在 GBE 处理的样品中没有发生(图 5.16)。进一步对样品进行横截面的电子背向散射衍射(EBSD)检查,确定了奥氏体、磁铁矿、Fe-Cr 尖晶石和赤铁矿(Tan 和 Allen,2005)。GBE 处理的样品显示出更高比例的赤铁矿(具有较小的晶粒尺寸)与外层中的少量磁铁矿混合,以及内层的尖晶石相与被鉴定为奥氏体的相混合。应变分布(通过每个 EBSD 数据点测量值与其相邻值之间的局部平均取向差获得,不包括任何更高的角度边界(>5°))显示,对于收到的样品,应变累积接近内层和外层之间的界面,而 GBE 处理的样品的应变分布相对均匀。相对应变强度作为氧化物尺度上的位置函数,明显可以看到 GBE 处理后内部和外部氧化物之间的应变变化显著减少。研究表明,GBE 处理样品上较低程度的氧化物剥落是由于

① 一种利用过度塑性变形在整个合金体中形成细粒度微观结构的技术。

相对各向同性的氧化物结构和较大的赤铁矿含量。GBE 也被证明可以提高 617 合金的耐腐蚀性(Tan 等,2008b)。Nezakat 等(2014)对 316L 进行了热机械处理(在室温下进行多步交叉轧制,然后在 1050℃下每毫米厚度退火 5min),而后评估其耐腐蚀性。将处理过的样品暴露于 SCW(600℃,25MPa,DO=150μg/kg)中 100h、300h 和 1000h。结果表明,纹理对抗氧化性有显著影响,具有{103}纹理和{110}纹理的样品分别表现出最差和最好的抗氧化性。同时,还发现热机械加工过程导致高能晶界分数减少,进而提高了氧化层和奥氏体之间的附着力。

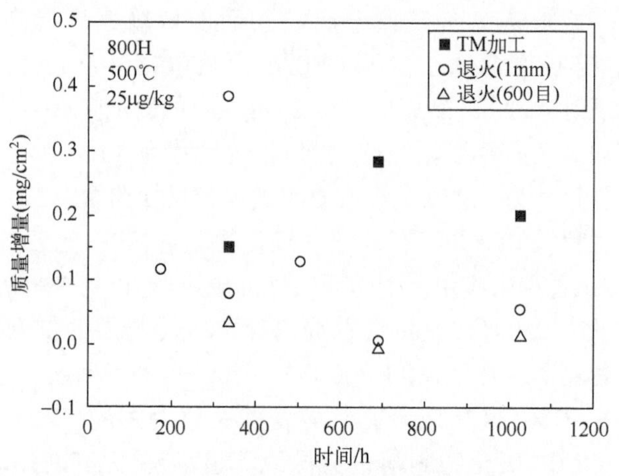


图 5.16 暴露于 500℃的 800H 合金的质量增量随时间的变化关系。TM 指热机械加工条件。在热处理之前,将退火样品抛光至 1mm 和 600 目的粗糙度。TM 样品的热处理之前被抛光至 1mm 的粗糙度

(引自文献 Allen, T. R., Chen, Y., Ren, X., Sridharan, K., Tan, L., Was, G. S., West, E., Guzonas, D. A., 2012. Material performance in supercritical water. In: Konings, R. J. M. (Ed.), Comprehensive Nuclear Materials, vol. 5. Elsevier, Amsterdam, pp. 279-326)

Ren 等(2010)使用有效扩散系数模拟了晶粒尺寸对 Cr 扩散的影响,该系数考虑了通过晶格和晶界的扩散系数、晶粒形状、晶粒尺寸和晶界宽度(估计为 0.5nm)。研究人员使用了 12CrMoV 钢的晶格和晶界扩散系数值(成分与他们研究中使用的材料接近)。通过假设 Cr 在晶粒及其边界之间的转移可以忽略不计,计算了 Cr 沿晶格和晶界的通量(图 5.17)。对于测试中的晶粒尺寸,发现沿晶界的(晶格+晶界)Cr 扩散百分比从原始条件下的 3%增加到晶粒细化条件下的 31%。使用类似的方法,Kim 等(2015)得出的结论是,为了确保在奥氏体钢上形成连续的 $Cr_2O_3$ 层,晶粒度需要小于 8μm,这与实验数据非常一致。

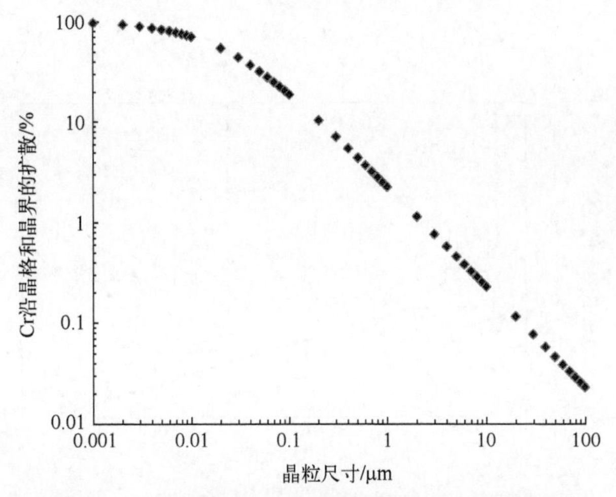


图 5.17　F/M 钢沿晶界(晶格加晶界)Cr 扩散的百分比

(引自文献 Ren, X., Sridharan, K., Allen, T. R., 2010. Effect of grain refinement on corrosion of ferriticemartensitic steels in supercritical water environment. Mater. Corros. 61,748-755)

## 5.3.3 水化学

水化学因素包括 pH 值、添加剂和杂质浓度以及氧化剂的质量分数和辐射分解产物。由于水、添加剂或杂质与合金中特定元素的相互作用,水的化学效应被认为是合金特有的。杂质对 SCWR 条件下腐蚀的影响尚未进行研究。

### 5.3.3.1　pH

在近临界区,pH 值的影响可以参照 Pourbaix 图(图 5.18)来理解①。许多相关系统的 Pourbaix 图已由不同的作者提出,铁-水(Kriksunov 和 Macdonald, 1997;Saito 等,2006;Cook 和 Olive,2012a)、铬-水(Saito 等,2006;Cook 和 Olive, 2012b)、镍-水(Saito 等,2006;Cook 和 Olive,2012c)系统的关系图已被报道;用于高温的 Pourbaix 图的一个缺点是,它们通常基于在较低温度下获得的热力学数据的推断。注意到在 400℃、25MPa 条件下,SCW 的密度($167kg/m^3$)已经超出了修订后的 HKF 模型已验证的范围。第 4 章已经讨论热力学数据的附加限制。Kritzer 等(1999)讨论了 Pourbaix 图的其他局限性,具体如下。

(1) 无法预测氧化物的形态(如多孔氧化物的防护性较差)。

① Pourbaix 图是一种物种优势度(由热力学决定)随 pH 值和电化学电位变化的图。电化学电位与吉布斯自由能的关系为 $\Delta G=-nFE$。

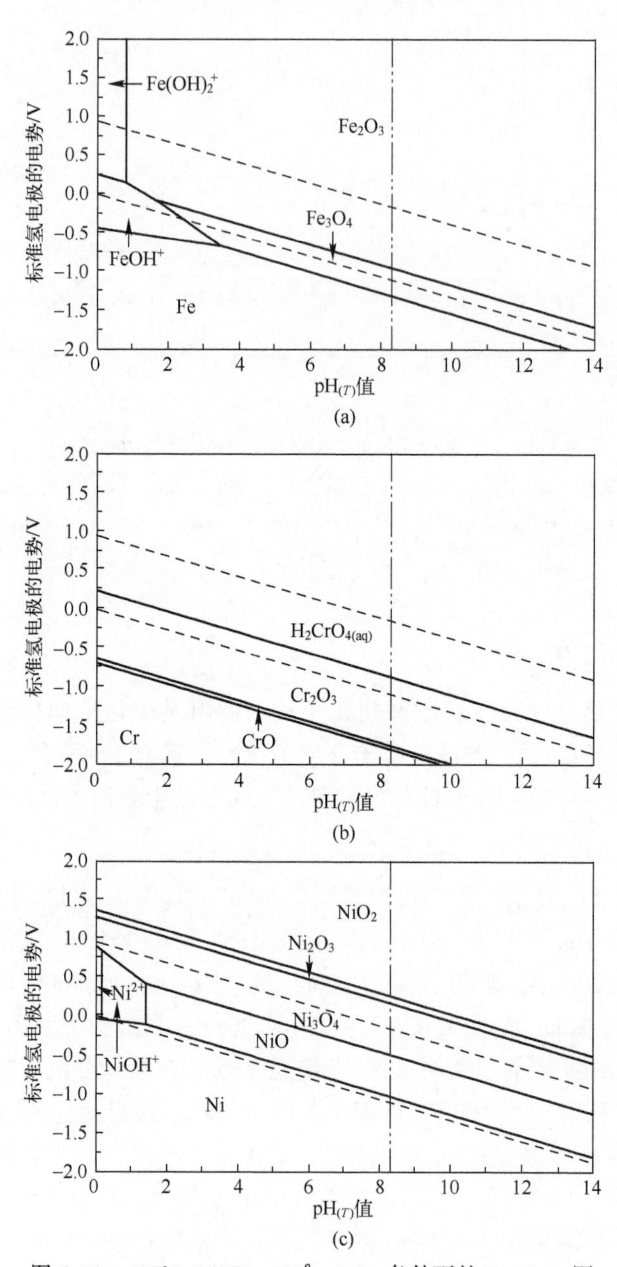


图 5.18　400℃、25MPa、$10^{-8}$mol/kg 条件下的 Pourbaix 图

(a) Fe;(b) Cr;(c) Ni。

(2) 未考虑动力学因素,如一个热力学不稳定的氧化物向稳定缓慢转换的速度。

(3) 无法解释局部的 pH 值影响(如坑、裂缝),必须单独建模,其化学参数可能很难衡量。

SCW 中使用 Pourbaix 图的另一个困难是缺乏标准化的 pH 值刻度,正如 Kriksunov 和 Macdonald(1995b)所讨论的,他们提出了一种实用的 pH 值刻度(而且,正如作者所指出的,有点武断)。该 pH 值是根据可用的高温数据,即 HCl 和 NaOH 的离解常数和水的活性得出的。

Mitton 等(2000)用热力学预测比较了 Kritzer 等(1998)关于 pH 对金属释放影响的试验结果,支持了 SCWO 的发展。在 350℃、24MPa 的条件下,金属的析出量随着 pH 值的增加而减少,各种析出元素的相对比例也发生了变化,与中性 pH 下合金的相对比例接近,结果与热力学预测相吻合。Guzonas 等(2010)报道在几个建议的 SCWR 给水化学的条件下(含氧水、脱氧水、水和氨、联氨),450℃时的腐蚀差别不大(图 5.19)。他们发现,在含氨的脱氧水化学反应中,金属的释放量和氧化层厚度都更高,氧化层厚度的增加与氨质量分数的增加有很好的相关性。Glebov 等(1976)报道了在压力为 28MPa、温度在 380~520℃范围内、pH = 9.5 ~ 10.5 相较于 pH = 6.5 ~ 7.5 时,4 种合金(12Cr1MoV、EI756、Cr18Ni12Ti 和 12Cr2Mo1)的腐蚀速率降低。图 5.20 显示了在pH=10 和 pH=7,以及温度为 350℃ 和 400℃ 稳态条件下,Fe 和 Ni 从 316 SS 中的释放情况。

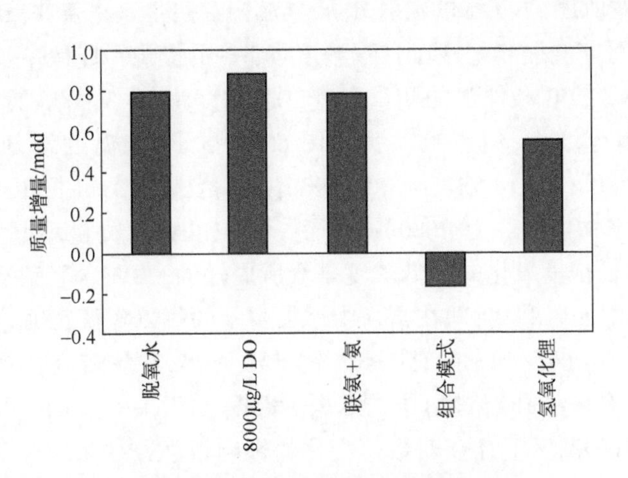


图 5.19 304 SS 在不同给水化学条件下不锈钢在超临界水中的质量增量

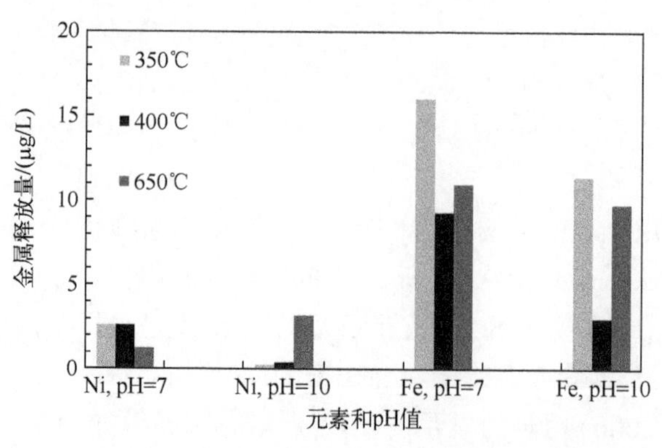


图 5.20　pH = 7 和 pH = 10、25MPa 时在含有 LiOH 的超临界水中，从 316 SS 释放 Fe 和 Ni 到测试溶液中的量

一般而言，pH 值对 SCW 近临界区腐蚀的影响与 Pourbaix 图中典型的热力学预测一致。在温度为 550℃、压力为 25MPa 的条件下，离子的离解很少，pH 值对腐蚀速率影响不大（见 4.2 节）。在这些较高的温度下，Pourbaix 图没有意义，而 Ellingham 图①则更相关。Qiu 和 Guzonas（2012）讨论了这两种方法，并说明了它们在近临界区域的收敛性。进入核心入口"蒸发器"区域的给水 pH 值可能会对腐蚀和腐蚀产物沉积产生一定的影响，当冷却剂加热到 400℃ 以上时，这种影响会减小。



#### 5.3.3.2　溶解氧

如第 4 章所述，由于水的辐射分解，堆芯内会产生氧化条件（DO 浓度为 1～20mg/kg）而可能会被氧化，因此 DO 浓度对腐蚀的影响一直是许多研究的主题。一般来说，在 SCWR 条件下的测试发现，DO 浓度为 0～8mg/kg 时，增加 DO 浓度，腐蚀速率也增大，但影响不大。图 5.21 显示，在温度为 500℃、压力为 25MPa 的条件下，当 DO 浓度为 10～300μg/kg 范围内增加时，HCM12A 的质量增量减小；当 DO 浓度为 2000μg/kg 时，HCM12A 的质量增量变大（Allen，2012）。这种低浓度氧化的最低浓度是众所周知的，并在含氧和复合水化学策略（第 4 章）中加以利用，即在给水中添加少量的氧以降低腐蚀速率。图 5.22 以暴露时间为函数，比较了铁素体钢 T91 和 T92 暴露在类似 SCW 的环境（500℃，但有一些 DO 浓度）中的数据。数据之间的一致性相对较好，表明在这些相对较低的浓度下，DO 对数据变化的影响很小，DO 浓度为 8000μg/kg 时

---

① Ellingham 图是在特定浓度的反应物（如 $p(O_2)$）下，氧化物稳定性与温度的函数关系图。纵坐标的吉布斯自由能变化对应 Pourbaix 图（$\Delta G=-nFE$）中的电极电位。

可能会产生积极效应。在奥氏体钢中观察到类似的现象。从图 5.7 可以看出，在 DO 浓度为 8mg/kg 时，316 和 316L 在 500℃ 和 550℃ 时的质量增量对 DO 浓度的依赖性较弱。图 5.23 显示，将 DO 浓度提高 40 倍，导致 304 SS 在 SCW（温度 400～550℃、25MPa）暴露 500h 后重量减轻至 1/2。

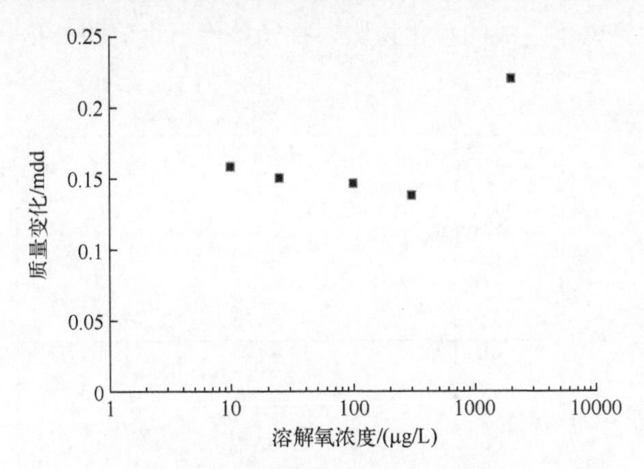


图 5.21　500℃、25MPa 条件下 HCM12A 暴露于不同 DO 浓度的超临界水中 1026h 后质量的变化

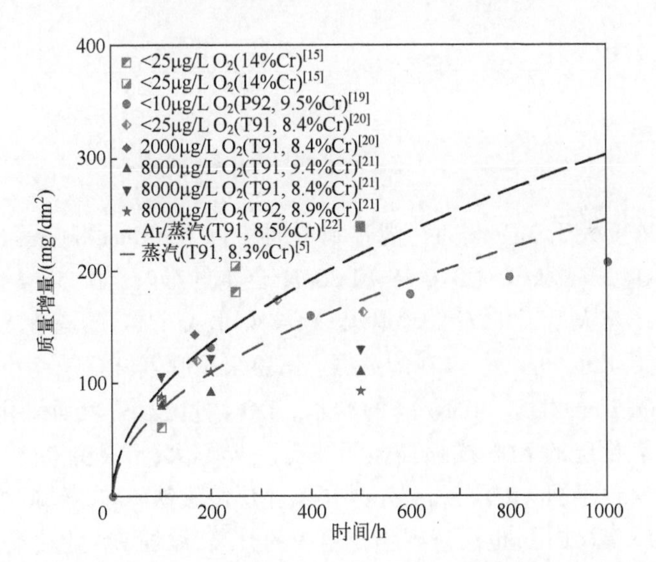


图 5.22　T91（实心符号）和 T92（半实心符号）暴露于 500℃ 和不同浓度 DO 的超临界水中质量随时间的变化而增加

Tanet 等（2006b、2007）研究了 HCM12A（500℃、25MPa、1μm 的粗糙度）在 DO 浓度为 25μg/kg 和 2mg/kg 的条件下暴露 576h 后的氧化情况，结果显示 DO

浓度对氧化层的组成影响不大,氧化层厚度和增重略有增加,但对氧化层的形态影响显著,氧化层厚度的相对差异随着暴露时间的增加而减小。暴露在较高DO浓度下的样品显示 $Cr_2O_3$ 晶体优先垂直于表面和赤铁矿的外层。Ren等(2007)发现,625 合金和 718 合金在 500℃和 600℃,DO 浓度为 2mg/kg 的SCW 中在脱气条件下生成的氧化物比未脱气条件下生成的氧化物少,且点蚀程度较低。

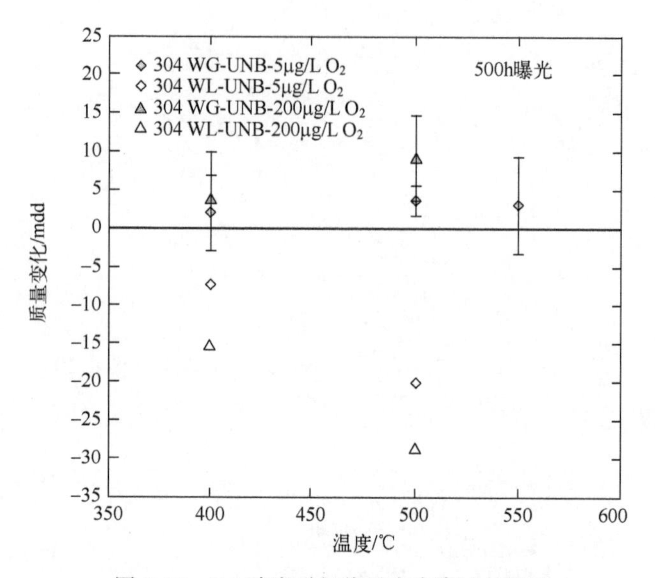


图 5.23　DO 浓度对超临界水中腐蚀的影响

在 DO 浓度大于 2mg/kg 时,观察到不同的行为。Choudhry 等(2015)研究了高浓度 DO(20mg/kg)对 316 SS 和 800H 合金的影响。图 5.24 所示为合金800H(在水冷淬火后于 1177℃时退火,颗粒尺寸 ASTM4)暴露在 SCW(650℃、25MPa、流速 0.1mL/min)中 140s 的数据。试验采用脱氧(DO 浓度小于 10μg/L)和高浓度 DO(20mg/L 和 40mg/L)的给水。DO 浓度越高,表面氧化物越致密。虽然横断面氧化层的 EDS 线扫描显示成分差异不大,但暴露在 20mg/kg SCW的样品上的氧化层的 XRD 分析显示外层氧化层由磁铁矿和赤铁矿组成。

Choudhry 等(2015)也注意到随着温度的升高,添加 DO 的效果也发生了变化。当温度到达 700℃时,800H 合金在 DO 浓度为 20mg/kg 时的金属释放量低于在除氧水中的释放量,但是 750℃、DO 浓度为 20mg/kg 条件下导致高金属释放量。Liu(2013)研究了添加不同质量分数的 Si(0.6%~2.2%)和 Mn(0.6%~1.8%)的 Fe12Cr 合金在 SCW(500℃、25MPa、8mg/kg、5% $H_2O_2$)中的氧化模型。他们发现在 SCW 中,氧化动力学从 DO 浓度为 8mg/kg 时的抛物线变化为高浓度 DO 时的 0.7～0.8 的指数形式,在 1000h 后质量增加了 1 倍。两种氧化物在整体结构上相似(均为双层氧化物),但在含 5% $H_2O_2$的 SCW 中形成的氧化物含量比 8mg/kg 时的更均匀、更连续。DO 浓度高时形成的外层是赤铁矿,而不是低浓度时发现的磁铁矿。Leistikow(1972)发现了当 800 合金暴露于 SHS(620℃、0.1MPa、流速 2～4cm/s)、DO 浓度为 6mg/kg 时,质量下降(Cr 的溶解使金属损失了 32%到蒸汽相),DO 浓度为 1mg/kg 时,质量随时间增加而增加,只有 8%的 Cr 溶解。Watanabe 等(2002)、Watanabe 和 Daigo(2005)报道了在 400℃、30MPa、含 0.01mg/kg $H_2SO_4$的 SCW 中,随着合金中 Cr 质量分数的增加,Fe 和 Ni 基合金重量损失减少。但在高得多的氧浓度下,这种关系就不成立了。在 800mg/kg 的氧浓度下,316 SS 的耐腐蚀性有所提高,而大多数 Ni 基合金在较高的 DO 浓度下腐蚀速率增大。Fe 基合金耐蚀性的提高是由于表面形成了保护性的 $Fe_2O_3$氧化物,而 Ni 基合金耐蚀性的降低是由于可溶性 Cr(Ⅵ)的形成,这与 Fe 和 Cr 的 Pourbaix 图一致。

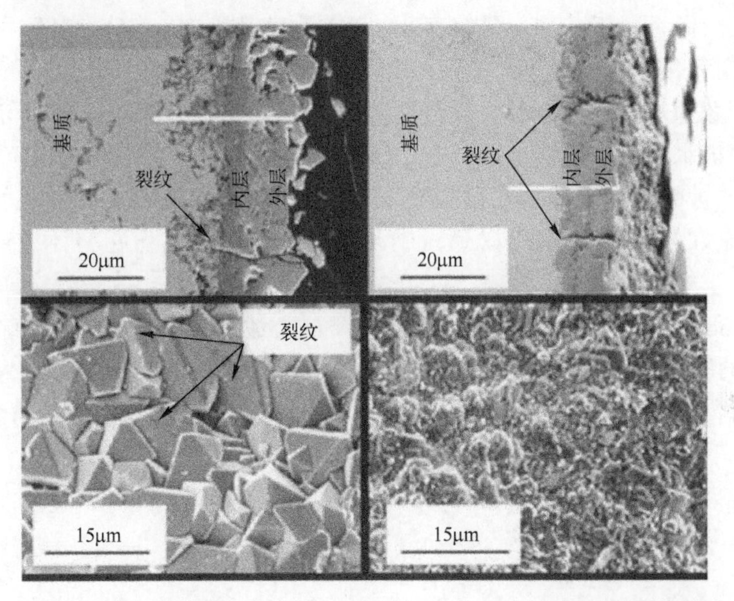


图 5.24　合金 800H 在超临界水(650℃、25MPa、DO = 20mg/kg(右)和脱氧给水(DO<10μg/kg)(左))中形成的氧化层截面(上)和表面形貌(下)(Choudhry 等,2016)

(引自文献 Choudhry, K. I., Mahboubi, S., Botton, G. A., Kish, J. R., Svishchev, I. M., 2015. Corrosion of engineering materials in a supercritical water cooled reactor: characterization of oxide scales on Alloy 800H and stainless steel 316. Corros. Sci. 100, 222-230)

如第 4 章所述,Cr(Ⅳ)类物质在 SCW 中是可溶的(挥发的),并导致保护性的富 Cr 氧化物的溶解。Fujiwara 等(2007)通过化学气相沉积法测量在温度为

350℃和450℃，不同DO浓度条件下，沉积在Pt衬底上的$Cr_2O_3$薄膜的溶解度。结果表明，当温度为450℃、DO浓度小于400μg/kg时，$Cr_2O_3$很稳定。Dong等(2012)研究了$Cr_2O_3$陶瓷在含有不同浓度DO的SCW中的稳定性，测量了在测试溶液中溶解Cr的质量分数。XPS分析了一个暴露在600℃、DO浓度为168g/kg的SCW中25h的样品，结果显示只有$Cr^{6+}$粒子。他们还注意到，一些暴露在高DO环境中的样品表面有氧化钼结晶沉淀。氧的影响与$Cr_2O_3$样品的晶粒尺寸有关，对晶粒尺寸为10~20μm的样品影响最低，对晶粒尺寸为0.2μm的样品影响最高。

在气相氧化的讨论中，Hauffe(1976)指出，如果在表面吸附氧气是速率限制的，那么气相氧化在很大程度上依赖于氧分压(相当于DO浓度)，然而，如果氧的固态扩散是速率限制的，那么依赖性就会弱得多。然而，即使在低压下，有水存在的金属氧化也不同于干燥空气中的氧化。在高密度SCW中的氧化机理远比在低密度SCW(SHS)中复杂，因为水分子的体积密度相差了2~3个数量级，而且最近的MD研究表明，水分子密度在$Fe(OH)_2$表面(与水接触的最外层氧化物表面的模型)高于整体水分子(第4章)。每个暴露的OH基团(表面覆盖)所吸附的水的数量随着温度的升高而减少。虽然由于密度较低，表面被水覆盖的比例减小(图4.26)，但即使在密度最低的情况下，合金表面也主要被吸附的水覆盖。模拟还表明，水是在纳米孔的表面发现的，而且，重要的是在超临界温度下，$O_2$通过纳米尺寸的裂缝扩散比在整体溶液中扩散更容易。



Saunders等(2008)指出，与干燥条件下的类似过程相比，水蒸气存在时生成氧化物的所有特性都发生了改变。反应的第一阶段(吸附和离解)受氧化物性质的控制，数据表明，表面缺陷有利于气体分子的离解。吸附在氧化物表面的水解离为

$$H_2O \longleftrightarrow O^{2-}+2H^+ \tag{5.1}$$

或

$$H_2O \longleftrightarrow H^++OH^- \tag{5.2}$$

在高温下，$O_2$是由水的热分解形成的，所以即使是一个名义上除氧的系统，在高于600℃的温度下，也会含有μg/kg量级浓度的氧。在曝气系统中，增加的$O_2$提供额外的氧化剂。更重要的是，在SCWR核心中，水的辐射分解除了产生短暂的自由基氧化和还原物质外，还能产生高浓度的$O_2$和$H_2O_2$。氧与铁的总体反应为

$$2Fe+2H_2O+O_2 \longleftrightarrow 2Fe(OH)_2 \tag{5.3}$$

$$4Fe(OH)_2+2H_2O+O_2 \longleftrightarrow 4Fe(OH)_3 \tag{5.4}$$

其他合金成分也有类似的反应。金属氢氧化物在低浓度下会溶解(蒸发)到SCW冷却液中(第4章)或在表面进行脱水反应,以形成金属氧化物。

$O^{2-}$和$OH^-$都能通过氧化物扩散(Galerie,2001),而且对表面氧化物的分析表明,$O^{2-}$、$OH^-$、$O_2$和$H_2O$均可通过固态扩散($O^{2-}$和$OH^-$)或通过氧化物中的裂纹和孔隙($O_2$和$H_2O$)渗透到氧化层中。由式(5.1)和式(5.2)中产生的氢所形成$H_2$进入溶液,或通过氧化物和金属扩散,最终释放到环境中。Rodriguez(2014)以及Rodriguez和Chidambaram(2015)用XPS法研究了氧化层中$O^{2-}$、$OH^-$和吸附的$H_2O$。Yamamura等(2005)使用XPS报告了在SCW(450℃、50MPa、$pH_{25}$值中性)中的304 SS上形成的氧化物存在的CrOOH,CrOOH占氧化物中Cr的90%。Was(2004)报道了在暴露于SCW的304 SS和316L SS上形成的氧化层中存在Fe和Cr的氢氧化物。氧化铁和氢氧化物的含量相似,他们认为,氧化物中存在的大部分Cr都是以氢氧根的形式存在的。Galerie等(2001)注意到$OH^-$可以很容易地填满通常由$O^{2-}$离子填充的氧化物,并以类似的方式通过氧化物进行运输。

Zhu等(2016)的工作对DO在腐蚀中的作用提供了一些见解。他们将P92抛光至1mm,暴露在600℃、25MPa、2mg/kg的同位素标记的氧($^{18}O_2$纯度大于95%)但未标记的水($H_2^{16}O$)中100h,(二次离子质谱法SIMS)深度剖面监测氧化物中$^{18}O$和$^{16}O$的含量。在这个温度下,与水平衡的$O_2$的浓度是3mg/kg。不考虑任何表面效应,水的浓度比氧的浓度大约高6个数量级。$^{16}O/^{18}O$的比例约为500,在整个氧化物中相对稳定,在金属-氧化物界面附近和过渡区的比例略有增加。作者认为,氧的主要作用是增加氧在氧化物中的电势,增加阳离子的扩散。Jacob等(2002)使用类似的同位素标记系统研究了水蒸气对粉末冶金Cr(PM)在800℃和1000℃时的氧化作用。样品暴露在Ar和20% $O_2$的混合物中,水含量既不是很低($P_{H_2O}<13Pa$)也不是很高($P_{H_2O}<1kPa$)。在“湿”气流中,氧的2/3来源于水,虽然它只占总气流的1/20(以分压为基础)。他们发现,无论是Cr还是氧,氧化物的缺陷结构都没有发生变化。他们注意到,SIMS的数据显示,氧化物中含有少量的$OH^-$,但未被定量。

由于水的辐照分解会导致高浓度的氧化剂,然而很少有关于辐照过的SCW腐蚀的研究。Karasawa等(2004)报道了几种Fe基合金(304、316和310 SS)、Ni基合金(600合金、625合金、X-750和哈氏合金C276)、Ti和Ti基合金(含21.2%(质量分数)V、3.7%(质量分数)Al和0.08%(质量分数)Fe)。研究了3种水化学条件:2000μg/L $O_2$、50μg/L $H_2$、脱气。

他们发现,结果取决于合金的成分。辐照改变了拉曼光谱测定的表面氧化物。对于不锈钢,他们发现质量增量可以用$WG=(kT)^{0.25}$来描述,对于Ni和Ti

基合金他们发现可以用 $WG=(kT)^{0.5}$ 来描述数据。他们的结论是,OH·自由基或 $H_2O_2$ 氧化了表面,对于 316 SS 的表面,这促进了镍铁氧体的形成。图 5.25 表明,在辐照条件下,DO 浓度为 400μg/kg 时 316 SS 表面所形成薄膜与未辐照的表面膜相比,Cr 被耗尽。Bakai 等(2015)研究发现,在辐照后的 SCW 中,由于水的辐照分解,Cr 的释放增加(图 4.23)。

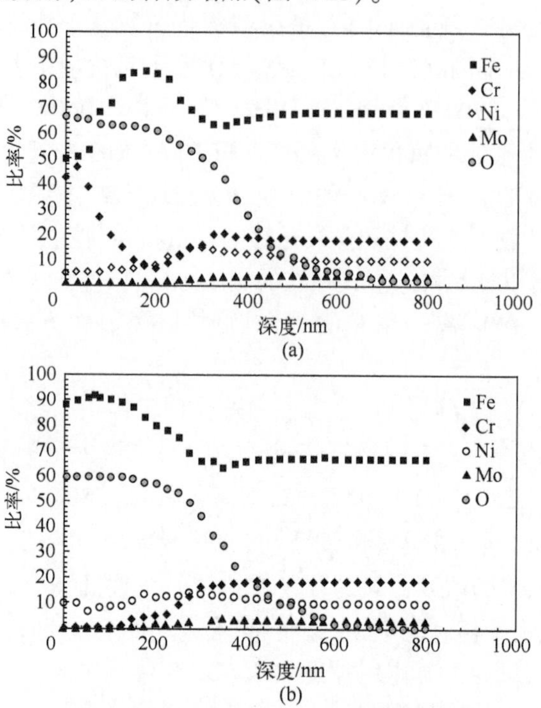


图 5.25　辐照对 316 SS 在超临界水中腐蚀的影响
(a) 无辐照;(b) 有辐照($3\times10^4$Gy/h)。
(引自文献 Karasawa, H., Fuse, M., Kiuchi, K., Katsumura, Y., 2004. Radiolysis of supercritical water. In:5th Int. Workshop on LWR Coolant Water Radiolysis and Electrochemistry, 2004 October 15, San Francisco, USA)

### 5.3.3.3　超临界水压力/密度

大多数支持 SCWR 开发的腐蚀测试是在恒压下进行的,特别是在区域 1 和区域 2("蒸发器"区域),如 SCW 密度随着温度的变化而变化。虽然试验很简单,而且直接适用于 SCWR(它将在基本恒定的压力下工作),但是这种类型的测试使数据解释变得复杂,因为 SCW 的特性在临界点附近发生了显著的变化(第 4 章)。很少有系统地研究腐蚀速率对压力或密度的依赖关系,一方面是因为试验的困难,另一方面是因为 SCWR 在基本恒定的压力下工作。Montgomery 和

Karlsson(1995)、Wright 和 Dooley(2010)的数据表明,压力对氧化层厚度或质量的影响很小。压力效应具有重要的实际意义,因为在较低压力下进行测试比在25MPa 压力下进行试验上更简单,以及这种测试在高温($T$>500℃)下易于观察所筛选材料的优点。在这种温度下,氧化物的溶解度较低,析出度也较低,在较高的水密度下会发生析出,但对氧化物层厚度影响不大。由于在核蒸汽过热开发项目中获得了包括反应堆内测试的大量 SHS 腐蚀数据,忽略压力依赖性将使这些数据对 SCWR 发展更有意义。

Macdonald(2004)提出了在反应发生时,压力对反应速率的 3 种可能的影响:①对活化能的影响;②对反应物体积浓度的影响;③对酸、碱解离的影响。如果反应速率是由被动式薄膜控制的,则需要考虑其他因素。当金属在氧化物中的扩散占优势时,Wright 和 Dooley(2010)提出抛物线速率常数与 $P^x$($\lg K_p = x \cdot \lg P_T$)成比例,对于磁铁矿,$x$=0.11~0.15。Holcomb(2014)提出增加压力会提高形成保护氧化铬膜所需的临界 Cr 浓度。

根据第 4 章的讨论和本章已经提供的数据,将感兴趣的区域划分为①近临界区和②$T \gg T_c$。在接近临界的区域,在恒定的温度下改变压强可以使密度和介电常数等性质的值从“类液体”变为“类气体”。在较高的温度下,改变压强对其他性质的影响很小。因此,这两个区域的腐蚀对压力的依赖关系应该是不同的。

1) 近临界区域

Watanabe 等(2001)研究了 5 种合金在 400℃、含 0.01mol/kg $H_2SO_4$ 和 0.025mol/kg DO 的 SCW 中腐蚀对压力的依赖性(图 5.26)。一般来说,当密度在 150~600kg/m$^3$之间增加时,腐蚀产物(100h 后)随之增加。这一区域压力的影响与根据水的性质随密度的变化所做的预测是一致的。在 390℃恒温和变密度条件下,对 304 SS(Guzonas,2013)和 625 合金(Li 等,2009)的腐蚀进行了密度依赖性测试。随着 SCW 密度的增加,质量增量(金属损失和氧化物形成的量)减少(图 5.27),但是质量损失(腐蚀)几乎保持不变(一种腐蚀指标),这反映了随着 SCW 密度的增加,氧化膜溶解度的增加(Guzonas,2013)。如第 4 章所讨论的,用 $\lg S \approx \lg \rho$ 给出了溶解度对密度的合理近似(图 4.13),而对温度的依赖性相对较弱。Guzonas(2013)假设外层氧化层的质量(厚度)与 $S$ 成反比,导出了质量变化的对数与 SCW 密度之间的简单关系,即

$$\lg(\mathrm{WC}) = -k \cdot \lg\rho + C \tag{5.5}$$

式中:$k$ 和 $C$ 为常数。

结果表明,该表达式与去除垢后的质量损失数据吻合较好。

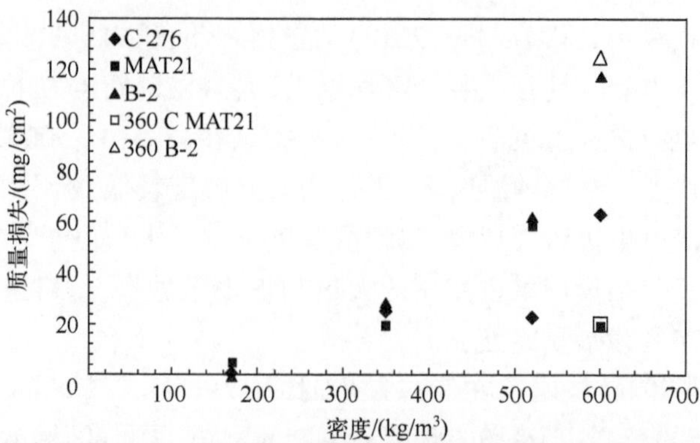


图 5.26　5 种合金在 400℃、含 0.01mol/kg $H_2SO_4$ 和 0.025mol/kg DO 的 SCW 中的腐蚀对压力的依赖性

(引自文献 Watanabe, Y., Kobayashi, T., Adschiri, T., 2001. Significance of Water Density on Corrosion Behavior of Alloys in Supercritical Water. NACE Corrosion 2001, Paper 01369)

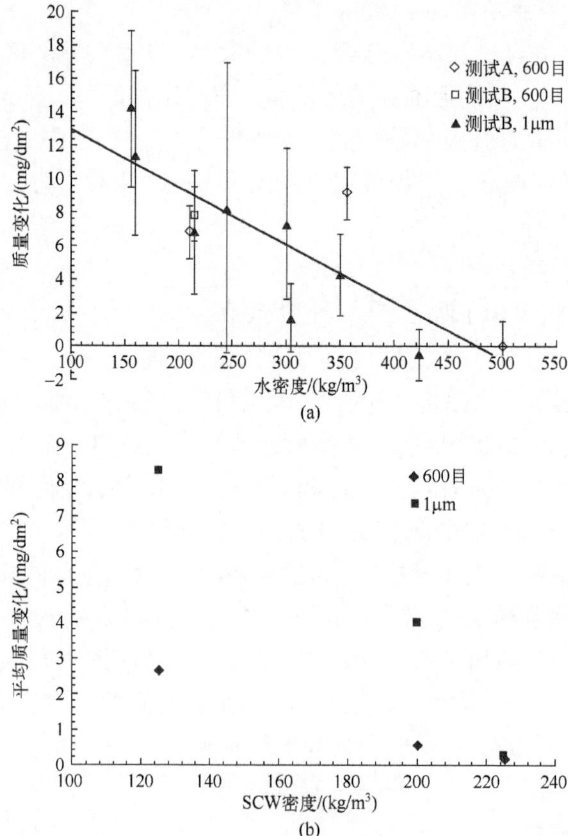


图 5.27　304 不锈钢(上)和 625 合金(下)在超临界水(SCW)下的平均质量变化随 SCW 密度的变化

(引自文献 Guzonas 等(2013)(图(a)), Li 等(2009)(图(b))。图(a)中的垂直误差条表示权重变化平均值的标准差)

2) 高温区域

Cowen 等(1966)报道了 4 种合金(Nimonic PE 16、20/25/Nb、FV 555 和 800 合金)暴露在 700℃,压力分别为 6.9MPa、24.1MPa 和 34.5MPa 的 SCW 中的质量增量。从图 5.28 中可以看出,随着压力的增大,800 合金质量增加,不同压力下的增量差异随时间缓慢增大。尽管在长时间暴露的情况下,PE 16 和 20/25/Nb 在 $P>P_c$ 处的质量增量比 $P<P_c$ 处要高,大约与 800 合金的相对含量相同,但 3 种合金的测试结果没有系统性的差异。在所有情况下,压力影响都很小,3500h 后 800 合金在 34.5MPa 时的质量增量仅为 6.9MPa 的 1.75 倍左右。图 5.29 所示为 3 种合金(20/25/Nb、600 合金和 800 合金)暴露于含少量 DO 和 $H_2$ 的 SCW 中 2000h,在温度分别为 600℃、650℃ 和 700℃ 时,24.1MPa 下的质量增量与 6.9MPa 下的质量增量之比。大多数样品被磨成 180 目,一些样品被冷轧。600 合金受压力的影响最大,质量增量比例随温度的升高而增大;700℃时,24.1MPa 下的质量增量是 6.9MPa 下的 9.3 倍。800 合金和 20/25/Nb 的质量增量比例随温度 $T$ 的增加仅略有增加。

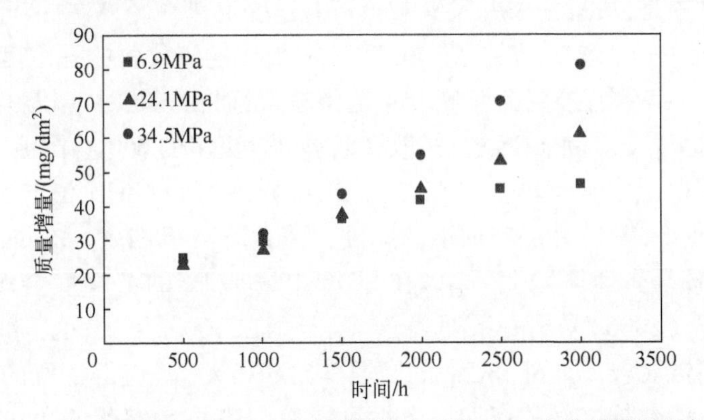


图 5.28 合金 800 在 700℃时 3 种不同压力下的 SCW 中的质量增量与时间的关系

(引自文献 Cowen, H. C., Longton, P. B., Hand, K., 1966. Corrosion of Stainless Steels and Nickel-Base Alloys in Supercritical Steam. U. K. Atomic Energy Authority TRG Memorandum 3399(C))

Bischoff(2011)和 Bischoff 等(2012)报道了 T91、NF616、HCM12A、9Cr ODS 和 14Cr ODS 在 500℃和 600℃的 SCW 中以及 500℃的蒸汽中的腐蚀结果。在 SCW 试验中采用了循环回路,并在循环回路中加入了 10mg/kg 和 25mg/kg 的 DO。蒸汽腐蚀试验是在压力为 10.8MPa 的静态脱气高压釜中进行的。由于 Fe 的扩散发生了变化,导致 10.8MPa 下的腐蚀少于 25MPa,在较低的压力下,质量增量和氧化层厚度降低到 1/2 至 2/3,这可能是由于样品表面的水密度不同造

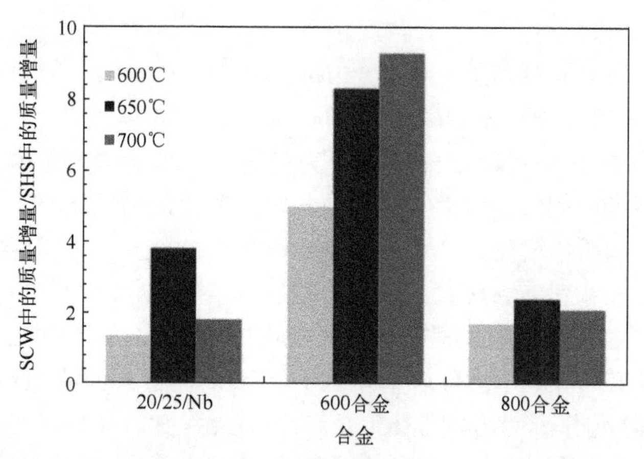


图 5.29　700℃时 3 种合金在 24.1MPa 时的质量增量与在 6.9MPa 时的质量增量之比
(引自文献 Cowen, H. C., Longton, P. B., Hand, K., 1966. Corrosion of Stainless Steels and Nickel-Base Alloys in Supercritical Steam. U. K. Atomic Energy Authority TRG Memorandum 3399(C))

成的。不幸的是,质量损失并没有被报道。速度定律在 SCW 中接近抛物线,而在 SHS 中近乎是 3 次方的关系。在 SCW 中形成的氧化物的孔隙度较大,尤其是内层。在 SCW 中,外层孔隙度分布范围较广,而在 SHS 中,孔隙度分布在内层与外层的交界面。在 SHS 中,扩散层似乎包含更多的氧化物沉淀。SHS 样品在内部和扩散层中含有较多的 $Cr_2O_3$。SCW 样品的 Fe/Cr 比值与 SHS 样品无显著差异,但 SCW 的外层与内层的厚度比值始终高于 SHS。Bischoff 指出,如果 Fe 的迁移是受速率限制,那么在 SCW 中增加 Fe 的溶解度会增加腐蚀速率。Jeong 等(2005)采用一系列 Zr 合金,分别在 500℃下的 SHS 和 SCW 进行了类似的腐蚀比较。此外,SCW 的腐蚀率比 SHS 略高,值得注意的是,在两种压力下获得的数据之间的密切一致性可以使 SHS 成为替代 SCW 进行材料筛选的替代物。



Yi 等(2001)在使用脱气水的静态高压釜中研究了不同压力对 4 种钢(HCM12A、NF616、HCM2S 和 STBA24)腐蚀的影响。这些样品片是用 1500 目的碳化硅磨成的。他们的结论是,抛物线速率常数在 600℃时随着压力的增加降低 1/2,但在 700℃时几乎没有影响。图 5.30 所示为 Yi 等(2001)与 Bischoff 等(2011)和 Bischoff 等(2012)对 NF616 进行研究的数据对比;虽然 Bischoff 等工作中的速率定律更接近于 3 次,但它被认为是比较有用的。Bischoff 的数据高于 Yi 等(2001)在 500℃的数据集所预期的 $K_p$ 值,这是因为 Bischoff 对样品表面进行了抛光处理,这表明表面抛光对 $K_p$ 的影响比压力更大。

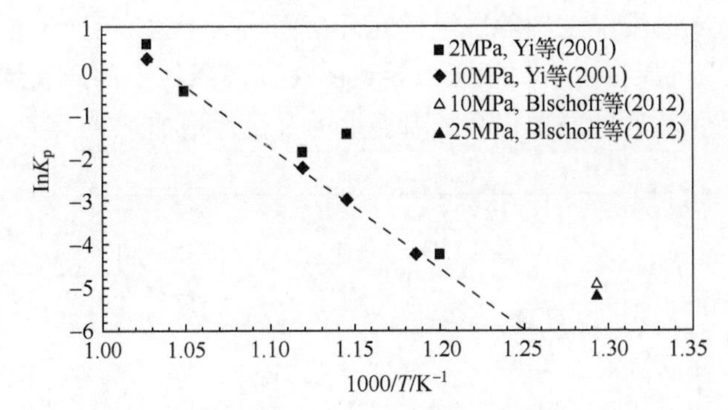


图 5.30　描述了 NF616 在超临界水中的腐蚀过程中，抛物线速率常数 $K_p$ 的对数随温度变化的关系

（引自 Yi 等（2001）和 Bischoff 等（2012）的研究。在 Yi 等（2001）的研究中，样品为采用 1500 目的 SiC 研磨制造的，并将其暴露在除氧水中。Bischoff 等（2012）将样品抛光至 0.05mm 使标记物能够发生沉积。实线为用线性最小二乘法拟合的 Yi 等（2001）在 10MPa 条件下的数据）

Gómez-Briceno 等（2013）研究了压力对 316L SS、600 合金和 625 合金氧化的影响。他们将这 3 种合金放置在温度分别为 400℃ 和 500℃、DO＝8mg/kg 的 SCW 中 780h。试验样品表面已磨光至 600 目。在温度为 400℃ 时，30MPa 下的 316L SS 的质量增量在 200h 后基本不变，而在 25MPa 下的质量增量随时间增加。在 500℃ 时，30MPa 时的质量增量约为 25MPa 时的 50%，在 25MPa 时形成的氧化物（富铬内氧化物、富铁外层及下扩散带）与其他作者在类似试验条件下对该材料的报告一致，在 30MPa 时形成的氧化层更薄，扩散区更厚。对于合金 600，压力在两种温度下的不确定度都很小。

Holcomb 等（2016）检验了 18-8 奥氏体不锈钢和 Ni 基合金在氧化过程中的抛物线速率常数与压力的相关性的试验数据。在 700℃ 条件下，压力对于 Ni 基合金氧化过程的影响能够被 $\lg K_p \propto P_T$ 更好地描述，而不是基于缺陷传输模型预测的 $\lg K_p \propto \lg P_T$。对于细粒度的奥氏体钢，$\lg K_p \propto \lg P_T$ 与数据符合得更好，而对于粗粒度的奥氏体钢，抛物线速率常数和压力之间的关系还没有被发现。

Choudhry 等（2016）报告了 800H 合金在 25MPa 时金属的释放浓度随温度（500℃、550℃、650℃、700℃ 和 750℃）变化的数据。Cr 和 Ni 的稳态浓度在 500～550℃ 之间升高，然后随着温度的进一步升高而降低。根据文献测量，随着温度的升高，腐蚀速率增加，Cr 和 Ni 的减少是由于氧化物溶解度的变化。他们发现 lg（金属浓度）和 lg（SCW 密度）之间有很好的相关性（图 5.31），这表明之前所观察到的浓度下降大部分可以归因于 SCW 密度的变化，尽管这种影响的幅度很小。这一结论与图 4.10 所示的计算溶解度一致。图 4.10 表明，在 500℃ 以

上，NiO 和 $Cr_2O_3$的溶解度均随着温度的升高而增大。这表明，SCW 密度变化的影响虽小，但仍可能在由于实际压力导致的在高温条件下的质量增加现象中发挥作用。预计除垢后的减重仍会随着温度的升高而增加。

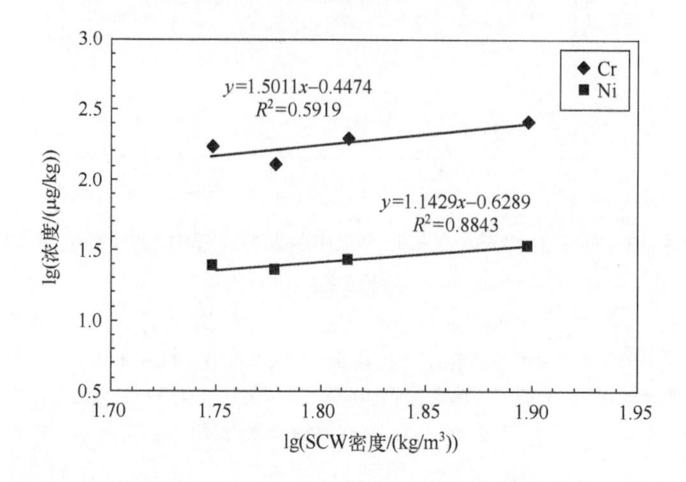


图 5.31 Ni 和 Cr 从 800H 合金中析出的浓度与超临界水(SCW)密度的对数的函数(Choudhry 等,2016)



虽然数据显示，在 $T \gg T_c$处，腐蚀速率没有表现出很强的密度(压力)依赖关系，但由此形成的氧化层的形貌确实如此(如 Bischoff,(2011)之前讨论过的工作)。Sanchez(2014)、Li 等(2015)和 Huang 等(2015b)报道了 4 种合金(304 SS、310 SS、A-286 和合金 625)在 625℃、8MPa、29MPa(静态高压釜)和 0.1MPa(管式炉流动蒸汽)条件下，在 SCW 中暴露 1000h 的研究结果。将样本研磨至 600 目。3 种 Fe 基合金的增重(氧化层厚度)在密度为 20kg/m³(8MPa)时达到峰值，而 625 合金的质量增量略有下降(图 5.32)。Sanchez(2014)报道了 625 合金和 310 SS在 8MPa 和 29MPa 条件下经过 3000h 后的质量增量差异不大，而 A-286 的质量增量较大。625 合金在任何密度下的质量增量几乎没有差异，而 304 SS 和A-286 在0.1MPa 时的质量增量。是在 8MPa 和 29MPa 时的 1/100 至 1/10。且发现 304 合金在 8MPa 条件下经过 1000h 后被氧化物完全覆盖，但在 29MPa 下仅被部分覆盖。在 0.1MPa 条件下暴露的样品仅被一层薄薄的氧化物部分覆盖。310 SS 形成了一层富 Cr 的具有尖晶石结构的氧化物层，且底层再结晶的 γ 层会在 Cr 中被耗尽(图 5.33 和表 5.1)。在不同压力下，尖晶石结构氧化物的晶格参数基本相同；内部氧化物的组成大致相似(表 5.1)，而外部氧化物的组成则与压力(密度)有关。

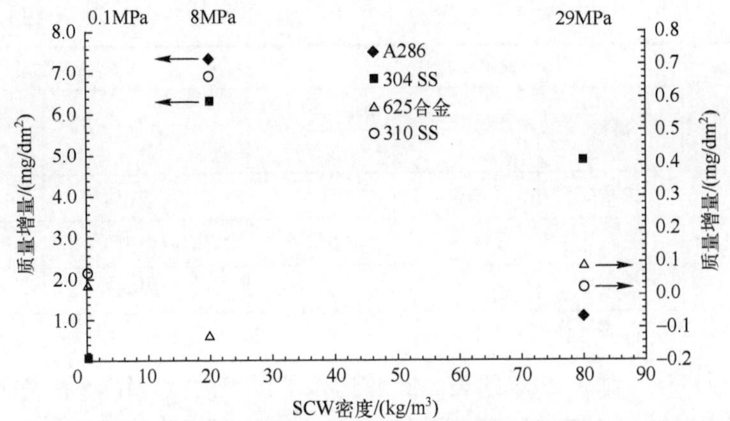


图 5.32　4 种合金在 625℃下暴露 3000h 后质量增量与密度(压力)的关系(注意 625 合金和 310 SS 的点在 $80kg/m^3$的密度下重叠)

(引自文献 Li, W., Woo, O. T., Guzonas, D., Li, J., Huang, X., Sanchez, R., Bibby, C. D., 2015. Effect of pressure on the corrosion of materials in high temperature water. In: Characterization of Minerals, Metals, and Materials. John Wiley & Sons Inc., Hoboken, NJ (Chapter 12))

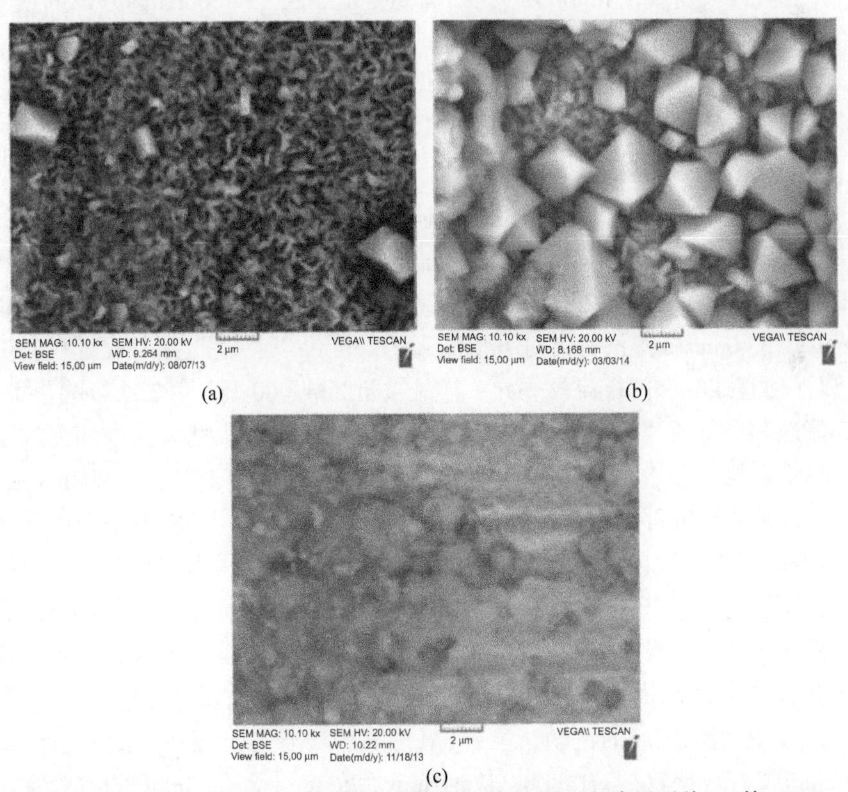


图 5.33　625℃时 310 SS 在不同条件下经过 1000h 辐照后的表层形貌(Li 等,2015)
(a)29h;(b)8h;(c)0.1MPa。

表 5.1 310 SS 在不同压力下的表面组成和结构(Li 等,2015)

| 压力/MPa | 表面氧化层 | | | 再结晶晶粒层 | |
|---|---|---|---|---|---|
| | 结构 | 厚度/μm | | 结构 | 厚度/μm |
| | | 外部 | 内部 | | |
| 29 | 尖晶石 | 0.5 | 0.4 | $\gamma w/tM_{23}C_6$ | 2 |
| 8 | 尖晶石 | 4.2 | 3.3 | $\gamma w/tM_{23}C_6$ | 8 |
| 0.1 | 尖晶石 | 0.2① | | $\gamma w/tM_{23}C_6$ | 1 |

① 不包括连续的二氧化硅薄层。

在本研究中,对 A-286 的表面区域进行了详细的微观结构的表征(Huang 等,2015b)。在 0.1MPa 时,形成了连续的 $Cr_2O_3$层,该层由接近连续的 $SiO_2$层和富 Ti 层构成。在 29MPa 时,形成了一个几乎连续的富 Cr 层,被一层 $Fe_2O_3$覆盖;在这些区域,氧化物较薄,且在氧化物下方有 1~2mm 深的再结晶层。在有较厚的氧化物的区域,可以发现氧化已经深入到衬底,且在内部氧化区域发现了$(Cr、Mo、Fe)_2O_3$相。在 8MPa 时,表面完全被一层 15~20mm 的外部的铁氧化物和一层 15~20mm 的部分氧化的内层所覆盖。

### 5.3.4 流量



Ruther 和 Greenberg(1964)报道了,在 650℃、42kg/cm、DO = 0.05mg/kg 的条件下,将蒸汽速度从 30m/s 变化为 91m/s,对 304 SS 和 406 SS 在受 5 周的辐照内产生的金属损失影响不大。Lukaszewicz 等(2013)研究了 T23、T92、TP347HFG 和 Super 304 在 650℃ 和 700℃ 条件下的脱氧去离子水中 4mm/s、16mm/s 和 40mm/s 蒸汽流量对氧化的影响。样品以特定的方向放置在测试炉中,以评估直接流和间接流的差异。样品被研磨成 600 目。T23 的影响很大,在最高流速和最低流速之间损伤显著增加。对于 T92,蒸汽流量的增加促进了最外层赤铁矿层的形成,形成的氧化物有更多的间隙或空隙。对 T347HFG 而言,表面上直接流动的部分形成了最厚、黏附性最强的氧化物。随着暴露时间的增加,保护氧化物被破坏并形成结节。Super 304 的变化是类似的,但其氧化物较薄,结节较小。图 5.34 所示为所研究的不同流量下 Super 304 的质量变化关系。作者将观察到的流量效应归因于表面较高的 $p(O_2)$,这是由于表面水热分解产生的 $H_2$被去除的结果。奥氏体合金上氧化物的分解是由于合金中 Cr 的耗竭引起的,这是由于在较高的流动速度下,氧化物的增长速度较快造成的。Knodler 和 Straub(2014)比较了 P92、FB2、Waspalloy 和 625 合金在 5000h 的实验室腐蚀试验中暴露于 SHS 的抛物线速率常数(F/M 钢为 650℃,Ni 基合金为 750℃),与在630℃时(F/M钢)和725℃时(Ni基合金)将样本置于运行中的发电厂的旁路线路中得到的数值(19MPa)。在实验室和工厂中暴露的F/M钢的速率常数差别不大,但在工厂中暴露的Ni基合金的速率常数比在实验室中测量的要高得多。

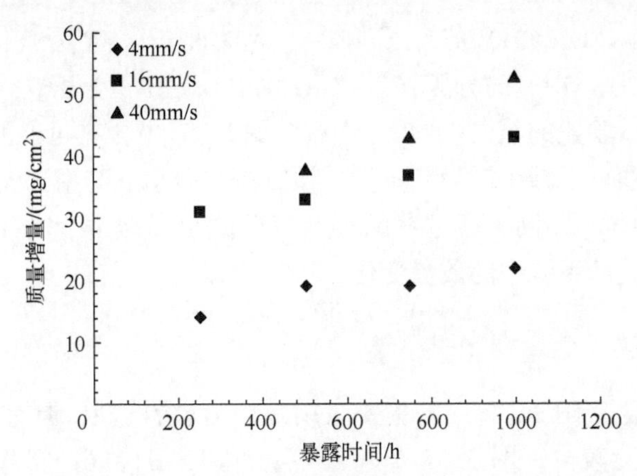


图5.34 在700℃过热蒸汽中Super 304的质量增量与在3种流量下暴露时间的关系(图中给出的值)

(引自文献 Lujaszewicz,M.,Simms,N. J.,Dudziak,T.,Nicholls,J. R.,2013. Effect of steam flow rate and sample orientation on steam oxidation of ferritic and austenitic steels at 650 and 700℃. Oxid. Met. 79,473-483)

Zhang等(2016)分别在流动和静态条件下研究了SCW中T22和P92的氧化特性,其中温度分别为550℃和600℃、压力为25MPa、DO浓度为2mg/kg、进口电导率为0.1μS/cm。结果表明,流动系统的速率常数接近抛物线,而静态系统的速率常数更接近立方。并在查阅其他文献资料的基础上,还提出了流动会促进这些合金上赤铁矿的形成。作者将这种行为上的差异归因于流动冷却剂造成的腐蚀,进而产生对$H_2$的去除,其改变了表面的氧势。Guzonas等(2016b)在一项国际实验室中的比较试验中,考察了质量增量与高压釜刷新时间的关系(图2.5)。参与试验的几个测试循环在测试前后监测了DO浓度;一个实验室报告说,在进口和出口之间系统损失约为600μg/kg,并假设是由于腐蚀片和包括高压釜在内的回路高温部分的腐蚀。通过对静态高压釜和流动回路中暴露的剥落腐蚀样本数据的比较,表明了这一点:腐蚀产物在静态高压釜中的迁移并不是一个限制因素,因为在静态高压釜中测试的每种合金的质量损失都比在流动回路中测试的合金质量损失大得多。假设在回路内较高的流速下,氧化层剥落可能较大,从而导致较低的增重。然而,尽管对暴露于气流中的样本进行的检查显示有氧化层剥落的证据,但从除垢后的重量损失中计算出的质量增量低于测量的质量增量,这与氧化层剥落造成的高氧化层损失不一致。

Holcomb(2009)研究了高温蒸汽中的色度蒸发,提出了以下方程:

$$k_e = \mathrm{Sh}_{\mathrm{ave}} \frac{D_{\mathrm{AB}} M_0}{LRT}(P_0 - P'_0) \tag{5.6}$$

式中:$k_e$为蒸发速率;$\mathrm{Sh}_{\mathrm{ave}}$为平均舍伍德数;$D_{\mathrm{AB}}$为 $CrO_2(OH)_2$与溶剂气体的气体扩散系数;$M_0$为 $CrO_2(OH)_2$的分子质量;$L$ 为曲面长度;$R$ 为气体常数;$T$ 为温度(K);$P_0$为 $CrO_2(OH)_2$的分压;$P'_0$为 $CrO_2(OH)_2$的体积分压。

如果进入试验段的蒸汽(或工厂中有关的管道)中含有足够的 $CrO_2(OH)_2$,则包含分压差项可以减少或抑制蒸发。该方法在概念上与用于模拟腐蚀产物沉积的方法(第 4 章)相同。通过考虑 Cr 在表面的固态扩散速率和蒸发速率,可以确定合金在流动系统中的质量变化。

## 5.3.5 传热

尽管堆芯合金在 SCER 中的主要应用是作为燃料包层,但有关传热对腐蚀速率影响的研究较少。图 5.35 是来自 Gaul 和 Pearl(1961)的数据。在 593~704℃之间进行了传热试验,并在 621℃进行了等温试验。试验分别在 6.9MPa、20mg/kg DO 和 2.5mg/L $H_2$的条件下进行;测试样品在暴露前进行了制备。在 4000h 的暴露时间内,除垢减重几乎没有差别,但在传热试验(质量损失份额为 10%~50%)中,与等温试验(非常小的质量增量)相比,系统损失了更多的氧化物(通过比较质量增量和除垢减重)。Pearl 等(1967)报道了合金 625 的类似结



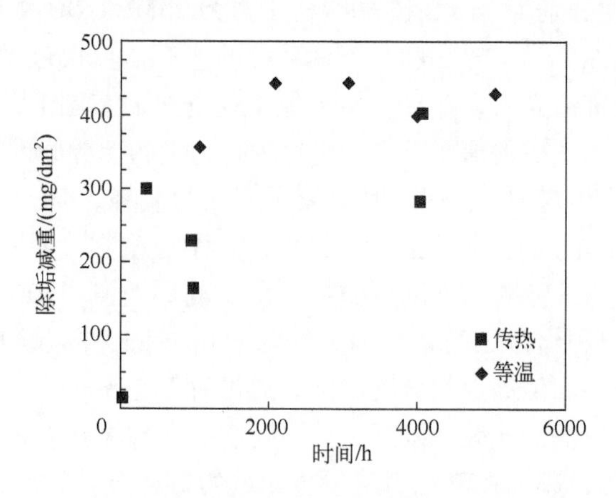


图 5.35 温度为 593~704℃(传热)和 621℃(等温)时,在 6.9MPa、20mg/kg DO 和 2.5mg/L $H_2$的超临界水中,304 SS 的除垢减重与暴露时间的关系

(引自文献 Gaul,G. G. ,Pearl,W. L. ,1961. Corrosion of Type 304 Stainless Steel in Simulated Superheat Reactor Environments. General Electric Company Report GEAP-3779)

果(除垢减重差异不大,但传热表面系统的氧化损失更大)。两种合金的冷却剂中较高的氧化物损失对活动性运输有重要意义。高温下800合金在738℃下的辐照试验(Comprelli 等,1969)表明,800合金的腐蚀渗透与反应堆外的数据基本一致,但比预测的氧化损失要高。

### 5.3.6 老化

SCWR 设计的一个基本挑战是,正如第3章所讨论的,燃料包层温度峰值(加拿大 SCWR 的设计高达850℃)导致的热老化(扩散和二次相沉淀,特别是晶界)使微观结构演化缓慢。到目前为止,许多测试都是在相对较短的持续时间内使用预制材料进行的,而金属间相(如χ、η和σ相)的形成需要较长时效时间的影响却很少受到关注。在 SCWO 发展的早期研究中,Daigo 等(2006)研究了在400℃、30MPa、含0.01mol/kg $H_2SO_4$的 SCW 中,热老化对 Ni 基合金的影响。夏比冲击韧性和耐腐蚀性降低,高 Cr 合金的降解程度显著。Jiao 等(2014)研究了热老化对在 SCW(550℃、25MPa、DO=8mg/kg)中轧制退火(MA)和热处理(TT)316L 和 310S 的耐蚀性的影响。经过这些处理产生了预期的金属间相($M_{23}C_6$、χ、η和σ相)。结果表明,热老化对材料的耐蚀性影响不大,这可能是由于在晶界和晶内形成的析出物具有不连续的性质造成的。

### 5.3.7 辐照



由于腐蚀是一种界面现象,很难将水的辐照分解效应与金属表面和氧化层的辐照效应区分开来。在反应堆内辐照(不在 SCW 中)后,合金暴露于700℃以下的 SCW 中(Higuchi 等,2007),总体腐蚀没有增加;然而,在这项工作中,冷却剂辐照(水辐照分解)的效果还无法被评估。Zr-1%Nb 和 Ni-Cr 基合金在9.76MeV/6.23kW 电子束照射下,在伪临界点附近的水中放置497h,在辐照下的氧化速率增强(Bakai 等,2015)。正如5.3.5节所指出的,800合金在高达738℃的高温下的辐照试验(Comprelli 等,1969)表明,腐蚀渗透总体上与在反应堆外的数据一致,但高于预测的氧化损失。如第4章所述,Ishigure 等(1980)报道称,γ辐照增加了304 SS 在亚临界水中溶解铁的释放。

## 5.4 氧化物形态

在 SCW 中形成的表面氧化物的形态和组成可能是复杂的,这些特征受到5.3节中讨论的所有关键变量的影响。一般来说,在水中,氧化物的结构取决于合金金属成分的氧化物(或氢氧化物)的热力学稳定性(如 Pourbaix 或 Ellingham

图所描述的那样)以及金属离子和氧化剂在形成的氧化物层中的固相扩散速率。Kofstad(1988)提供了对在高温气体中的氧化物生长的热力学和动力学方面的一个很好的概述。在预期 SCWR 堆芯的条件下,所形成的氧化物的预期热力学稳定性被预测为在低 DO 浓度下的 $Al_2O_3>TiO_2>SiO_2>MnO>Cr_2O_3>Fe_3O_4>NiO$。在较高的 DO 浓度下(根据 5.3 节提供的试验数据,DO 浓度大于 2mg/kg),似乎形成了 Cr(Ⅵ)氢氧化物,并且在 SCW 中相对溶解,赤铁矿的形成比磁铁矿更有利。金属扩散速率为 Fe>Ni>Cr。

### 5.4.1 铁素体-马氏体钢

对于在暴露于 SCW 中的 F/M 钢上形成的氧化物,目前存在大量非常详细的研究(Jang 等,2005b;Ren 等,2006;Ampornrat 和 Was,2007;Hwang 等,2008;Ampornrat,2011;Tan 等,2010;Bischoff,2011;Liu,2013;Li,2012),且不同合金的试验结果基本一致。所述氧化物具有层状结构,其外为 $Fe_3O_4$层,内为具有尖晶石结构的$(FeCr)_3O_4$层,其最内为扩散层。有时可在内层观察到 $Cr_2O_3$,这取决于合金的温度和 Cr 含量,并可能形成或不形成连续层。外层主要由粗柱状晶粒组成,而内层由极细的等轴晶组成。扩散层的结构和厚度随温度的变化而变化。Bischoff(2011)指出,NF616 中的氧化物沉淀物倾向于沿板条边界形成。他指出,在 500℃时,氧化层厚度随着暴露时间的增加而增加,但氧化层形貌保持相对不变。



在 500℃时,内部的氧化物被观察到具有好的多孔性,并被分为靠近内氧化物/外氧化物界面的相对致密的一层,与在内层/扩散层界面附近的一层较为多孔的。在 500℃的蒸汽中,孔隙率很小,但双层形貌类似。HCM12A 在 600℃时,扩散层不像 NF616 那样均匀,这似乎是由微观结构决定的。外层的多孔性较强,孔隙率随时间增大,而内部氧化物孔隙率随时间减小。在 500℃时,氧化物更加均匀,在整个暴露时间内,其形态相对稳定。类似的内部氧化物分为多孔层和非多孔层。在 500℃下暴露的 14Cr ODS 样品中,扩散层较大(略大于内层),许多氧化物沉淀沿铁素体晶界生长方向形成长枝晶,且内层和外层都是多孔的。

标记试验表明,内部氧化物/外部氧化物间相对应于原始合金表面,即氧化物是同时从内氧化金属界面和外氧化金属界面生长的。Bischoff(2011)和 Bischoff 等(2012)描述了将 Pd 标记置于腐蚀片上,利用光刻法研究原始水界面位置的试验结果。由于在光刻曝光过程中,表面粗糙度会影响焦点和照度,所以所有的样品,包括没有标记的样品,都被抛光到 0.05mm 的粗糙度。将标记好的和未标记的样品分别置于 SCW(500℃、25MPa、约 1000h)和 SHS(500℃、

10.8MPa、约1350h)中,取出横切并利用SEM确定标记位置。在这两种压力下,在外-内层界面观察到标记物(图5.36),证实了SCW中F/M钢的氧化是通过Fe的向外迁移形成外层和氧的向内迁移形成内层同时发生的。

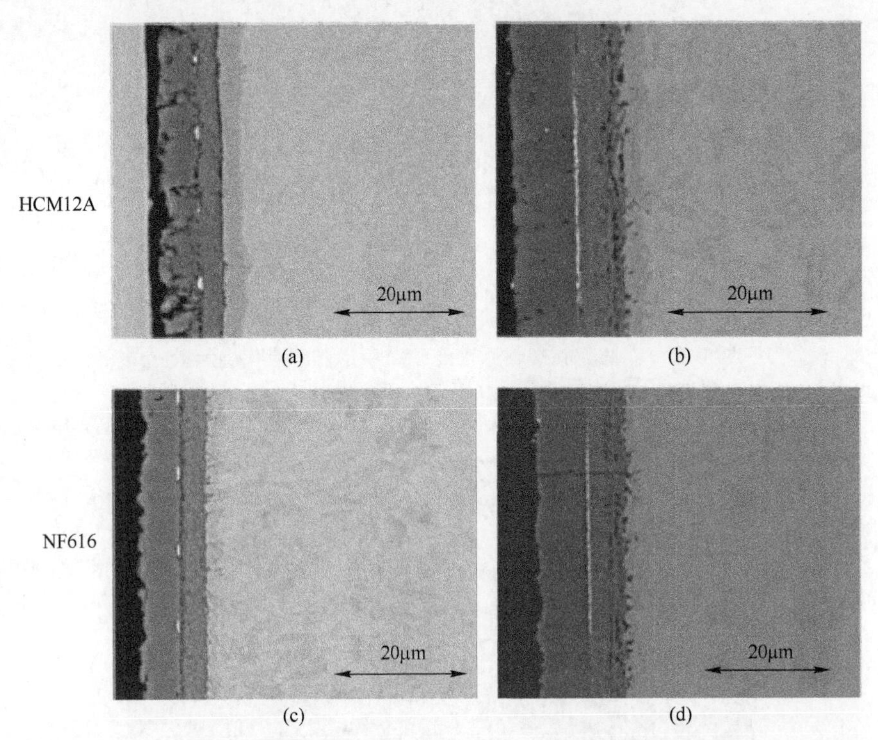


图5.36 在((a)和(c))蒸汽及((b)和(d))SCW中,((a)和(b))HCM12A及((c)和(d))NF616在500℃下连续6周形成的有标记的氧化层的SEM图像(Bischoff等,2012)(标记物特征可以被看作薄的、明亮的白色,大致位于氧化层的中间)

稳态氧化过程如图5.37所示,外部氧化物由含有孔隙的柱状颗粒组成,内部较细颗粒的氧化物含有孔隙和孔洞,并在扩散区的晶界处发生氧化。$O_2$和$H_2O$的向内扩散和部分Fe的向外扩散(作为离子或各种水解产物,见图4.9)是通过孔隙形成的,$O^{2-}$和$OH^-$的向内扩散和部分Fe(作为离子)的向外扩散是通过固态扩散形成的。这些不同工艺的相对重要性尚不清楚,可能会随着温度、Cr浓度、表面粗糙度、晶粒尺寸、SCW密度和微量合金元素浓度等因素而变化。

在SCW中,所有F/M钢的腐蚀研究的一个重要特征是孔隙率的存在(图5.38),孔隙的大小和密度随温度的变化而变化。Tomlinson和Cory(1989)注意到9Cr-1Mo钢在501℃(流量为50mL/min)的Ar和蒸汽混合物中暴露10~20h后形成的氧化物中存在孔隙。他们认为孔隙主要分布在晶界和晶界三相点上。

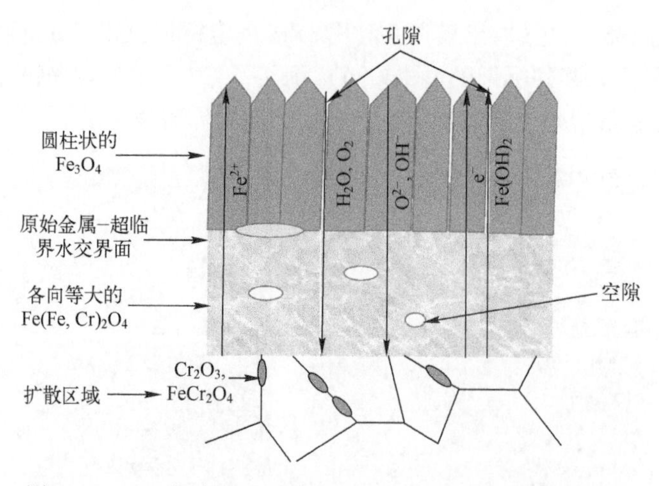


图 5.37　F/M 钢在超临界水(SCW)中形成的氧化层示意图。这些孔洞被证明是直的,尽管一个相互连接的网络更有可能。

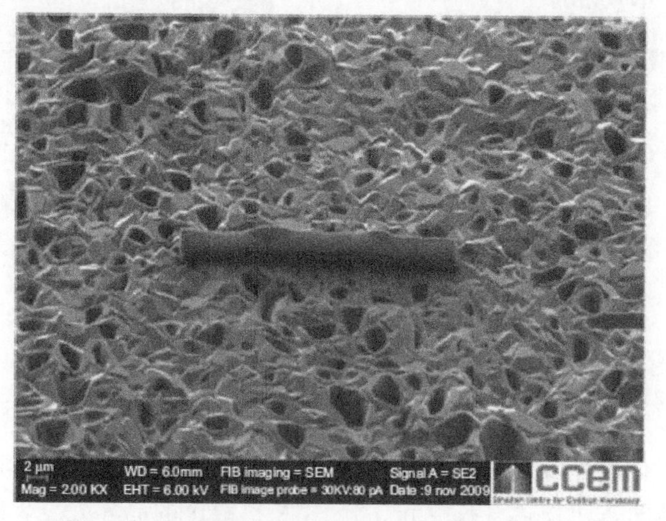


图 5.38　14Cr 合金表面氧化层暴露于超临界水中,表面孔隙率大(图像中心的细长条带是一层薄的保护涂层,用于保护 TEM 样品制备的最表层。标尺:2μm)(Artymowizc 等,2010)



Cook 和 Miles(2012)指出,将 DO 浓度从 5mg/kg 增加到 200mg/kg 会增加孔隙率。Zhang 等(2016)发现,在 550℃和 600℃条件下 SCW 暴露 1000h 后,孔隙度“愈合”(在扫描电镜的 SEM 图像中无法再观察到)。他们还观察到,在更高的温度下形成的气孔更少。Li(2012)指出,在模型 Fe-9Cr 和Fe-9Cr-1.5X(X=Si、Al、Mn、V 或 Ti)合金上形成的孔隙率表现出对表面粗糙度的依赖性,在金刚石刀片切割表面上的孔隙率最低。我们注意到,表面粗糙度小于 1mm 的氧化物具有很高的多孔性,这表明富 Cr 内层保护层的形成速度是很重要的。还发现,在 F/M 钢马氏体晶粒上形成的氧化物比在铁素体晶粒上形成的氧化物的多孔性更强,这可能是由于晶界缺陷含量的差异造成的。这些数据表明,孔隙度与外部氧化物的初始形成有关,当氧化物达到稳态结构时,孔隙度消失。

Cook 等(2010)认为孔隙是热液成因①。一种可能的水热机制是表面上高度缺陷的磁铁矿晶体快速初始形成,然后在水化的表层中溶解和再沉淀,导致了观察到的“愈合”。由于表面水的密度随着温度的升高而减小,在较高的温度下孔隙会减少。Zhu 等(2016)认为气孔的形成和愈合与晶界和散粒扩散速率的差异有关。Tan 和 Allen(2009)提出孔隙度是由于局部腐蚀造成的,局部腐蚀是由于吸附了从其他回路表面运输到表面的 $Cu_2S$ 颗粒造成的。循环水的化学分析发现了 6μg/kg Cu、36μg/kg Mn 和可测量浓度的 Ni 和 Mo。他们发现,将 HCM12A 打磨至 1mm 的粗糙度后,将其暴露在 SCW(500℃、25MPa、流量约 1m/s)下,约 1000 以下的孔隙深度约为 1.5mm,有的达到约 4mm(表面磁铁矿膜的 40%)。他们指出,如果这些孔隙相互连通,那么腐蚀速率将显著提高。

## 5.4.2 奥氏体钢

对暴露于 SCW 的奥氏体合金上形成的氧化物的研究显示了一种由外磁铁矿层或混合磁铁矿层组成的层状结构和具有脊状结构$(Fe、Cr、Ni、M)_3O_4$(M 为其他微量元素)或刚玉型氧化物 $M_2O_3$(M=Fe、Cr、Mn)的内层。Payet(2011)使用含同位素标记氧的水来研究氧化层的生长。将 316L SS 暴露于 SCW(600℃、25MPa)中 2 个周期,初始氧化期为 $H_2{}^{16}O$,第 2 个氧化期为 $H_2{}^{18}O$。在 SCW 暴露后,使用 SIMS 深度剖面对样品进行横断面分析,以确定 $^{16}O$ 和 $^{18}O$ 剖面。在金属/氧化物和氧化物/SCW 界面都观察到 $^{18}O$,表明其与 F/M 钢一样,氧化物层同时向外和向内生长,内部氧化物在金属/氧化物界面生长,外层氧化物在氧化物/SCW 界面生长(图 5.39)。

合金的微观组织对氧化物的形貌有很大的影响。结果发现,20Cre25Ni 合金 NF709 的近表面奥氏体晶粒易于发生晶内氧化,形成内部氧化垢,其中,氧化物晶粒表现出原奥氏体晶粒取向所施加的优先取向(Chen 等,2008)。值得注意的是,在 SCW 中,在 F/M 钢表面形成的氧化层中广泛观察到孔隙率(图 5.38),而在奥氏体合金中没有观察到。球墨组织的氧化物在接触 SCW 的奥氏体钢中普遍存在;Chen 等(2005、2007a)报道了在 D9 SS 上形成的氧化层在 500℃下的

① A. Anderson(圣弗朗西斯泽维尔大学)也对暴露于 SCW 表面上氧化物形态的热液性质发表了类似的评论。

SCW 暴露后呈现出表面不均匀的结节状结构。结节间距为 5~10mm，且与合金中奥氏体晶粒大小相当。在 SCW（625℃、25MPa、500h、静态高压釜）中，800HT 合金表面形成的氧化物被发现在氧化基层的顶部存在氧化物结节（Mahboubi, 2014；Mahboubi 等，2015）（图 5.40）。这些结节富含 Cr、Fe、Mn 和 O，而底层的氧化坑则富含 Cr、Mn 和 O。基体氧化层（130nm 厚）富含 Cr、Mn、O 和少量的 Fe。基体氧化层（633nm 厚）以下区域的 Cr 和 Mn 含量降低，Fe 和 Ni 含量增加。在氧化基层下面的薄层（62nm 厚）富含 Ti、Si 和 O。

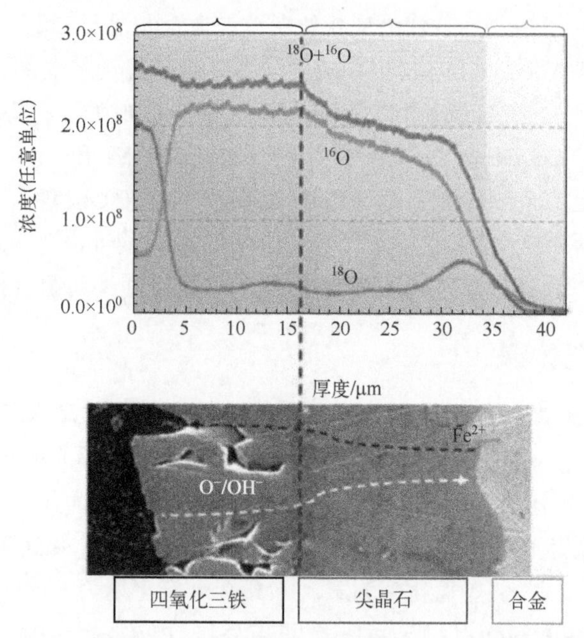


图 5.39　316L SS 氧化层中 $^{16}O$ 和 $^{18}O$ 剖面在 600℃、25MPa、$H_2{}^{16}O$ 条件下的第一氧化期（760h）的演变，以及在 600℃、25MPa、$H_2{}^{18}O$ 条件下的第二氧化期（305h）的演变

（引自文献 Sarrade, S., Féron, D., Rouillard, F., Perrina, S., Robin, R., Ruizc, J. C., Turc, H. -A., 2017. Overview on corrosion in supercritical fluids. J. Supercrit. Fluids 120, 335-344）

图 5.41 概述了 316L SS（16%（质量分数）Cr）、800HT 合金（20.6%（质量分数）Cr）和 33 合金（33.4%（质量分数）Cr）在 25MPa、SCW、550℃ 静高压下的氧化层结构。较高的 Cr 块含量促使形成了刚玉型 $M_2O_3$（M = Fe、Cr 和 Mn）的外部基础层。已经证明，$M_2O_3$ 层的组成取决于合金的 Cr 含量和温度：在 33 合金上形成的层的 Fe 含量明显少于在 800HT 合金上形成的层的 Fe 含量（Mahboubi, 2014）。

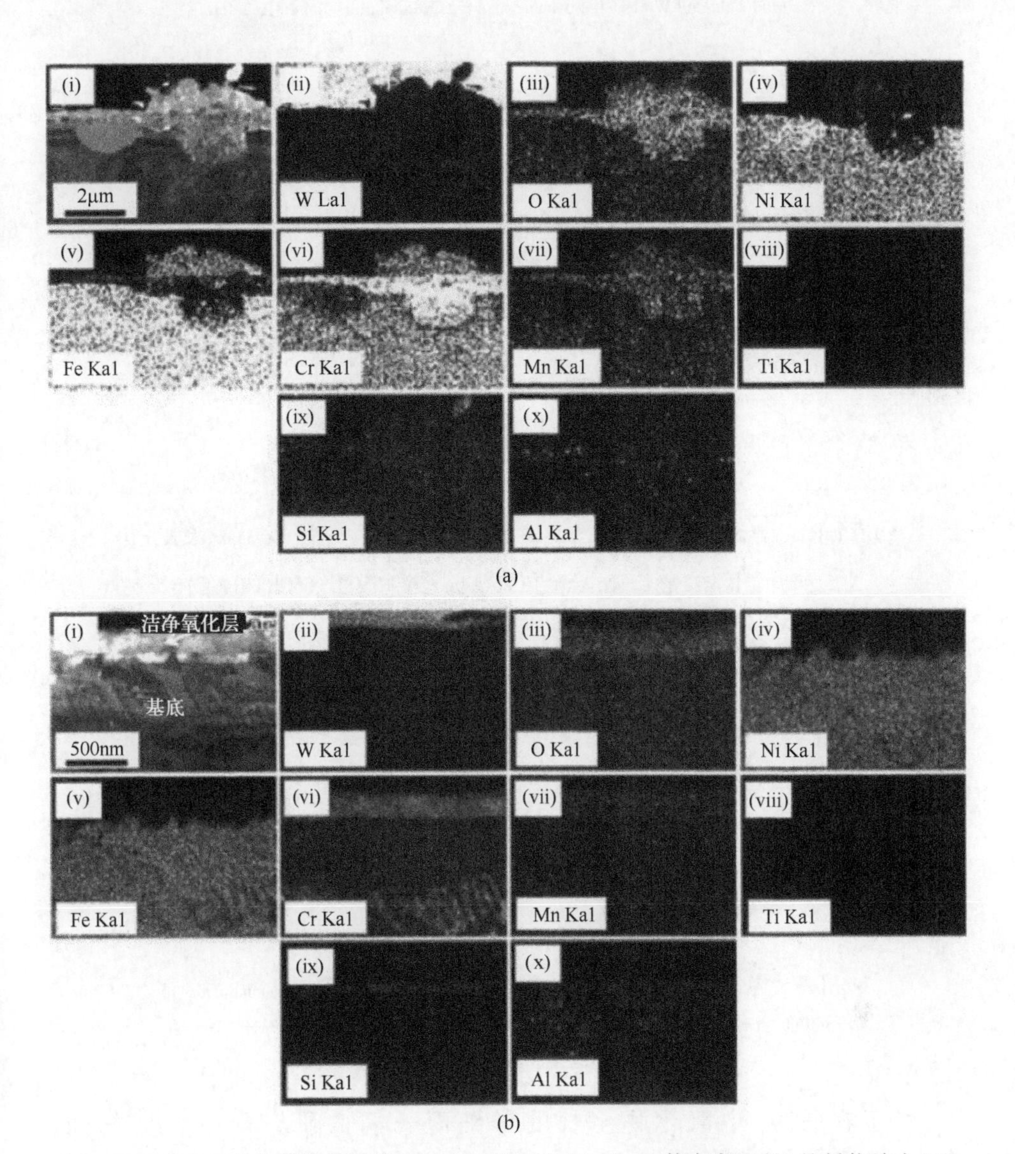


图 5.40　800HT 合金薄片暴露于 SCW(625℃、25MPa、500h、静态高压釜)的低倍放大(i) BF-STEM 图像和(ii)~(x)相关的 STEM/EDS 图((a)典型区域;(b)清洁氧化区域)(Mahboubi,2014)

图 5.42 是 Zhang 等(2009a)提出的结节形成机理示意图。由于 Cr 原子沿晶界和穿过晶内的扩散速率差异较大(如 5.3.2 节所述(图 5.17)),Cr 原子在晶界上的快速扩散可以为金属-氧化物界面提供足够的 Cr,以补偿晶界以上氧化膜中 Cr 的消耗,而在晶粒内部的表面扩散则较慢。因此,晶粒内部上方的氧化膜在 Cr 和非防护性物质中被耗尽,这促进了氧化剂向基底金属扩散,并在膜

下形成镜像结节。来自基底金属的 Fe 原子扩散到膜上，与氧或水发生反应，在膜的上方形成氧化的 $Fe_3O_4$ 岛，如图 5.40 所示。

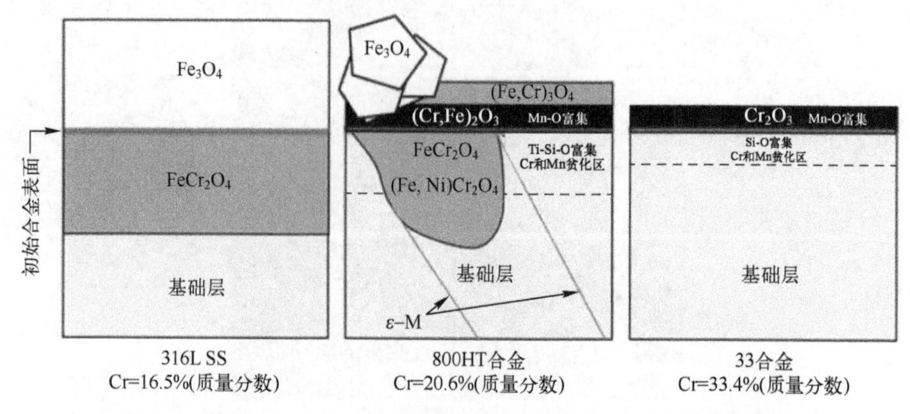


图 5.41 奥氏体 Fe-Cr-Ni 合金在 25MPa SCW、550℃、500h 的静态高压釜中暴露后对其氧化垢结构和成分的总结（Mahboubi, 2014）

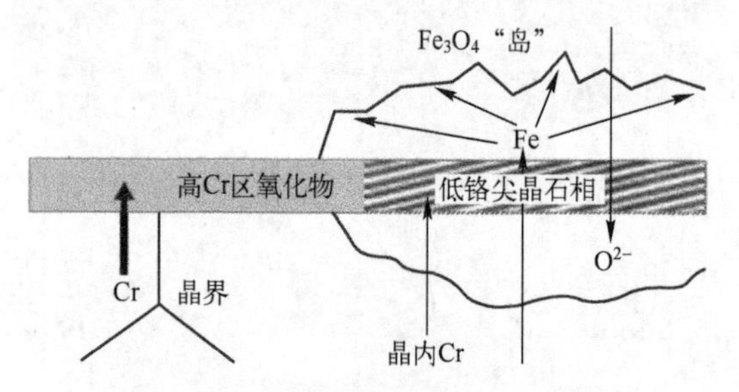


图 5.42 奥氏体钢结节机理的示意图

（引自文献 Zhang, L., Zhu, F., Tang, R., 2009a. Corrosion mechanisms of candidate structural materials for supercritical water-cooled reactor. Front. Energy Power Eng. China 3(2), 233-240）

关于在 SCW 中氧化物生长的详细而长期的研究很少。Behnamian 等(2017)在 500℃下对 316L SS 进行了长期(20000h)小容器试验。暴露 500h 后，外层氧化物大约为 5mm 厚，由与厚度大致相同直径的晶体组成。将暴露时间增加到 5000h，导致氧化层厚度增加，晶粒直径增大（图 5.43）。暴露 20000h 后，外氧化层厚度为 20~25mm，内氧化层厚度为 13~53mm。暴露 $10^4$h 后，外层氧化层呈现出连续但多孔的形态。在包膜内氧化层检测到少量微裂纹，可能是在 SCW 暴露过程中包膜内表面氧化和应力强度增加所致。氧化层厚度和质量变化均遵循抛物线速率定律（厚度测量，$n=0.427$；质量变化，$n=0.455$）。所观察到的孔隙率是由于 $Fe_3O_4$层空位浓度足够高时空位塌缩成孔隙所致。图 5.44 图示了 316L SS 氧化的长期演化过程。

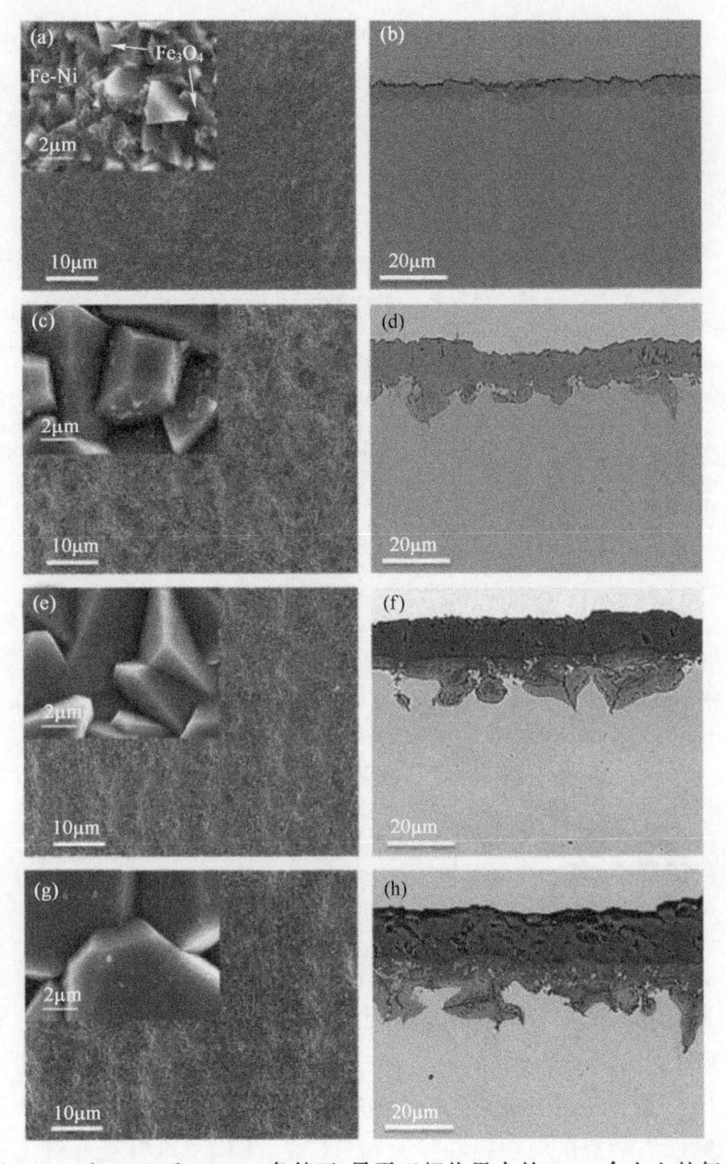


图 5.43　在 500℃和 25MPa 条件下，暴露于超临界水的 316L 合金上的氧化膜的俯视图和截面 SEM 显微照片：((a),(b))500h；((c),(d))时间 5000h；((e),(f))10000h；((g),(h))20000h

(引自文献 Behnamian, Y., Mostafaei, A., Kohandehghan, A., Zahiri, B., Zheng, W., Guzonas, D., Chmielus, M., Chen, W., Luo, J. L., 2017. Corrosion behavior of alloy 316L stainless steel after exposure to supercritical water at 500℃ for 20,000h. J. Supercrit. Fluids 127,191-199)

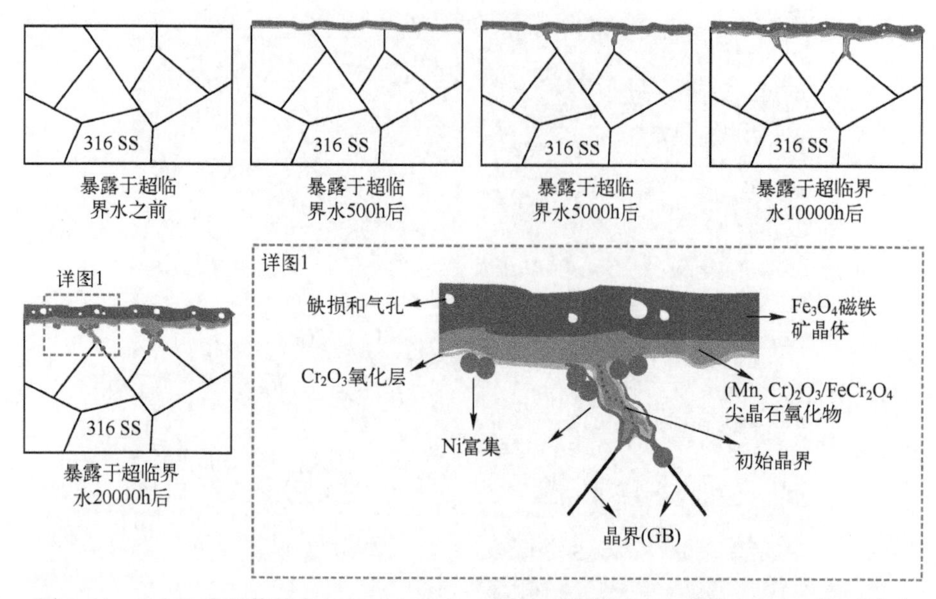


图 5.44　316 SS 在超临界水(SCW)(500℃、25MPa、胶囊样品)中长达 20000h 的不同时间的氧化示意图

(引自文献 Behnamian, Y., Mostafaei, A., Kohandehghan, A., Zahiric, B., Zheng, W., Guzonas, D., Chmielus, M., Chen, W., Luo, J. L., 2017. Corrosion behavior of alloy 316L stainless steel after exposure to supercritical water at 500℃ for 20,000h. J. Supercrit. Fluids 127, 191-199)



## 5.4.3 镍基合金

除 625 合金外,在 SCWR 条件下对 Ni 基合金的研究要少于 Fe 基合金。在大多数研究中,这种合金与奥氏体钢相比具有相当高的耐腐蚀性。在 Ni 基合金上发现的氧化物与其他合金类相比较薄;625 合金和 690 合金在 450℃和 23MPa 的 SCW 条件下暴露 483h 后,其氧化物厚度约为 50nm(Guzonas 等,2007)。即使在 600℃下,质量变化通常也很小,这使得测量变得困难,并导致数据会显著分散。而质量损失经常被观察到。对于 625 合金,人们发现其表面氧化物是由典型尺寸约为 1mm 的氧化物颗粒构成的,其均匀的氧化层由几十纳米范围内的非常细小的氧化物颗粒组成。不同直径的凹坑通常出现在样品表面(Was 等,2006;Fujisawa等,2006;Guzonas 等,2007;Chang 等,2012;Ren 等,2007),可能起源于金属碳化物包裹体(Ren 等,2007)。当样品暴露于相对较高的氧含量或较高的 SCW 温度时,点蚀会被抑制。据报道,在 $T$>600℃条件下,经长时间暴露后会出现晶间腐蚀(Ren 等,2007)。

通过在超临界水中腐蚀在625合金上形成的表面氧化物具有双层结构，该双层结构由富Ni/Fe的外层和富Cr的内层组成。Li等(2009)报告说，在450℃、23MPa的超临界水中暴露483h后，形成了薄的(约50nm)富含Cr的表面氧化物，其厚度与超临界水密度无关，尽管随着超临界水密度从$125kg/m^3$增加到$200kg/m^3$(图5.27)，在较低超临界水密度下观察到的重量增加变为质量损失。在450℃的超临界水中暴露483h后，在625合金和690合金上形成的氧化物的拉曼光谱(Guzonas等，2007)主要是$NiFe_2O_4$。Zhang等(2009b)将625合金暴露于超临界水(500℃、25MPa)中1000h，发现$NiFe_2O_4$的外层较为粗糙，$NiFe_2O_4$的内层是细晶粒。此外，还观察到结节腐蚀，这归因于基体中γ′团簇的局部腐蚀。将625合金在600℃下的超临界水暴露1026h后，确定的主要相为化学计量为$Ni(Fe,Cr)_2O_4$的尖晶石相，并确定了其他两种氧化物为$Cr_{1.3}Fe_{0.7}O_3$和NiO(Ren等，2007)。625合金表面的俄歇电子能谱显示了超临界水在500℃和600℃下暴露后的双重氧化物结构(Ren等，2007)，由Ni/富Fe外层和富Cr尖晶石内层组成，底层具有扩散层。将温度升至600℃会形成带有细颗粒的均匀氧化物表面，而晶界则由形貌升高的氧化物勾勒出轮廓。张等(2009b)指出，随着C-276在超临界水中的暴露时间延长，沉淀相(存在于合金中用于增强)会被氧化，破坏外部保护性氧化物层并在表面形成微裂纹，从而增加腐蚀。

### 5.4.4 涂层



许多团队(Khatamian，2013；Guzonas等，2005；Wills等，2007；Zheng等，2008；Hui等，2011；Huang和Guzonas，2013；van Nieuwenhove等，2013)已经研究了使用涂层改善超临界水冷反应堆中的材料性能，尤其是加拿大超临界水冷反应堆概念，其包壳峰值温度更高。例如，可以将具有良好抗蠕变和辐射破坏性能的基体合金与对一般腐蚀和SCC具有高抗性的表面层进行合金化(Zheng等，2008)。如果表面层可以在制造堆内组件之后产生，如在成形和连接操作之后，则可以避免可成形性和可焊性问题。还有一些组件可能会从隔热涂层的应用中受益。

Khatamian(2013)探索了锆合金Zircaloy-2、Zircaloy-4、Zr-2.5Nb、Zr-1Nb和Zr-Cr-Fe的Cr涂层，而Hui等(2011)证明了$ZrO_2$涂层在室温下在锆合金上的成功沉积。在这两种情况下，超临界水中基材的耐腐蚀性都得到了显著提高。Guzonas等(2005)的研究结果表明，可以使用溶胶-凝胶基底的技术在碳钢和不锈钢表面上沉积相当坚固的氧化锆膜。Wills等(2007)研究了使用常压等离子体射流在超临界水冷反应堆中使用的各种表面上形成涂层的方法，尽管不能生产出合适的坚固涂层。已经有许多用于加拿大超临界水冷反应堆概念的

MCrAlY 型涂层的研究(Huang 和 Guzonas,2013);如果需要 YSZ 腐蚀或隔热涂层,尤其是对除了燃料包层以外的堆芯内组件,则需要这些类型的材料作为基材和上部氧化锆涂层之间的结合层。在超临界水中对 NiCrAlY 和 FeCrAlY 进行的长期测试表明,两种材料的质量变化都非常小,Ni-20Cr-5Al 在 550℃ 的超临界水中暴露 6300h 后,质量变化仅为 0.009mg/dm$^2$(Huang 和 Guzonas,2014)。TiAlN、CrN 和 $ZrO_2$涂层通过物理气相沉积法沉积在 316L 不锈钢和 600 合金基体上(Van Nieuwenhove 等,2013);尽管 TiAlN 和 $ZrO_2$涂层在超临界水中的性能不能令人满意,但 CrN 涂层提供了稳定的保护性腐蚀屏障。超临界水冷反应堆的涂层应用仍是研究的热门领域,但是要使这种混合材料用于燃料包壳,还需要进行大量的开发工作。因此,在为加拿大超临界水冷反应堆概念的"Mark I"版本最终选择燃料包层材料时,涂层没有纳入考虑范围内。

其他形式的改进的表面组分已经被测试过。为了改进 9Cr ODS 合金的表面,采用了 Y 表面处理。在经过 Y 表面处理的 9Cr ODS 合金上形成了富钇层,将磁铁矿层分为两部分(Tan 等,2007;Chen 等,2007c)。外磁铁矿层是多孔的,而内磁铁矿层是细颗粒的。Y 主要集中在一个中心层中,这表明该层是由于 Y 氧化物的高热力学稳定性而在暴露的初始阶段形成的。随后,由于下面细小的 $Fe_3O_4$晶粒的生长,该层被向外驱动。XRD 分析表明 Y 主要以 $YFeO_3$的形式存在。作者得出的结论是,虽然 $YFeO_3$颗粒是阳离子扩散的有效局部扩散屏障,但粒子间区域却为阳离子提供了其他扩散路径。



## 5.5 氧化物增长动力学

从超临界水冷反应堆设计人员的角度来看,研究腐蚀的任何程序所需要的关键输出是一种预测部件使用寿命结束时的总金属损耗(或金属渗透)的方法,以及与该预测相关的不确定性。

通常,腐蚀动力学(金属损失、质量增量、氧化物厚度或其他参数随时间变化)分为对数、抛物线或线性 3 种形式。这些分类来自对理想模型的理论考虑。在实际的合金中,这些分类并不总是与数据相符,并且速率定律可能随时间或温度而变化。Kofstad(1988)和 Hauffe(1976)对可能遇到的各种速率定律及其机理基础进行了详尽的讨论。

对数形式的速率定律为

$$x = k\lg \cdot \lg(t+t_0)+C \tag{5.7}$$

反对数形式的速率定律为

$$x^{-1} = C-k\lg \cdot \lg t \tag{5.8}$$

在超临界水冷反应堆和超临界化石燃料电厂腐蚀文献中,常将抛物线动力学假定为默认值,如果反应物通过生长的氧化膜的扩散是限速步骤,则所期望的形式为

$$x^2 = k_p t + C \tag{5.9}$$

当速率确定步骤是受反应物供应限制,允许反应物快速扩散到反应前沿或通过恒定厚度的阻挡层扩散的多孔氧化物的存在限制的表面反应时,可能会产生线性动力学形式,即

$$x = k_{lin} t + C \tag{5.10}$$

尽管需要注意,但使用双对数图(lg$x$ 与 lg$t$ 比较)可以帮助确定要遵循的动力学类型以及速率随时间的任何变化,即

$$\lg x = \frac{1}{m}\lg t + C \tag{5.11}$$

$m$ 的值(1、2 或 3)分别表示线性、抛物线或 3 次方动力学。$m$ 的非整数值通常在实际系统中找到。

氧化物生长模型通常假设氧化物层结构已经建立,忽略了从裸露表面形成氧化物的初始阶段。Choudhry 等(2016)研究了 DO 浓度为 20mg/L 时超临界水中 800H 合金腐蚀初始阶段氧、氢和溶解金属浓度的变化。图 5.45 显示了在 650℃下随时间变化的 $O_2$ 和 Fe、Cr 和 Ni(800H 合金的主要成分)的浓度,以及所收集样品中 $O_2$、Mn、Al、Cu 和 Ti 的浓度。在 550℃下获得了相似的数据。在这两个温度下,每种物质的时间分布可分为 4 个阶段(Ⅰ、Ⅱ、Ⅲ和Ⅳ),该阶段由反应堆出口处的 DO 浓度确定。第Ⅰ阶段对应于一个初始阶段,在此期间出口处测得的 $O_2$浓度小于 300μg/L。第Ⅱ阶段是出口处 $O_2$浓度短暂的非常迅速增加的时期。在第Ⅲ阶段,出口处的 $O_2$浓度逐渐增加。在第Ⅳ阶段,出口处的 $O_2$浓度达到给水的 $O_2$浓度并保持恒定,这表明氧化的稳态速率很低。

第Ⅰ阶段中,在 650℃下,在废水中观察到相对较高的 Fe、Ni 和 Al 浓度以及非常低的溶解 $O_2$浓度(约 300μg/L)。合金的氧化消耗了几乎所有添加的 $O_2$。从氧气消耗量计算,所释放的 Fe、Ni 和 Al 的量仅占形成氧化物总量的一小部分(约 1/500)。在此期间,冷却剂中的 Fe 浓度约为 $6\times10^{-10}$m,与 Olive(2012)报告的磁铁矿的计算溶解度($1\times10^{-11}$m)合理吻合。第Ⅰ阶段,Fe 的释放量与其在合金中的含量大致相同,并且 Ni 的释放量略少,而 Mn 和 Al 的释放量明显多于基于合金成分的预期释放量。第Ⅰ阶段在 550℃下持续约 90h,在 650℃下持续 300h。

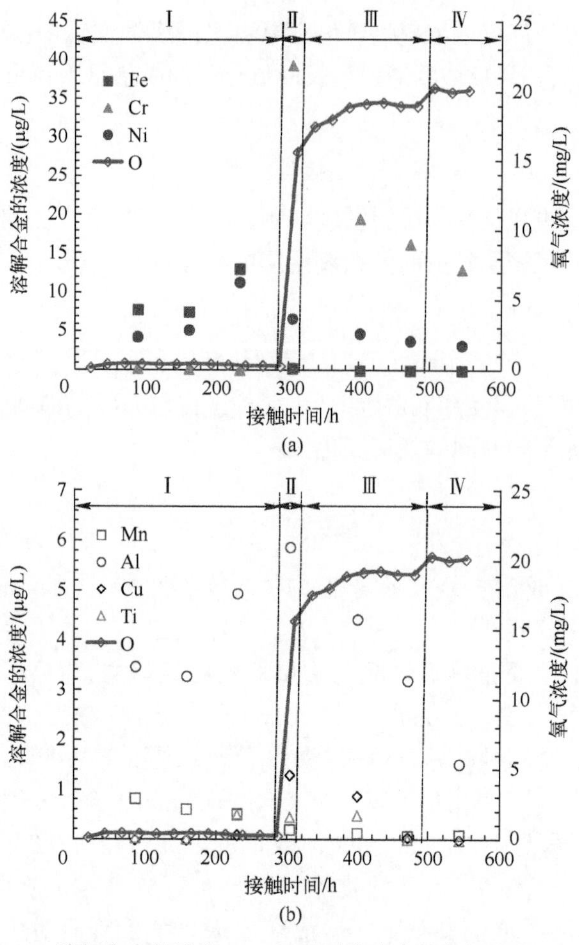


图 5.45 溶解合金的浓度(μg/L)以及废水中的氧气浓度(mg/L)随接触时间的变化(温度、压力和流速分别保持恒定在 650℃、25MPa 和 0.1mL/min)(Choudhry 等,2016)
(a) 溶解 Fe、Cr、Ni;(b) 溶解 Mn、Al、Cu 和 Ti。



第Ⅱ阶段,在 650℃下持续不到 24h,在 550℃下持续约 4h。在这段时间内,两种温度下出口处的氧气浓度迅速增加至约 17mg/L。第Ⅱ阶段,在 650℃下,Fe 的释放突然减少,而 Cr 的释放迅速增加。基于在合金中的浓度 Cr、Al、Cu 和 Ti 的释放量比预期要大得多。同时观察到 Cr 释放量增加,还发现 Mn 释放量大大减少。在第Ⅲ阶段中,各种元素的相对释放量与第Ⅱ阶段中观察到的相似。Al、Mn、Ti 和 Cu 的浓度随暴露时间而缓慢降低。

在第Ⅲ阶段中,出口处的氧气浓度随时间增加的速率在 500℃、550℃和 650℃时大致相同。在第Ⅲ阶段中 Cr 的行为在 550℃和 650℃下有所不同,在

550℃下随时间增加而在650℃下随时间减少。

在第Ⅳ阶段，基于在合金中的浓度，Cr和Al释放量仍然高于预期。稳态下的释放份额在550℃和650℃下是相同的。基于这些观察，建立了初始阶段的氧化物增长模型（图5.46）。

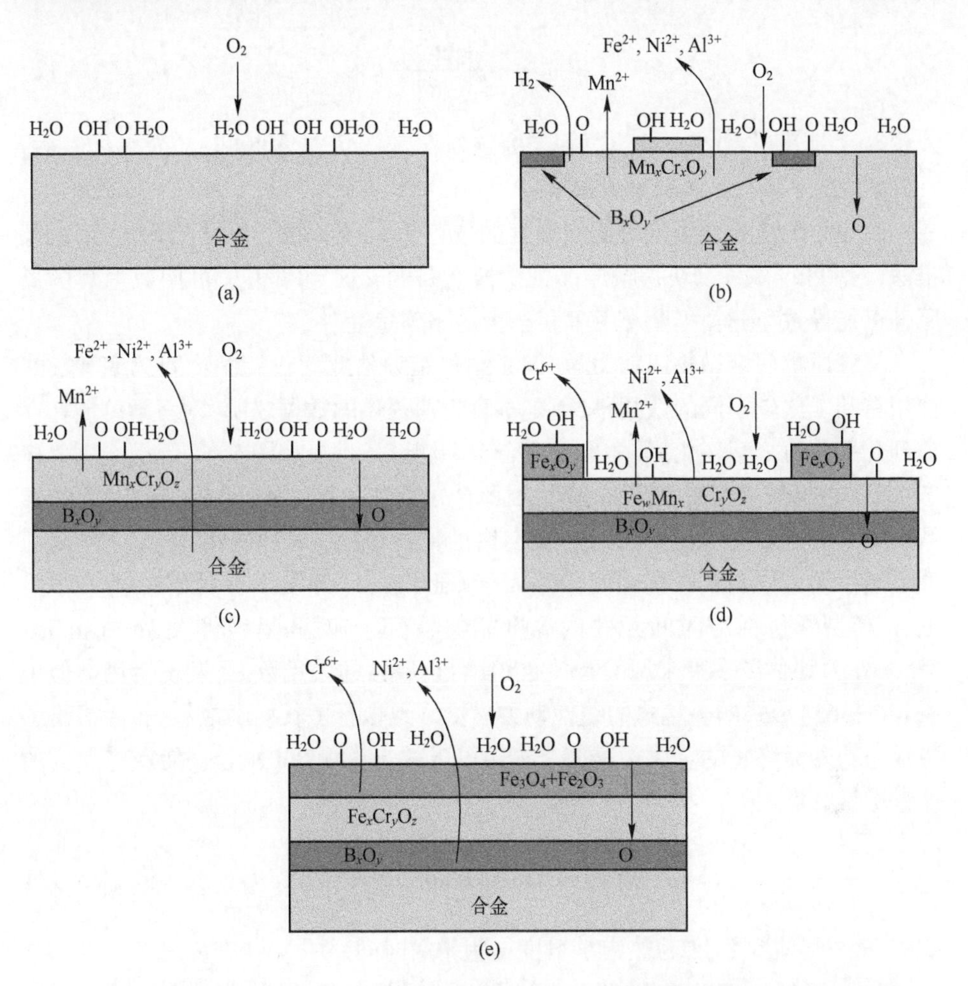


图5.46　在含氧水中800H合金上氧化物形成的模型（氧化物层的厚度不成比例。根据金属释放数据，B可能是Ti（即内层是$TiO_2$）。注意，虽然释放到溶液中的金属物质被写为离子，但它们最有可能是不带电的离子对（如$Fe(OH)_2$、$Al(OH)_3$））（Choudhry等，2016）

Novotny等（2013）报道了温度为500℃和550℃、压力为25MPa、流量为15～17L/h（高压釜刷新时间约6min）、DO浓度为1985～2020μg/kg、入口电导率小于0.1μS/cm的测试条件下，DO相似的瞬态行为。出口水电导率从测试开始时的

0.4μS/cm 缓慢降低到 0.12μS/cm,并且在测试的初始阶段高压釜出口 DO 浓度从 0 稳定增加到约 1800μg/kg。此后,在其余的加热过程中,它几乎保持恒定在(1800±10)μg/kg。

从现象学上讲,预计腐蚀速率(CR)具有以下函数形式,即

$$\mathrm{CR}=k\cdot\exp\left(-\frac{E_a}{RT}\cdot t^x\right) \tag{5.12}$$

式中:$k$ 为速率常数;$E_a$为活化能;$T$ 为绝对温度;$t$ 为暴露时间;$R$ 为气体常数;$x$(在式(5.11)中为 $1/m$)通常在 0~1 之间。

在独立的试验中测量时间和温度的依存关系,可以得出执行所需的两次外推法(时间和温度)以获得报废腐蚀渗透所需的 $x$、$E_a$ 和 $k$ 值。可用数据集的局限性可能导致与此类推断相关的相当大的不确定性。

尽管通常假定抛物线动力学,但必须仔细评估此假设。抛物线速率常数的使用有利于比较不同的数据集,但是本章先前提供的数据表明,动力学可能在立方到几乎线性之间变化,具体取决于合金和测试条件。如图 5.45 所示,建立"稳态"氧化物生长可能需要数百小时,因此从短期测试(如 500h)得出的动力学参数可能具有有限的价值。大量的长期腐蚀数据表明,SHS 中 800H 合金的腐蚀遵循线性动力学,如 Pearl 等(1965)。Brush(1965)开发了一种腐蚀动力学模型,该模型既包括与时间无关的温度相关项,又与时间呈线性关系。Guzonas 等(2016a)评估了多种来源的有关 800H 合金腐蚀的大量数据,发现使用类似于 Brush(1965)形式的表达式的线性动力学更好地拟合了长期腐蚀数据,并得出了更保守的寿命终止的金属渗透值。提出以下式来预测 800H 合金使用寿期内的质量增量 $\Delta W$,即



$$\Delta W=(0.6T-302)+17\exp\left(\frac{51000}{R}\cdot T\right)\cdot t \tag{5.13}$$

式中:$R$ 为气体常数;$T$ 为热力学温度;$t$ 为暴露时间。

对于 347 SS、310 SS 和 625 合金可以推导得到相似的形式。

Steeves 等(2015)使用几年来获得的数据以及 Guzonas 和 Cook(2012)的腐蚀模型得出超临界水中 625 合金腐蚀的动力学参数。数据是在 400℃、500℃、550℃和 600℃下获得的;观察到重量损失,在每个温度下重量损失基本上与时间呈线性关系,即

$$\mathrm{WL}=323.4\exp(-8684T^{-1})\cdot t \tag{5.14}$$

作为对比,Pearl 等(1967)报道了 625 合金的腐蚀速率定律,即

$$WL=2427\exp(-3037T^{-1})+1.07\times10^{6}\exp(-9944T^{-1})\cdot t \tag{5.15}$$

其在流量为174~217kg/h、压力为6.9MPa的条件下，在SHS中的回路测试得出。为了模拟水的放射分解，在回路中保持20mg/kg的氧气和2.5mg/kg的溶解氢。Steeves等(2015)获得的活化能(72.2kJ/mol)和Pearl等(1967)(81.7kJ/mol)的相对比较一致。Pearl等(1967)的模型包括在低于500℃的温度下获得的数据；省略较低温度数据的重新分析得出$E_a$值为74.4kJ/mol。修订后的腐蚀速率定律源自Pearl的数据，有

$$WL=260\exp(-8950T^{-1})\cdot t \tag{5.16}$$

且与式(5.14)非常一致。

## 5.6 机理与建模

由于材料性能(如负载能力)以及高温腐蚀和耐EAC的局限性，可能需要用于超临界水冷反应堆开发的改进材料。但是，只有证明可以在期望环境中运行之后，才接受新的材料作为选项并将其引入到常规操作中，这对于核级材料来说是一个非常缓慢的过程。为了成功地为给定的组件引入新材料，候选材料必须满足一系列与成本、制造、组装等相关的性能标准，然后与现有材料相比要表现出足够的改进才能获得认可。尽管标准可能很明确，与现有解决方案的竞争也很明确，但要实现显著的改进可能会充满挑战。改善特定性能可能需要不成比例的高水平的昂贵合金，或者由于同时需要的性能相互冲突，即，由于块状合金中较高的Ni含量对中子经济性产生不利影响，从而提高了总体耐蚀性。因此，推进超临界水冷反应堆材料的开发需要理论、建模和可靠的试验数据的结合。

超临界水冷反应堆开发人员需要模型来预测关键参数，例如，在组件寿命期内不同时间的金属渗透率或氧化膜厚度。预测合金一般腐蚀行为的模型在复杂性上有所不同，从纯粹的经验模型(基于将假定的速率定律拟合到测得的腐蚀速率数据)到基于腐蚀机理的数学公式的确定性模型。最终，所有模型都必须通过试验测量进行验证，并且必须能够预测时间和温度相关性、水化学的影响、氧化物结构和层组成以及微量合金元素的影响。

考虑一些复杂程度不同的建模选项，以根据耐腐蚀性预测候选超临界水冷反应堆材料的预期使用寿命。以下各节将讨论这些用于预测超临界水中腐蚀速率的方法。

### 5.6.1 经验和现象学模型

在亚临界水和高密度超临界水中的腐蚀是一种电化学过程,涉及独特的氧化和还原半反应,只要阳极和阴极位点之间既有电子接触又有电解接触,就可以分开。低密度超临界水的低离子电导率不利于这种分离,并且腐蚀机理变得类似于气相氧化。为了说明临界点附近水的物理特性的变化,Guzonas 和Cook(2012)通过明确地将直接 CO 作为并行过程添加到 EO 中,扩展了 Guan 和 Macdonald(2009)的现象学模型①。相对腐蚀速率 $R$ 由下式给出,即

$$R=k_{\mathrm{EO}}\exp\left(-\frac{E_{\mathrm{EO}}}{RT}\right)C_{\mathrm{H}}^{m}C_{\mathrm{O_2}}^{n}+k_{\mathrm{CO}}\exp\left(-\frac{E_{\mathrm{CO}}}{RT}\right)C_{\mathrm{O_2}}^{p} \tag{5.17}$$

式中:$k_{\mathrm{EO}}(E_{\mathrm{EO}})$和 $k_{\mathrm{CO}}(E_{\mathrm{CO}})$分别为 EO 和 CO 机制的异质速率常数(活化能);$T$ 为热力学温度;$C_{\mathrm{O_2}}^{n}$和 $C_{\mathrm{O_2}}^{p}$为氧浓度;$C_{\mathrm{H}}^{m}$为氢离子浓度;$m$、$n$ 和 $p$ 为相对于每种组分的反应顺序。

图 5.47 显示了 EO 和 CO 成分以及总的相对腐蚀随温度的变化。EO 组分的 Arrhenius 温度相关性导致 $R$ 随着温度升高到 $T_c$ 而升高,此后超临界水密度的快速下降降低了反应物的利用率,导致 $R$ 下降。当温度远高于 $T_c$ 时,CO 机理占主导地位,$R$ 再次显示 Arrhenius 温度行为。$R$ 接近 $T_c$ 时的局部最大值可以通过预测得到。



该模型(图 5.47)的预测结果可以与亚临界水和过热蒸汽中 304 SS 腐蚀的数据进行比较(Maekawa 等, 1968),这表明在 $T_c$ 附近存在类似的拐点(图 5.47)。$T_c$ 上下的斜率变化表明活化能不同。Steeves 等(2015)报道了超临界水中 800H 合金和 625 合金都在 $T_c$ 附近的局部最大值(图 5.48),并使用式(5.17)提取 EO 和 CO 组分的活化能。虽然这个现象学模型提供了对关于 $T_c$ 以上和 $T_c$ 以下腐蚀速率的温度依赖性的原因的见解,并且能够计算活化能,但是一个主要的缺点是无法预测时间依赖性。这种类型的模型可以与对时间依赖性的经验测量相结合(式(5.12))来解决这个问题。

如果有足够的数据,这种半经验模型对特定合金具有良好的预测能力。然而,这些模型不能用来预测新合金的行为。

---

① Guan 和 Macdonald(2009)引入了并行的 EO 和 CO 机制的概念,但没有建立一个包含这两个过程的模型。

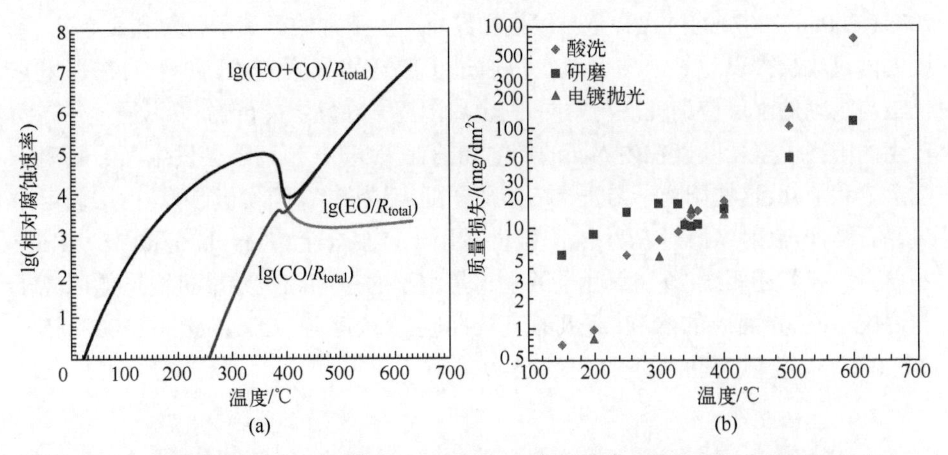


图 5.47 (a) 依据式(5.17)取值 $m$、$n=0.5$、$p=1$,在压力为 25MPa、$E_{EO}=50$kJ/mol、$E_{CO}=200$kJ/mol 条件下预测得到的相对腐蚀速率;(b) 304 SS 在亚临界水和过热蒸汽中的腐蚀数据。

(引自文献 Guzonas, D. A., Cook, W. G., 2012. Cycle chemistry and its effect on materials in a supercritical water-cooled reactor: a synthesis of current understanding. Corros. Sci. 65, 48-66)

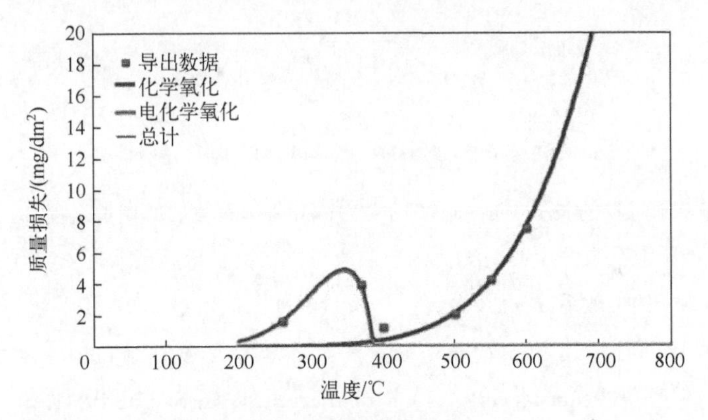


图 5.48 625 合金的试验腐蚀速率和模型腐蚀速率(Steeves 等, 2015)(见彩插)



## 5.6.2 确定性模型

Bojinov 等基于不锈钢在轻水反应堆(LWR)条件下氧化膜生长的模型,首先提出了 SCW 条件下腐蚀的确定性模型。混合传导模型(MCM)强调准稳态无源膜中离子和电子缺陷之间的耦合。该模型能够确定氧化层的主要动力学和输运参数,以计算在 SCW 条件下许多合金的氧化膜厚度随暴露时间的变化,并能深入了解合金元素对 SCW 抗氧化的影响因素。

根据 MCM(Betova 等, 2008、2009a、2009b; Penttila 等, 2011; Bojinov 等, 2001、

2005a、2005b、2007)通过在合金与氧化物界面产生正常的阳离子位置和氧空位，致密的内氧化层得以生长。然后氧空位通过扩散迁移到膜/电解质界面，在那里它们被与吸附的水反应消耗。与此同时，金属阳离子在该层的传输，可以通过外界面产生的阳离子空位及其在内界面的传输和消耗来实现，预计还会发生间隙阳离子的运输和消耗；这些阳离子与正常的阳离子亚点位交换(间化机制)。内层假定为尖晶石结构；X 射线衍射表明，这一假设适用于许多合金(Penttila 等，2011、2015a、2015b)。我们还假设，外层氧化层的生长是由于通过间隙位置输进内层的阳离子与内层/外层界面层的水和/或氧的直接反应(Ehlers 等，2006；Sun 等，2009b；Ren 等，2006)。这些过程的简化图见图 5.49。

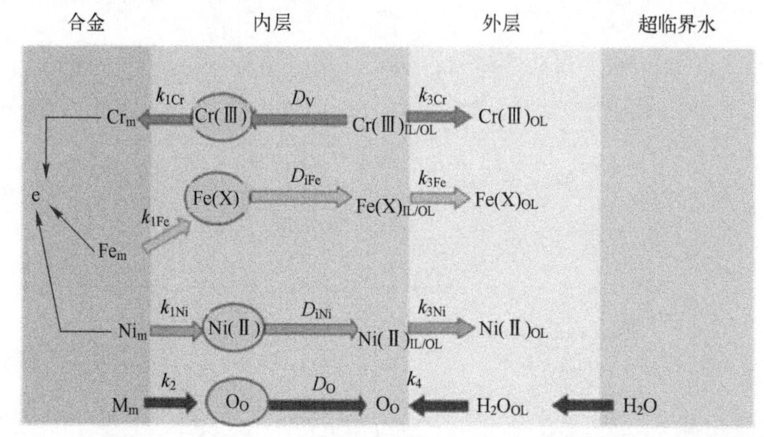


图 5.49 示意性的展示 3Fe-Cr-Ni 合金内部和外部氧化层的增长过程、$D_O$为内层氧空穴的扩散系数($cm^2/s$)，$D_V$为内层阳离子空穴的扩散系数($cm^2/s$)，$D_i$为内层间隙阳离子的扩散系数($cm^2/s$)，$k_{1j}$($j$=Fe,Ni,Mn,Si,Cu,Nb,Ti,Mo)为合金内表面间隙阳离子生成速率常数($mol \cdot cm^{-2} \cdot s^{-1}$)，$k_2$为通过氧化和注入氧空穴形成氧化物的速率常数($mol/(cm^2 \cdot s)$)，$k_{3j}$($j$=Fe,Cr,Ni,Mn,Si,Cn,Nb,Ti,Mo)为在内外层界面处外层氧化物生成速率常数($mol/(cm^2 \cdot s)$)

在自由腐蚀条件下生长的薄膜，需要通过内层传递电子，使金属组分的氧化与水或 DO 溶剂的还原相结合。MCM 假设离子点缺陷起电子给体或受体的作用(Bojinov 等，2005a)，电子输运与离子缺陷的输运相耦合。该模型不认为电子传导是速率限制，因为 SCW 钢在 500~700℃氧化过程中的原位电阻和电化学阻抗测量表明，氧化层的比电导率可与纯铁相媲美(Betova 等，2006)。

在 LWR、150℃以上条件下观察到的情况与在 SCW、500~700℃条件下的氧化过程相似(Betova 等，2008、2009a、2012；Bojinov 等，2005a)，氧化物中的大量缺陷不太可能支持高电场条件。从而得到 Fromhold 和 Cook(1967)的广义输运

方程的低场近似，用于 $j(j=\mathrm{I}$、O 和 V 等间隙阳离子、氧和阳离子空位）型点缺陷的通量，即

$$J_j(x,t)=-D_j\frac{\delta c_j(x,t)}{\delta x}-\frac{XF\boldsymbol{E}}{RT}D_jc_j(x,t) \tag{5.18}$$

式中：$c_j$为体积摩尔浓度（$mol/cm^3$）（$j$=Fe、Cr、Ni、Mn、Si、Cu、Nb、Ti、Mo）；$D_j$为内层扩散系数（$cm^2/s$）；$X$ 为氧化物中阳离子的价位（Ni 和 Mn 的 $X=2$，Cr 的 $X=3$，Fe 的 $X=2.67$）；$F$ 为法拉第常量；$\boldsymbol{E}$ 为内层电场强度（V/cm）；$R$ 为气体常数；$T$ 为温度（K）。

氧化物中的电场强度被认为与氧化物内部的距离无关（Bojinov 等，2005a；Macdonald，2012），由于电子态之间的带间隧穿，产生电荷分离以补偿电场强度的变化。

金属氧化物组分的浓度可以表示为其原子分数，其中 $V_{m,MO}$为层中相的摩尔体积，假设为尖晶石氧化物（$47cm^3/mol$）。而其他相（如 $Fe_2O_3$或 $Cr_2O_3$）在 SCW 的氧化过程中，这些相的摩尔体积差异约为 10%，且假设误差较小。

非平稳输运方程是由通量对给定薄膜组分浓度随时间变化的微分而得到的。为了获得内层的组成剖面，采用 Cranke-Nicolson 方法在初始条件和边界条件下求解方程组，该方法考虑了通过氧阴离子和铬离子空位的生长机制。采用类似于内层处理的方程组来计算外层的成分剖面。

测内层氧化层的生长作为暴露时间的函数，增长率（式（5.19））派生自点缺陷模型（PDM）（Macdonald，2011）和 MCM（Betova 等，2008、2009a；Bojinov 等，2007），有

$$L_{\mathrm{in}}(t)=L_{\mathrm{in}}(t=0)+\frac{1}{b}\ln\left[1+V_{m,\mathrm{MO}}k_2b\mathrm{e}^{-bL_{\mathrm{in}}(t=0)}t\right],\quad b=\frac{3\alpha_2F\boldsymbol{E}}{RT} \tag{5.19}$$

式中：$L_{in}(t)$为内层氧化物的厚度（cm）；$L_{in}(t=0)$为初始内层氧化物的厚度（cm）；$V_{m,MO}$为氧化层中相的摩尔体积（$cm^3/mol$）；$k_2$为通过氧化和注入氧空位形成氧化物的速率常数（$mol/(cm^2\cdot s)$）；$t$ 为时间（s）。

针对 SCW（Penttila 等，2011）外层氧化物的增长特性，基于近以扩散机制假设推导出来的抛物线定律被认为是不充分的，特别是对于高合金材料，因为在外层氧化物中观察到了显著的 Cr 和其他组分浓度，如 Ni，Al 等。由于氧化物的溶解度可以忽略，因此也假定不可能有溶解沉淀机制。因此，假定外层的生长是通过与用于内层输送间隙阳离子的水发生反应而发生。以双层膜生长模型为出发点（Sloppy 等，2013），外层生长被认为是由于通过内层的间隙阳离子的通量，即

$$\frac{\mathrm{d}L_{\mathrm{out}}(t)}{\mathrm{d}t}=-V_{m,\mathrm{MO}}J_{\mathrm{I}} \tag{5.20}$$

式中：$J_{\mathrm{I}}$为内层间隙阳离子的通量(mol/(cm$^2$·s))。

由于间隙晶格点的阳离子通量等于向氧化物中注入间隙阳离子后金属组分$j$的氧化速率常数，有

$$J_{\mathrm{I}}=-\sum_{j}k_{lj}y_{j,\mathrm{a}}\mathrm{e}^{-bL_{\mathrm{in}}(t)} \tag{5.21}$$

进一步得到以下微分方程：

$$\frac{\mathrm{d}L_{\mathrm{out}}(t)}{\mathrm{d}t}=V_{m,\mathrm{MO}}\sum_{j}k_{lj}y_{j,\mathrm{a}}\mathrm{e}^{-bL_{\mathrm{in}}(t)} \tag{5.22}$$

在合金/内层界面各反应传递系数相等的前提下，该方程是有效的。从0到$t$(氧化时间)和从0到$L_{\mathrm{out}}(t)$积分得到以下增长规律：

$$L_{\mathrm{out}}(t)=\frac{\sum_{j}k_{lj}y_{j,\mathrm{a}}}{k_2}(L_{\mathrm{in}}(t)-L_{\mathrm{in}}(t=0)) \tag{5.23}$$

通过比较模型预测和使用GDOES测量的氧化物成分的基本轮廓，验证了模型的有效性(图5.50)。在界面速率常数和扩散系数取值范围内，主要组分(Fe、Cr、Al)和次要组分(Mn、Si、Ni、Nb、Ti等)氧化物组分的大小和深度分布具有较好的一致性。



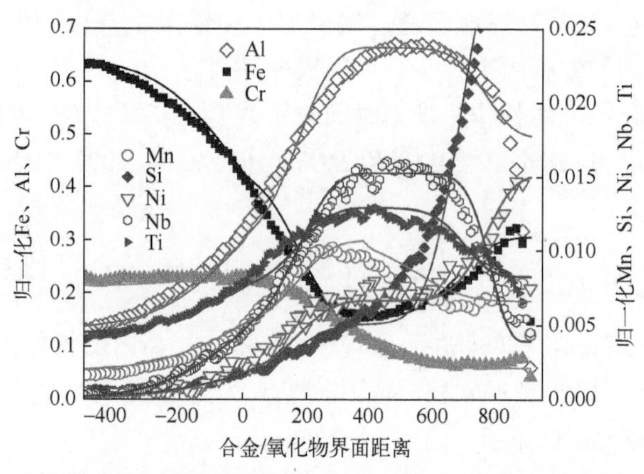


图5.50 测定(散点)和计算(实线)在MA956上形成的氧化物中金属成分的分数之间的比较(这些氧化物是在650℃下暴露600h后与合金/氧化物界面距离的函数(Penttila等,2015a、2015b)。金属组分的分数被归一化为金属组分的总浓度)

对每个深度剖面的敏感性研究表明，主要合金成分的速率常数和扩散系数的置信区间接近±10%。敏感性研究确定了以下参数对成分分布的影响最大：合金/内层界面的速率常数、在内层/水界面产生Cr空位的速率常数，以及内层中点缺陷的扩散系数。

得到的速率常数被用来预测氧化膜厚度作为曝光时间的函数。在 PM2000 和 MA956 上计算和测量的氧化物厚度(图 5.51(a)),以及 690 合金和 Sanicro 28(图 5.51(b))作为曝光时间的函数,对比发现函数预测曲线和测量数据之间具有良好的一致性,为所提出方法的有效性提供了额外的证据。

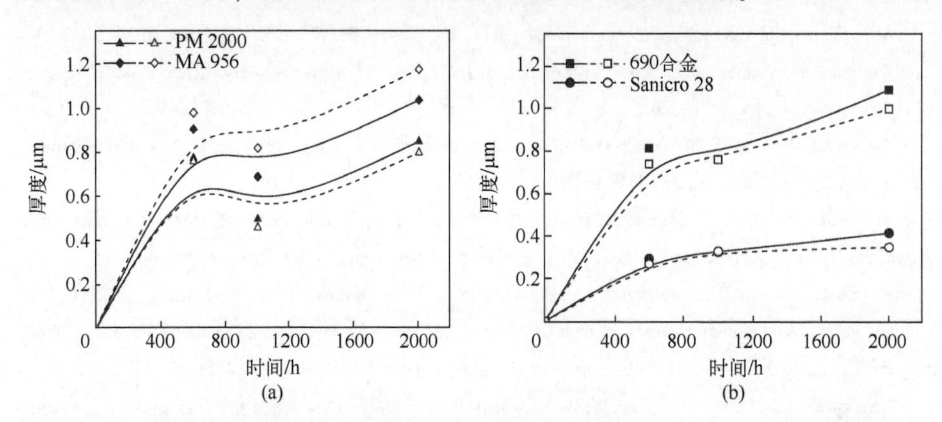


图 5.51 对比(a)PM 2000 和 MA 956 以及(b)690 合金和 Sanicro 28 在 650℃超临界水中的模型(开口符号)计算的氧化物厚度和辉光放电光谱深度剖面(封闭符号)估算的氧化物厚度(左)(Penttila 等,2015a、2015b)。

上述模型没有考虑溶剂的物理、化学性质。Guzonas(2013)建议使用 Cook 和 Olive(2011)提出的关于沉淀氧化物厚度的表达式(见第 4 章)来引入这种相关关系:

$$\frac{\mathrm{d}L_0}{\mathrm{d}t}=k_{\mathrm{OLFe}}(C_{\mathrm{O/S}}-C_{\mathrm{sat}}) \tag{5.24}$$

式中:$\mathrm{d}L_0/\mathrm{d}t$=外层厚度变化;$k_{\mathrm{OLFe}}$为动态沉积(溶解)常数(Penttila 等(2011)的符号);$C_{\mathrm{O/S}}$和 $C_{\mathrm{sat}}$为氧化物溶液浓度和饱和浓度。

将式(5.24)与外氧化层厚度表达式结合,利用 $C_{\mathrm{sat}}=K_{\mathrm{s}}\cdot\rho^n$得到氧化层厚度随 SCW 密度变化的表达式,即

$$\frac{\mathrm{d}L_0}{\mathrm{d}t}=D_{0,\mathrm{Fe}}\cdot\frac{\Delta y_{\mathrm{Fe}}}{L_0}+k_{\mathrm{OLFe}}(C_{\mathrm{O/S}}-K_{\mathrm{s}}\cdot\rho^n) \tag{5.25}$$

该方法与 Macdonald(1999)建立氧化层溶解对内层厚度影响的模型相似。当 $C_{\mathrm{O/S}}$较小时,如测试段进口杂质浓度较低,则氧化层厚度将取决于 SCW 密度。在低密度 SCW 中,除非氧化物溶解度高,否则氧化物的厚度将以式(5.5)~式(5.24)中的第一项为主,除非有氧化物剥落,重量增加是合理的腐蚀增加的表现。式(5.25)还介绍了添加流量依赖关系的可能性。

## 参考文献

Abe, F., Yoshida, H., 1985. Corrosion behaviours of heat-resisting alloys in steam at 800℃ and 40 atm pressure. Z. fur Metllkd. 76, 219-225.

Allen, T. R., Chen, Y., Ren, X., Sridharan, K., Tan, L., Was, G. S., West, E., Guzonas, D. A., 2012. Material performance in supercritical water. In: Konings, R. J. M. (Ed.), Compre-hensive Nuclear Materials, vol. 5. Elsevier, Amsterdam, pp. 279-326.

Ampornrat, P., Was, G. S., 2007. Oxidation of ferritic-martensitic alloys T91, HCM12A and HT-9 in supercritical water. J. Nucl. Mater. 371(1-3), 1-17.

Ampornrat, P., 2011. Determination of Oxidation Mechanisms of Ferritic/martensitic Alloys in Supercritical Water (Ph. D. thesis). Nuclear Engineering and Radiological Sciences, Uni-versity of Michigan.

Artymowicz, D., Andrei, C., Huang, J., Miles, J., Cook, W. G., Botton, G. A., Newman, R. C., 2010. Preliminary study of oxidation mechanisms of high Cr steels in supercritical water. In: 2nd Canada-China Joint Workshop on Supercritical Water-cooled Reactors (CCSC-2010) Toronto, Ontario, Canada, April 25-28, 2010.

Artymowicz, D., Andrei, C., Cook, W. G., Newman, R. C., 2014. Influence of the cold work on the supercritical water oxidation of Fe14Cr1.5Si model alloy. In: Canada-China Confer-ence on Advanced Reactor Development (CCCARD-2014), Niagara Falls, Canada, April 27-30, 2014.

Asay, R. H., 1990. Plant Tests of Surface Preconditioning for Steam Generators. EPRI Report NP-6616.

Bakai, O. S., Guzonas, D. A., Boriskin, V. M., Dovbnya, A. M., Dyuldya, S. V., 2015. Combined effect of irradiation, temperature, and water coolantflow on corrosion of Zr-, Ni-Cr-, and Fe-Cr-based alloys. In: 7th International Symposium on Supercritical Water-cooled Reactors (ISSCWR-7), March 15-18, 2015, Helsinki, Finland. Paper ISSCWR7-2012.

Behnamian, Y., Mostafaei, A., Kohandehghan, A., Zahiri, B., Zheng, W., Guzonas, D., Chmielus, M., Chen, W., Luo, J. L., 2017. Corrosionbehavior of alloy316Lstainless steel after exposure to supercritical water at 500℃ for 20,000h. J. Supercrit. Fluids 127, 191-199.

Betova, M. I., Bojinov, M., Kinnunen, P., Lehtovuori, V., Peltonen, S., Penttilä, S., Saario, T., 2006. Composition, structure, and properties of corrosion layers on ferritic and austenitic steels in ultrasupercritical water. J. Electrochem. Soc. 153(11), B464-B473.

Betova, I., Bojinov, M., Karastoyanov, V., Kinnunen, P., Saario, T., May 2012. Effect of water chemistry on the oxide film on alloy 690 during simulated hot functional testing of a pressurised water reactor. Corros. Sci. 58, 20-32.

Betova, I., Bojinov, M., Kinnunen, P., Lehtovuori, V., Penttila, S., Saario, T., 2007. Surface film electrochemistry of AISI316 stainless steel and its constituents in supercritical water. In: Proc. ICAPP 2007, Nice, France, May 13-18, 2007, Paper 7123.

Betova, I., Bojinov, M., Kinnunen, P., Lundgren, K., Saario, T., 2008. Mixed-conduction model for stainless steel in a high-temperature electrolyte: estimation of kinetic parameters of inner layer constituents. J. Electrochem. Soc. 155(2), C81-C92.

Betova, I., Bojinov, M., Kinnunen, P., Lundgren, K., Saario, T., 2009a. Influence of Zn on the oxide layer on AISI 316L(NG) stainless steel in simulated pressurised water reactor coolant. Electrochim. Acta 54(3),1056-1069.

Betova,I.,Bojinov,M.,Kinnunen,P.,Lundgren,K.,Saario,T.,2009b. Akinetic model of the oxide growth and restructuring on structural materials in nuclear power plants. In:Hagy,F. H.(Ed.),Structural Materials and Engineering. Nova Science Publishers,New York,pp. 91-133.

Berge,P.,1997. Importance of surface preparation for corrosion control in nuclear power stations. Mater. Perform. 36(11),56-62.

Bischoff,J.,2011. Oxidation Behavior of Ferritic-martensitic and ODS Steels in Supercritical Water(Ph. D. dissertation in Nuclear Engineering). The Pennsylvania State University,Department of Mechanical and Nuclear Engineering.

Bischoff,J.,Motta,A. T.,Eichfeld,C.,Comstock,R. J.,Cao,G.,Allen,T. R.,2012. Corrosion of ferritic-martensitic steels in steam and supercritical water. J. Nucl. Mater. xx,xx.

Bojinov,M.,Fabricius,G.,Kinnunen,P.,Laitinen,T.,Mäkelä,K.,Saario,T.,Sundholm,G.,2001. Electrochemical study of the passive behaviour of Ni-Cr alloys in a borate solution-a mixed conduction model approach. J. Electroanal. Chem. 504(1),29-44.

Bojinov,M.,Kinnunen,P.,Lundgren,K.,Wikmark,G.,2005a. A mixed-conduction model for the oxidation of stainless steel in a high-temperature electrolyte. Estimation of kinetic parameters of oxide layer growth and restructuring. J. Electrochem. Soc. 152(7),B250-B261.

Bojinov,M.,Heikinheimo,L.,Saario,T.,Tuurna,S.,2005b. Characterisation of corrosion films on steels after long-term exposure to simulated supercritical water conditions. In:Pro-ceedings of the American Nuclear Society—International Congress on Advances in Nuclear Power Plants 2005,ICAPP'05,pp. 1799-1807.

Bojinov,M.,Galtayries,A.,Kinnunen,P.,Machet,A.,Marcus,P.,2007. Estimation of the parameters of oxide film growth on nickel-based alloys in high temperature water elec-trolytes. Electrochim. Acta 52(26),7475-7483.

Boyd,W. K.,Pray,H. A.,1957. Corrosion of stainless steels in supercritical water. Corrosion 13(6),33-42.

Brar,C.,Davis,W.,Mann,E.,Ridley,C.,Wirachowsky,B.,Rangacharyulu,C.,Guzonas,D.,Leung,L.,2015. GEN IV SCWR cladding analysis project:nickel content in SCWR cladding material. In:Proceedings 5th Int. Youth Conference on Energy,Pisa,Italy,May 27-30,2015,pp. 103-110.

Brush,E. G.,1965. Corrosion rate law considerations in superheated steam. Nucl. Appl. 1(3),246-251.

Chang,K.-H.,Huang,J.-H.,Yan,C.-B.,Yeh,T.-K.,Chen,F.-R. Kai,J.-J.,2012. Corrosion behavior of alloy 625 in supercritical water environments. Prog. Nucl. Energy 1-12.

Chen,Y.,Sridharan,K.,Allen,T. R.,2005. Corrosion Behavior of NF616 and D9 as Candidate Alloys for Supercritical Water Reactors. Corrosion 2005. NACE International,Houston,TX. Paper 05391.

Chen,Y.,Sridharan,K.,Allen,T. R.,2007a. Corrosion of Candidate Austenitic Stainless Steels for Supercritical Water Reactors. Corrosion 2007,Nashville,TN. NACE International,Houston,TX. Paper 07408.

Chen,Y.,Sridharan,K.,Allen,T. R.,2007b. Corrosion of 9Cr oxide dispersion strengthened steel in supercritical water environment. J. Nucl. Mater. 371,118-128.

Chen,Y.,Sridharan,K.,Kruizenga,A.,Tan,L.,Allen,T. R.,2008. Corrosion behavior of austenitic alloys in supercritical water. Corros. Sci.

Chen, Y., Sridharan, K., Allen, T. R., 2007c. Effect of yttrium thin film surface modification on corrosion layers formed on 9Cr ODS steel in supercritical water". In: Allen, T. R., Busby, J. T., King, P. J. (Eds.), 13th Environmental Degradation of Materials in Nuclear Power Systems 2007. Canadian Nuclear Society.

Cho, H. S., Kimura, A., Ukai, S., Fujiwara, M., 2004. Corrosion properties of oxide dispersion strengthened steels in supercritical water environment. J. Nucl. Mater. 329-333, 387-391.

Choudhry, K. I., Mahboubi, S., Botton, G. A., Kish, J. R., Svishchev, I. M., 2015. Corrosion of engineering materials in a supercritical water cooled reactor: characterization of oxide scales on Alloy 800H and stainless steel 316. Corros. Sci. 100, 222-230.

Choudhry, K. I., Guzonas, D. A., Kallikragas, D. T., Svishchev, I. M., 2016. On-line monitoring of oxide formation and dissolution on alloy 800H in supercritical water. Corros. Sci. 111, 574-582.

Comprelli, F. A., Busboom, H. J., Spalaris, C. N., 1969. Comparison of radiation damage studies and fuel cladding performance for Incoloy-800. In: Irradiation Effects in Structural Alloys for Thermal and Fast Reactors, ASTM STP 457. American Society for Testing and Ma-terials, pp. 400-413.

Cook, W., Miles, J., Li, J., Kuyucak, S., Zheng, W., 2010. Preliminary analysis of candidate alloys for use in the CANDU-SCWR. In: 2nd Canada-China Joint Workshop on Super-critical Water-cooled Reactors (CCSC-2010) Toronto, Ontario, Canada, 2010 April 25-28.

Cook, W. G., Olive, R. P., 2011. Prediction of crud deposition in a CANDU-SCWR core through corrosion product solubility and transport modelling. In: Proceedings of the 5th Interna-tional Symposium on Supercritical Water-Cooled Reactors (ISSCWR5), March 13-16, 2011. Canadian Nuclear Society, Vancouver, Canada. Toronto, Ontario. Paper P83.



Cook, W. G., Olive, R. P., 2012a. Pourbaix diagrams for the iron-water system extended to high-subcritical and low-supercritical conditions. Corros. Sci. 55, 326-331.

Cook, W. G., Olive, R. P., 2012b. Pourbaix diagrams for chromium, aluminum and titanium extended to high-subcritical and low-supercritical conditions. Corros. Sci. 58, 291-298.

Cook, W. G., Olive, R. P., 2012c. Pourbaix diagrams for the nickel-water system extended to high-subcritical and low-supercritical conditions. Corros. Sci. 58, 284-290.

Cook, W., Miles, J., 2012. Effect of dissolved oxygen on the corrosion and oxide morphology of materials in SCW. In: 3rd Canada-China Joint Workshop on Supercritical Water-cooled Reactors (CCSC-2012), Xi'an, China, April 18-20, 2012. Paper 12063.

Cowen, H. C., Longton, P. B., Hand, K., 1966. Corrosion of Stainless Steels and Nickel-Based Alloys in Supercritical Steam. U. K. Atomic Energy Authority TRG Memorandum 3399(C).

Cox, B., 1973. Accelerated Oxidation of Zircaloy-2 in Supercritical Steam. Atomic Energy of Canada Limited Report AECL 4448.

Crank, J., Nicolson, P., 1947. A practical method for numerical evaluation of solutions of partial differential equations of the heat conduction type. Math. Proc. Camb. Philos. Soc. 43(1), 50-67.

Daigo, Y., Watanabe, Y., Sugahara, K., Isobe, T., 2006. Compatibility of nickel-based alloys with supercritical water applications: aging effects on corrosion resistance and mechanical properties. Corrosion 62(2), 174-181.

Dong, Z., Chen, W., Zheng, W., Guzonas, 2012. Corrosion behavior of chromium oxide based ceramics in supercritical water (SCW) environments. Corros. Sci. 65, 461-471.

Dong, Z., Liu, Z., Li, M., Luo, J.-L., Chen, W., Zheng, W., Guzonas, D., 2015. Effect of ultrasonic impact peening on the corrosion of ferritic-martensitic steels in supercritical water. J. Nucl. Mater. 457, 266-272.

Dong, Z., Li, M., Behnamian, Y., Liu, Z., Luo, J.-L., Chen, W., Zheng, W., Guzonas, D., 2014 June. Corrosion Behavior of Ferritic-martensitic Steels in High Temperature Hydrothermal Environments. NACE Northern AreaWestern Conference. Alberta, Edmonton, pp. 27-30.

Eberle, F., Kitterman, J. H., 1968. Scale Formation on Superheater Alloys Exposed to High Temperature Steam. In: Lien, G. E. (Ed.), Behavior of Superheater Alloys in High Temperature, High Pressure Steam. The American Society of Mechanical Engineers, New York.

Ehlers, J., Young, D. J., Smaardijk, E. J., Tyagi, A. K., Penkalla, H. J., Singheiser, L., Quadakkers, W. J., 2006. Enhanced oxidation of the 9% Cr Steel P91 in water vapour containing environments. Corros. Sci. 48(11), 3428-3454.

Fromhold Jr., A. T., Cook, E. L., 1967. Diffusion currents in large electric fields for discrete lattices. J. Appl. Phys. 38, 1546-1551.

Fry, A., Osgerby, S., Wright, M., 2002. Oxidation of Alloys in Steam Environmentse-A Review. National Physical Laboratory Report MATC(A)90.

Fujisawa, R., Nishimura, K., Nishida, T., Sakaihara, M., Kurata, Y., Watanabe, Y., 2006. Corrosion behavior of nickel-based alloys and type 316 stainless steel in slightly oxidizing or reducing supercritical water. Corrosion 62(3), 270-274.

Fujiwara, K., Watanabe, K., Domae, M., Katsumura, Y., 2007. Stability of Chromium Oxide Film Formed by Metal Organic Chemical Vapor Deposition in High Temperature Water up to Supercritical Region. Corrosion 2007, March 11-15, Nashville, Tennessee.

Galerie, A., Wouters, Y., Caillet, M., 2001. The kinetic behavior of metals in water vapour at high temperatures: can general rules be proposed. Mater. Sci. Forum 369-372, 231-238.

Gaul, G. G., Pearl, W. L., 1961. Corrosion of Type 304 Stainless Steel in Simulated Superheat Reactor Environments. General Electric Company Report GEAP-3779.

Gilewicz-Wolter, J., Zurek, Z., Dudała, J., Lis, J., Homa, M., Wolter, M., 2006. Diffusion rates of $^{51}Cr$, $^{54}Mn$ and $^{59}Fe$ in $MnC_2O_4$ and $FeCr_2O_4$ spinels. Adv. Sci. Technol. 46, 27-31.

Glebov, V. P., Antikain, P. A., Taratuta, V. A., Mamet, A. P., Kurilenko, E. P., 1976. Investigation of the influence of pH of the medium at supercritical pressure and of hydrogen concentration in it on the intensity of oxidation of different types of steel. Teploenerggetika 23, 77-81.

Gómez-Briceño, D., Blázquez, F., Sáez-Maderuelo, A., 2013. Oxidation of austenitic and ferritic/martensitic alloys in supercritical water. J. Supercrit. Fluids 78, 103-113.

Gong, M., Zhou, Z., Hu, H., Penttilä, S., Toivonen, A., 2015. Corrosion behavior of $^{14}Cr$ ODS ferritic steels in a supercritical water. In: 7th International Symposium on Supercritical Water-Cooled Reactors (ISSCWR-7), 2015 March 15-18, Helsinki, Finland. Paper ISSCWR7-2059.

Gu, G. P., Zheng, W., Guzonas, D., 2010. Corrosion database for SCWR development. In: 2nd Canada-China Joint Workshop on Supercritical Water-cooled Reactors (CCSC-2010) Toronto, Ontario, Canada, 2010 April 25-28.

Guan, X., Macdonald, D. D., 2009. Determination of corrosion mechanisms and estimation of electrochemical kinetics of metal corrosion in high subcritical and supercritical aqueous systems. Corrosion 65(6), 376-387.

Guzonas, D. A., Wills, J. S., McRae, G. A., Sullivan, S., Chu, K., Heaslip, K., Stone, M., 2005. Corrosion-resistant coatings for use in a supercritical water CANDU reactor. In: Proceedings of the 12th Int. Conf. on Environmental Degradation of Materials in Nuclear Power Systems—Water Reactors, Salt Lake City, August 14-18, 2005. TMS (The Minerals Metals and Materials Society).

Guzonas, D. A., Wills, J., Do, T., Michel, J., 2007. Corrosion of candidate materials for use in a supercritical water CANDU® reactor. In: 13th International Conference on Environmental Degradation of Materials in Nuclear Power Systems, pp. 1250-1261.

Guzonas, D., Wills, J., Dole, H., Michel, J., Jang, S., Haycock, M., Chutumstid, M., 2010. Steel corrosion in supercritical water: an assessment of the key parameters. In: 2nd Canada-China Joint Workshop on Supercritical Water-cooled Reactors (CCSC-2010), Toronto, Ontario, Canada, April 25-28, 2010.

Guzonas, D. A., Cook, W. G., 2012. Cycle chemistry and its effect on materials in a supercritical water-cooled reactor: a synthesis of current understanding. Corros. Sci. 65, 48-66.

Guzonas, D. A., 2013. The physical chemistry of corrosion in a supercritical water-cooled reactor. In: 16th International Conference on the Properties of Water and Steam (ICPWS16), University of Greenwich, London, UK, September 1-5, 2013.

Guzonas, D. A., Bissonette, K., Deschenes, L., Dole, H., Cook, W., 2013. Mechanistic aspects of corrosion in a supercritical water-cooled reactor. In: ISSCWR-6: The 6th International Symposium on Supercritical Water-Cooled Reactors, March 03-07, 2013, Shenzhen, Guangdong, China.

Guzonas, D. A., Edwards, M. E., Zheng, W., 2016a. "Assessment of candidate fuel cladding alloys for the Canadian supercritical water-cooled reactor concept. J. Nucl. Eng. Rad. Sci. 2, 011016.

Guzonas, D., Penttilä, S., Cook, W., Zheng, W., Novotny, R., Sáez-Maderuelo, A., Kaneda, J., 2016b. The reproducibility of corrosion testing in supercritical water-results of an international round robin exercise. Corros. Sci. 106, 147-156.

Guzonas, D., Novotny, R., Penttilä, S., 2017. Corrosion phenomena induced by supercritical water in generation IV nuclear Reactors. In: Yvon, P. (Ed.), Structural Materials for Generation IV Nuclear Reactors. Woodhead Publishing.

Hauffe, K., 1976. Gas-solid reactions-oxidation. In: Hannay, N. B. (Ed.), "Treatise on Solidstate Chemistry". Plenum Press, New York (Chapter 8).

Higuchi, S., Sakurai, S., Ishida, T., 2007. A study of fuel behavior in an SCWR core with high power density. In: Proc. ICAPP'07, Nice, France, May 13-18, 2007. Paper 7206.

Holcomb, G. R., Alman, D. E., 2006. The effect of manganese additions on the reactive evaporation of chromium in Nie-Cr alloys. Scr. Mater. 54, 1821-1825.

Holcomb, G. R., 2009. Steam oxidation and chromia evaporation in ultrasupercritical steam boilers and turbines. J. Electrochem. Soc. 156, C292-C297.

Holcomb, G. R., 2014. High Pressure Steam Oxidation of Alloys for Ultrasupercritical Conditions, vol. 82, pp. 271-295.

Holcomb, G. R., Carney, C., Doğan, Ö. M., 2016. Oxidation of alloys for energy applications in supercritical $CO_2$ and $H_2O$. Corros. Sci. 109, 22-35.

Hsu, R. H., Rankin, W. N., Summer, M. E., 1994. Characterizing Pre-polished Type 304L Stainless Steel (U). Westinghouse Savannah River Company Report WSRC-MS-94-0437.

Hudson, M. J. B., Ocken, H., 1996. Dose reduction measures at the Millstone Pointe-2 PWR. In: Water Chemistry of Nuclear Reactor Systems 7, vol. 2. BNES, p. 9.

Huang, X., Guzonas, D., 2013. Summary of Recent Coating Material Development and Testing in SCW. In: 6th International Symposium on Supercritical Water-cooled Reactors ISSCWR-6, March 03-07, 2013, Shenzhen, Guangdong, China. Paper ISSCWR6-13012.

Huang, X., Guzonas, D., 2014. Characterization of Ni-20Cr-5Al model alloy in supercritical water. J. Nucl. Mater. 445, 298-307.

Huang, X., Guzonas, D., Li, J., 2015a. Characterization of Fe-20Cr-6Al-Y model alloy in supercritical water. Corros. Eng. Sci. Technol. 50, 137-148.

Huang, X., Li, J., Amirkhiz, B. S., Liu, P., 2015b. "Effect of water density on the oxidation behavior of alloy A-286 at 625℃—A TEM study. J. Nucl. Mater. 467, 758-769.

Hwang, S. S., Lee, B. H., Kim, J. -G., Jang, J., 2005. Corrosion behavior of F/M steels and high Ni alloys in supercritical water. In: Proceedings of Global 2005, Tsukuba, Japan, October 9-13, 2005. Paper 043.

Hwang, S. S., Lee, B. H., Kim, J. G., Jang, J., 2008. SCC and corrosion evaluations of the F/M steels for a supercritical water reactor. J. Nucl. Mater. 372, 177-181.

Hui, R., Cook, W., Sun, C., Xie, Y., Yao, P., Miles, J., Olive, R., Li, J., Zheng, W., Zhang, L., 2011. Deposition, characterization and performance evaluation of ceramic coatings on metallic substrates for supercritical water-cooled reactors. Surf. Coatings Technol. 205, 3512-3519.

Ishigure, K., Kawaguchi, M., Oshoma, K., 1980. The effect of radiation on the corrosion product release from metals in high temperature water. In: Proceedings of the 4th BNES Conference on Water Chemistry of Nuclear Reactor Systems, Bournemouth, UK. Paper 43.

Jacob, Y. P., Haanappel, V. A. C., Stroosnijder, M. F., Buscail, H., Fielitz, P., Borchardt, G., 2002. The effect of gas composition on the isothermal oxidation behavior of PM chromium. Corros. Sci. 44, 2027-2039.

Jiao, Y., Kish, J., Zheng, W., Guzonas, D., Cook, W., 2014. Effect of thermal ageing on the corrosion resistance of stainless steel type 316L exposed in supercritical water. In: Canada-China Conference on Advanced Reactor Development (CCCARD-2014), Niagara Falls, Canada, April 27-30, 2014.

Jeong, Y. H., Park, J. Y., Kim, H. G., Busby, J. T., Gartner, E., Atzmon, M., Was, G. S., Comstock, R. J., Chu, Y. S., Gomes da Silva, M., Yilmazbayhan, A., Motta, A. T., 2005. Corrosion of zirconium-based fuel cladding alloys in supercritical water. In: Proceedings of the 12th International Conference on Environmental Degradation of Materials in Nuclear Power Systems-Water Reactors, pp. 1369-1378.

Jang, J., Lee, Y., Han, C. H., Yi, Y. S., Hwang, S. S., Sik, S., 2005a. Effect of Cr content on supercritical water corrosion of high Cr alloys. Mater. Sci. Forum 475-479, 1483-1486.

Jang, J., Han, C. H., Lee, B. H., Yi, Y. S., Hwang, S. S., 2005b. Corrosion behavior of 9Cr F/M steels in supercritical water. In: Proceeding of ICAPP 05, Seoul, Korea, No. 5136.

Kaneda, J., Kasahara, S., Kuniya, J., Moriya, K., Kano, F., Saito, N., Shioiri, A., Shibayama, T., Takahashi, H., 2005. General corrosion properties of titanium based alloys for the fuel claddings in the supercritical water-cooled reactor. In: Proc. 12th International Conference on Environmental Degradation of Materials in Nuclear Power Systems-Water Reactors, pp. 1409-1419.

Karasawa, H., Fuse, M., Kiuchi, K., Katsumura, Y., 2004. Radiolysis of supercritical water. In: 5th Int. Workshop on LWR Coolant Water Radiolysis and Electrochemistry, 2004 October 15, San Francisco, USA.

Kasahara, S., Kuniyo, J., Moriya, K., Saito, N., Shiga, S., September 2003. General Corrosion of Iron, Nickel and Titanium Alloys as Candidate Materials for the Fuel Cladding of Supercritical-Water Cooled Power Reactor, GENES4/ANP2003, 2003, Kyoto, Japan, Paper 1132.

Khatamian, D., 2013. Corrosion and deuterium uptake of Zr-based alloys in supercritical water. J. Supercrit. Fluids 78, 132-142.

Kim, J.-H., Kim, B. K., Kim, D.-I., Choi, P.-P., Raabe, D., Yi, K.-W., 2015. The role of grain boundaries in the initial oxidation behaviour of austenitic stainless steel containing alloyed Cu at 700℃ for advanced thermal power plant applications. Corros. Sci. 96, 52-66.

Knodler, R., Straub, S., 2014. Steam oxidation of martensitic steels and Ni-base alloys: comparison of lab tests with operating power plants. Oxid. Met. 82, 113-122.

Kofstad, P., 1988. High Temperature Corrosion. Elsevier Applied Science, London.

Kritzer, P., 2004. Corrosion in high-temperature and supercritical water and aqueous solutions: a review. J. Supercrit. Fluids 29, 1.

Kritzer, P., Boukos, N., Dinjus, E., 1998. Corrosion of Alloy 625 in high temperature sulfate solutions. Corrosion 54, 689-699.

Kritzer, P., Boukos, N., Dinjus, E., 1999. Factors controlling corrosion in high-temperature aqueous solutions: a contribution to the dissociation and solubility data influencing corrosion processes. J. Supercrit. Fluids 15, 205-227.

Kriksunov, L. B., Macdonald, D. D., 1995a. Corrosion in supercritical water oxidation systems: a phenomenological analysis. J. Electrochem. Soc. 142, 4069.

Kriksunov, L. B., Macdonald, D. D., 1995b. Corrosion testing and prediction in SCWO environments. HTD. In: Proc. of the ASME Heat Transfer Division, vols. 317-2, p. 281. Kriksunov, L. B., Macdonald, D. D., 1997. Potential-pH diagrams for iron in supercritical water. Corros. Sci. 53, 605-611.

Leistikow, S., 1972. Isothermal steam corrosion of commercial grade austenitic stainless steels and nickel base alloys in two technical surface conditions. In: Proceedings of the Fourth International Congress on Metallic Corrosion, Amsterdam, Netherlands, September 7 - 14, 1969. National Association of Corrosion Engineers, Houston, TX, pp. 278-290.

Li, J., Guzonas, D., Zheng, W., Wills, J., Dole, H., Michel, J., Woo, O. T., 2009. Property of passive films form on alloys tested in SCW water. In: 4th Int. Symposium on Supercritical Water-cooled Reactors", Heidelberg, Germany, March 8-11, 2009. Paper 56.

Li, W., Woo, O. T., Guzonas, D., Li, J., Huang, X., Sanchez, R., Bibby, C. D., 2015. Effect of pressure on the corrosion of materials in high temperature water. In: Characterization of Minerals, Metals, and Materials. John Wiley & Sons Inc., Hoboken, NJ (Chapter 12).

Li, J., Huang, X., Zeng, Y. W., Woo, O. T., Guzonas, D. A., 2014. Microscopy investigation on the corrosion of canadian generation IV SCWR materials. In: 19th Pacific Basin Nuclear Conference (PBNC 2014) Hyatt Regency Hotel, Vancouver, British Columbia, Canada, August 24-28, 2014.

Li, W., 2012. Oxidation Resistance and Nanoscale Oxidation Mechanisms in Model Binary and Ternary Alloys Exposed to Supercritical Water (M. A. Sc. thesis). Dept. of Materials Science and Engineering, University of Toronto.

Li, Q., Zhou, B., Yao, M., Liu, W., Chu, Y., 2007. "Corrosion behavior of zirconium alloys in supercritical water at 550℃ and 25MPa". Rare Metal Mater. Eng. 36,1358-1361.

Liu,Z. ,2013. Corrosion Behavior of Designed Ferritic-martensitic Steels in Supercritical Water(M. Sc. thesis). Department of Chemical and Material Engineering, University of Alberta, Canada.

Longton,P. B. ,1966. The Oxidation of Iron- and Nickel-Based Alloys in Supercritical Steam; A Review of the Available Data. United Kingdom Atomic Energy Authority TRG Report 1144.

Lillard,S. ,2012. Corrosion and compatibility. In: Konings,R. J. M. (Ed.), Comprehensive Nuclear Materials, vol. 5. Elsevier, Amsterdam, pp. 279-326.

Lujaszewicz,M. ,Simms,N. J. ,Dudziak,T. ,Nicholls,J. R. ,2013. Effect of steam flow rate and sample orientation on steam oxidation of ferritic and austenitic steels at 650 and 700℃ Oxid. Met. 79,473-483.

Macdonald,D. D. ,1999. Passivity—the key to our metals-based civilization. Pure Appl. Chem. 71(6), 951-978.

Macdonald,D. D. ,2004. Effect of pressure on the rate of corrosion of metals in high subcritical and supercritical aqueous systems. J. Supercrit. Fluids 30,375.

Macdonald,D. D. ,2011. The history of the point defect model for the passive state: a brief review of film growth aspects. Electrochim. Acta 56,1761-1772.

Macdonald,D. D. ,2012. The passive state in our reactive metals-based civilization. Arabian J. Sci. Eng. 37(5),1143-1185.

Maekawa,T. ,Kagawa,M. ,Nakajima,N. ,1968. Corrosion behaviors of stainless steel in high-temperature water and superheated steam. Trans. Jpn. Inst. Metals 9(2),130-136.

Mahboubi,S. ,September 2014. Effect of Cr Content on Corrosion Resistance of Fe-Cr-Ni Alloys Exposed in Supercritical Water(SCW)(M. A. Sc. thesis). Department of Materials Science and Engineering, McMaster University.

Mahboubi,S. ,Botton,G. A. ,Kish,J. R. ,2014. Microstructural characterization of oxide scales grown on austenitic Fe-Cr-Ni alloys exposed to supercritical water(SCW). In: 2014 Canada-China Conference on Advanced Reactor Development(CCCARD-2014), Niagara Falls, Canada, April 27-30, 2014. Paper CCCAD2014-017.

Mahboubi,S. ,Botton, G. A. ,Kish, J. R. ,2015. On the oxidation resistance of alloy 800HT exposed in supercritical water(SCW). Corrosion 71(8),992-1002.

Mitton,D. B. ,Yoon, J. H. ,Cline, J. A. ,Kim, H. S. ,Eliaz, N. ,Latanision, R. M. ,2000. The corrosion behaviour of nickel-base alloys in SCWO systems. Ind. Eng. Chem. Res. 39,4689-4696.

Montgomery,M. ,Karlsson,A. ,1995. Survey of oxidation in streamside conditions. VGB Kraftw. 75,235-240.

Motta,A. T. ,Yilmazbayhan,A. ,Gomes da Silva,M. J. ,Comstock,R. J. ,Was,G. S. ,Busby,J. T. ,Gartner, E. ,Peng,Q. ,Jeong,Y. H. ,Park,J. Y. ,2007a. Zirconium alloys for supercritical water reactor applications: challenges and possibilities. J. Nucl. Mater. 371(1-3),61-75.

Motta,A. T. ,Siwy,A. D. ,Kunkle,J. M. ,Bischoff,J. B. ,Comstock,R. J. ,Chen,Y. ,Allen,T. R. ,2007b. Microbeam synchrotron radiation diffraction and fluorescence study of corrosion layers formed on 9Cr ODS steel in supercritical water. In: Allen,T. R. ,Busby,J. T. ,

King,P. J. (Eds.), 13th International Conference on Environmental Degradation of Materials in Nuclear Power Systems. Canadian Nuclear Society, pp. 1501-1513.

Nezakat, M., Akhiani, H., Penttilä, S., Szpunar, J., 2014. Oxidation Resistance of Thermo-mechanically Processed Austenitic Stainless Steel 316L in Super Critical Water. PBNC2014-350.

Novotny, R., Janik, P., Nilsson, K. F., Siegl, J., Hausild, P., 2013. Pre-qualification of cladding materials for SCWR fuel qualification testing facility. In: 6th International Symposium on Supercritical Water-Cooled Reactors, March 3-7, 2013, Shenzhen, Guangdong, China. China: CGNPC, 2013. Paper ISSCWR6-13069.

Olive, R. P., 2012. Pourbaix Diagrams, Solubility Predictions and Corrosion-product Deposition Modelling for the Supercritical Water Reactor (Ph. D. thesis). University of New Bruns-wick, Department of Chemical Engineering.

Otsuka, N., Shida, Y., Fujikawa, H., 1989. Internal-external transition for the oxidation of Fe-Cr-Ni austenitic stainless steels in steam. Oxid. Met. 32, 13.

Payet, M., 2011. Corrosion en Eau Supercritique—Apportà la Compréhension des Mécanismes pour des Alliages Fe-Cr-Ni de Structure c. f. c. (Doctoral thesis) Ecole Doctorale Arts et Metiers ED 415, Paris.

Pearl, W. L., Brush, E. G., Gaul, G. G., Wozadlo, G. P., 1965. General corrosion of Incoloy 800 in simulated superheat reactor environment. Nucl. Technol. 1(3), 235-245.

Pearl, W. L., Brush, E. G., Gaul, G. G., Leistikow, S., 1967. General corrosion of Inconel 625 in simulated superheat reactor environment. Nucl. Appl. 3, 418-432.

Peng, Q., Gartner, E., Busby, J. T., Motta, A. T., Was, G. S., 2007. Corrosion behavior of model zirconium alloys in deaerated supercritical water at 500℃ Corrosion 63, 577-590.

Penttilä, S., Betova, I., Bojinov, M., Kinnunen, P., Toivonen, A., 2011. Estimation of kinetic parameters of the corrosion layer constituents on steels in supercritical water coolant conditions. Corros. Sci. 53, 4193-4203.



Penttila, S., Toivonen, A., 2013. Oxidation and SCC behaviour of austenitic and ODS steels in supercritical water. In: ISSCWR-6: The 6th International Symposium on Supercritical Water-cooled Reactors, March 03-07, 2013, Shenzhen, Guangdong, China. CGNPC, China. Paper ISSCWR6-13029.

Penttilä, S., Betova, I., Bojinov, M., Kinnunen, P., Toivonen, A., 2015a. Oxidation model for construction materials in supercritical water-estimation of kinetic and transport parameters. In: 7th Int. Symp. on Supercritical Water-cooled Reactors (ISSCWR-7), March 15-18, 2015, Helsinki, Finland. Paper #2064.

Penttilä, S., Betova, I., Bojinov, M., Kinnunen, P., Toivonen, A., 2015b. Oxidation model for construction materials in supercritical water-estimation of kinetic and transport param-eters. Corros. Sci. 100, 36-46.

Peraldi, R., Pint, B. A., 2004. Effect of Cr and Ni contents on the oxidation behavior of ferritic and austenitic model alloys in air with water vapor. Oxid. Met. 61(5/6), 463-483.

Potter, E. C., Mann, G. M. W., 1963. Mechanism of magnetite growth on low-carbon steel in steam and aqueous solutions up to 550 degrees C. In: 2nd Int. Congress on Metallic Corrosion, National Association of Corrosion Engineers, Houston.

Qiu, L., Guzonas, D. A., 2012. Prediction of metal oxide stability in supercritical water reactors—Pourbaix Versus Ellingham. In: The 3rd Canada-China Joint Workshop on Supercritical Water-cooled Reactors (CCSC-2012), Xian, China, April 25-28, 2012.

Ren, X., Sridharan, K., Allen, T. R., 2006. Corrosion of ferritic-martensitic steel HT9 in supercritical water. J. Nucl. Mater. 358(2-3), 227-234.

Ren, X., Sridharan, K., Allen, T. R., 2007. Corrosion behavior of alloys 625 and 718 in supercritical water. Corrosion 63, 603.

Ren, X., Sridharan, K., Allen, T. R., 2008. Effect of Shot Peening on the Oxidation of Ferritic-Martensitic Steels in Supercritical Water. Corrosion, Paper 08425.

Ren, X., Sridharan, K., Allen, T. R., 2010. Effect of grain refinement on corrosion of ferritice martensitic steels in supercritical water environment. Mater. Corros. 61, 748-755.

Robertson, J., 1991. The mechanism of high temperature aqueous corrosion of stainless steels. Corros. Sci. 32, 443-465.

Rodriguez, D., Merwin, A., Chidambaram, D., 2014. On the oxidation of stainless steel alloy 304 in subcritical and supercritical water. J. Nucl. Mater. 452, 440-445.

Rodriguez, D., Chidambaram, D., 2015. Oxidation of stainless steel 216 and Nitronic 50 in supercritical and ultrasupercritical water. Appl. Surf. Sci. 347, 10-16.

Ru, X., Staehle, R. W., 2013a. Historical experience providing bases for predicting corrosion and stress corrosion in emerging supercritical water nuclear technology: Part 1 —review. Corrosion 69(3), 211-229.

Ru, X., Staehle, R. W., 2013b. Historical experience providing bases for predicting corrosion and stress corrosion in emerging supercritical water nuclear technology: Part 2 —review. Corrosion 69(4), 319-334.

Ru, X., Staehle, R. W., 2013c. Historical experience providing bases for predicting corrosion and stress corrosion in emerging supercritical water nuclear technology: Part 3 —review. Corrosion 69(5), 423-447.

Ruther, W. E., Greenberg, S., 1964. Corrosion of steels and nickel alloys in superheated steam. J. Electrochem. Soc. 111, 1116-1121.

Ruther, W. E., Schlueter, R. R., Lee, R. H., Hart, R. K., 1966. Corrosion behavior of steels and nickel alloys in superheated steam. Corrosion 22(5), 147-155.

Saito, N., Tsuchiya, Y., Yamamoto, S., Akai, Y., Yotsuyanaji, T., Domae, M., Katsumura, Y., 2006. Chemical thermodynamic considerations on corrosion products in supercritical water-cooled reactor coolant. Nucl. Technol. 155, 105-112.

Sanchez, R. G., 2014. Corrosion Performance of Candidate Materials for Canadian Gen IV Supercritical Water Cooled Reactor(M. A. Sc. thesis). Dept. of Mechanical Engineering, Carleton University, Ottawa, Canada.

Sarver, J., 2009. The oxidation behavior of candidate materials for advanced energy systems in steam at temperatures between 650℃ and 800℃. In: 14th Int. Conf. on Environmental Degradation of Materials in Nuclear Power Systems, Virginia Beach, VA, August 23-27, 2009.

Sarver, J. M., Tanzosh, J. M., 2013. Effect of temperature, alloy composition and surface treatment on the steamside oxidation/oxide exfoliation behavior of candidate A-USC boiler materials. Babcock & Wilcox Technical Paper BR-1898. In: Proceedings of the Seventh International Conference on Advances in Materials Technology for Fossil Power Plants, Waikoloa, Hawaii, USA, October 22-25, 2013.

Sawochka, S., Leonard, M., 2006. Source Term Reduction: Impact of Plant Design and Chemistry on PWR Shutdown Releases and Dose Rates. Electric Power Research Institute Report 1013507.

Sarrade, S., Féron, D., Rouillard, F., Perrina, S., Robin, R., Ruizc, J. C., Turc, H. -A., 2017. Overview on corrosion in supercriticalfluids. J. Supercrit. Fluids 120, 335-344.

Saunders, S. M., Monteiro, M., Rizzo, F., 2008. The oxidation behavior of metals and alloys at high temperatures in atmospheres containing water vapour: a review. Prog. Mater. Sci. 53, 775-837.

Schulenberg, T., 2013. Material requirements of the high performance light water reactor. J. Supercrit. Fluids 77, 127-133.

Sloppy, J. D., Lu, Z., Dickey, E. C., Macdonald, D. D., 2013. Growth mechanism of anodic tantalum pentoxide formed in phosphoric acid. Electrochim. Acta 87(1), 82-91.

Spalaris, C. N., 1963. Finding a corrosion resistant cladding for superheater fuels. Nucleonics 21.

Steeves, G., Cook, W., Guzonas, D., 2015. Development of kinetic models for the long-term corrosion behaviour of candidate alloys for the Canadian SCWR. In: 7th International

Symposium on Supercritical Water-cooled Reactors (ISSCWR-7), March 15-18, 2015, Helsinki, Finland. Paper-2076.

Sun, M., Wu, X., Zhang, Z., Han, E. -H., 2009a. Oxidation of 316 stainless steel in supercritical water. Corros. Sci. 51(5), 1069-1072.

Sun, M., Wu, X., Han, E. -H., Rao, J., 2009b. Microstructural characteristics of oxide scales grown on stainless steel exposed to supercritical water. Scr. Mater. 61(10), 996-999.

Sun, C., Hui, R., Qu, W., Yick, S., 2009c. Progress in corrosion resistant materials for super-critical water reactors. Corros. Sci. 51, 2508-2523.

Svishchev, I. M., Carvajal-Ortiz, R. A., Choudhry, K. I., Guzonas, D. A., 2013. Corrosion behavior of stainless steel 316 in sub- and supercritical aqueous environments: effect of LiOH additions. Corros. Sci. 72, 20-25.

Tan, L., Allen, T. R., 2005. An EBSD study of grain boundary engineered Incoloy alloy 800H. Metall. Mater. Trans. A 36(7), 1921-1925.

Tan, L., Allen, T. R., 2009. Localized corrosion of magnetite on ferritic-martensitic steels exposed to supercritical water. Corros. Sci. 51, 2503-2507.

Tan, L., Sridharan, K., Allen, T. R., 2006a. The Effect of grain boundary engineering on the oxidation behavior of Incoloy alloy 800H in supercritical water. J. Nucl. Mater. 348, 263.



Tan, L., Yang, Y., Allen, T. R., 2006b. Oxidation behavior of alloy HCM12A exposed in supercritical water. Corros. Sci. 48, 3123.

Tan, L., Machut, M. T., Sridharan, K., Allen, T. R., 2007. Oxidation behavior of HCM12A exposed in harsh environments. J. Nucl. Mater. 371, 171-175.

Tan, L., Ren, X., Sridharan, K., Allen, T. R., 2008a. Effect of shot-peening on the oxidation of alloy 800H exposed to supercritical water and cyclic oxidation. Corros. Sci. 50, 2040-2046.

Tan, L., Ren, X., Sridharan, K., Allen, T. R., 2008b. Corrosion behavior of Ni-base alloys for advanced high temperature water-cooled nuclear plants. Corros. Sci. 50, 3056.

Tan, L., Ren, X., Allen, T. R., 2010. Corrosion behavior of 9-12% Cr Ferritic-martensitic steels in supercritical water. Corros. Sci. 52, 1520-1528.

Tomlinson, L., Cory, N. J., 1989. Hydrogen emission during the steamoxidation of ferritic steels: kinetics and mechanism. Corros. Sci. 29(8), 939-965.

Tsuchiya, Y., Saito, N., Kano, F., Ookawa, M., Kaneda, J., Hara, N., 2007. Corrosion and SCC Properties of Fine Grain Stainless Steel in Subcritical and Supercritical Pure Water. NACE 2007. NACE, Houston.

Van Nieuwenhove, R., Balak, J., Toivonen, A., Penttilä, S., Ehrnsten, U., 2013. Investigation of coatings, applied by PVD, for the corrosion protection of materials in supercritical water. In: 6th International Symposium on Supercritical Water-cooled Reactors (ISSCWR-6), March 03-07, 2013, Shenzhen, Guangdong, China. Paper ISSCWR6-13024.

Wills, J. S. C., Chiu, A., Guzonas, D., 2007. Deposition of anti-corrosion coatings by atmospheric pressure plasma jet. In:28th Annual Conference of the Canadian Nuclear Society, June 3-6, 2007, St. John, New Brunswick.

Was, G. S., Teysseyre, S., McKinley, J., 2004. Corrosion and Stress Corrosion Cracking of Iron- and Nickel-base Austenitic Alloys in Supercritical Water. NACE Corrosion 2004, Paper 04492.

Was, G. S., Allen, T. R., 2005. Time, temperature, and dissolved oxygen dependence of oxidation of austenitic and ferritic-martensitic alloys in supercritical water. In: Proc. International Congress on Advances in Nuclear Power Plants(ICAPP'05), Seoul, Korea, Paper 5690, May 15-19, 2005. Paper 5690.

Was, G. S., Ampornrat, P., Gupta, G., Teysseyre, S., West, E. A., Allen, T. R., Sridharan, K., Tan, L., Chen, Y., Ren, X., Pister, C., 2007. Corrosion and stress corrosion cracking in supercritical water. J. Nucl. Mater. 371, 176-201.

Was, G. S., Teysseyre, S., Jiao, Z., 2006. Corrosion of austenitic alloys in supercritical water. Corrosion 62, 989-1005.

Was, G. S., Teysseyre, S., 2005. Challenges and recent progress in corrosion and stress corrosion of austenitic alloys for supercritical water reactor core components. In: Proceedings of the 12th International Conference on Environmental Degradation of Materials in Nuclear Power Systems - water Reactors. The Minerals, Materials and Metals Society, Warrendale, PA, pp. 1343-1357.

Watanabe, Y., Kobayashi, T., Adschiri, T., 2001. Significance of Water Density on Corrosion Behavior of Alloys in Supercritical Water. NACE Corrosion 2001, Paper 01369.

Watanabe, Y., Shoji, K., Adschiri, T., Sue, K., 2002. Effects of Oxygen Concentration on Corrosion Behavior of Alloys in Acidic Supercritical Water. NACE Corrosion 2002, Paper 02355.

Watanabe, Y., Daigo, Y., 2005. Corrosion rate and oxide scale characteristics of austenitic alloys in supercritical water. Materials Science Forum, v 522 - 523, High - Temperature Oxidation and Corrosion 2005. In: Proceedings of the International Symposium on High-Temperature Oxidation and Corrosion 2005, 2006, pp. 213-220.

Wright, I. G., Tortorelli, P. F., 2007. Program on Technology Innovation: Oxide Growth and Exfoliation on Alloys Exposed to Steam. Electric Power Research Institute Report 1013666.

Wright, I. G., Dooley, R. B., 2010. A review of the oxidation behavior of structural alloys in steam. Int. Mater. Rev. 55, 129-167.

Xu, R., Yetisir, M., Hamilton, H., 2014. Thermal-mechanical behavior of fuel element in SCWR design. In: 2014 Canada-China Conference on Advanced Reactor Development, Niagara Falls, Ontario Canada, April 27-30, 2014. Paper CCCARD2014-005.

Yi, Y., Lee, B., Kim, S., Jang, J., 2006. Corrosion and corrosion fatigue behaviors of 9Cr steel in a supercritical water condition. Mater. Sci. Eng. A Struct. Mater. Prop. Microstruct. Pro-cess. 429, 161-168.

Yi, Y.-S., Watanabe, Y., Kondo, T., Kimura, H., Saito, M., 2001. Oxidation rate of advanced heat-resistant steels for ultra-supercritical boilers in pressurized superheated steam. J. Press. Vessel Technol. 123, 391-397.

Yamamura, T., Okuyama, N., Shiokawa, Y., Oku, M., Tomiyasu, H., Sugiyama, W., 2005. Chemical states in oxide films on stainless steel treated in supercritical water: factor analysis of X-ray photoelectron spectra. J. Electrochem. Soc. 152(12), B540-B546.

Yuan, J., Wu, X., Wang, W., Zhu, S., Wang, F., 2013. The effect of surfacefinish on the scaling behavior of stainless steel in steam and supercritical water. Oxid. Met. 79,541-551.

Zemniak, S. E. , Hanson, M. , Sunder, P. C. , 2008. Electropolishing effects on corrosion behavior of 304 stainless steel in high temperature, hydrogenated water. Corros. Sci. 50,2465-2477.

Zhang, L. , Zhu, F. , Tang, R. , 2009a. Corrosion mechanisms of candidate structural materials for supercritical water-cooled reactor. Front. Energy Power Eng. China 3(2),233-240.

Zhang, Q. , Tang, R. , Li, C. , Luo, X. , Long, C. , Yin, K. , 2009b. Corrosion behavior of Ni-base alloys in supercritical water. Nucl. Eng. Technol. 41,107-112.

Zhang, L. , Bao, Y. , Tang, R. , 2012. Selection and corrosion evaluation tests of candidate SCWR fuel cladding materials. Nucl. Eng. Des. 249,180-187.

Zhang, N. Q. , Zhu, Z. -L. , Xu, H. , Mao, X. -P. , Li, J. , 2016. Oxidation of ferritic and ferritic-martensitic steels in flowing and static supercritical water. Corros. Sci. 103,124-131.

Zheng, W. , Guzonas, D. , Li, J. , 2008. Novel approach to the development of in-core materials for a supercritical water cooled reactor. In: 16th Pacific Basin Nuclear Conference (16PBNC), Aomori, Japan, October 13-18, 2008. Paper ID P16P1413.

Zhou, Z. Y. , Lvov, S. N. , Wei, X. J. , Benning, L. G. , Macdonald, D. D. , 2002. Quantitative evaluation of general corrosion of Type 304 stainless steel in subcritical and supercritical aqueous solutions via electrochemical noise analysis. Corros. Sci. 44,841-860.

Zhu, Z. , Xu, H. , Jiang, D. , Mao, X. , Zhang, N. , 2016. Influence of temperature on the oxidation behavior of a ferritic-martensitic steel in supercritical water. Corros. Sci. 113,172-179.

Zhu, Z. , Xu, H. , Jiang, D. , Yue, G. , Li, B. , Zhang, N. , 2016. The role of dissolved oxygen in supercritical water in the oxidation of ferritic-martensitic steel. J. Supercrit. Fluids 108,56-60.



Zurek, J. , Young, D. J. , Essuman, E. , Hänsel, M. , Penkalla, H. J. , Niewolak, L. , Quadakkers, W. J. , 2008. Growth and adherence of chromia-based surfaces scales on Ni-base alloys in high and low $pO_2$ gases. Mater. Sci. Eng. A 477,259-270.

# 第6章 环境敏感开裂

## 6.1 引　言

环境辅助开裂(EAC)是一种由机械、化学和冶金相关因素的协同相互作用引起的复杂现象,如图6.1所示。EAC包含诸如应力腐蚀开裂(SCC),应变腐蚀开裂(SICC)以及腐蚀疲劳(CF)的环境敏感破坏机理。这些劣化模式可以根据应力的特性进行区分,典型的例子包括由常应力引起的SCC、单调递增应变引起的SICC以及循环应力引起的CF。不同的诱因间复杂的相互作用使得对EAC进行试验测量非常困难。因此,我们对超临界水冷反应堆(SCWR)条件下EAC知识的了解程度不如普通腐蚀条件下的深入。由于EAC的发生需要这3种诱因同时出现,在某些情况下可以通过仔细优化其余两个因素以补偿另一个欠佳的因素,如对水化学、设计以及残余应力的优化以补偿材料的不足。

图6.1在某种程度上过于简化了其中的影响因素,如Andresen和Ford(2011)中强调的,诸如冷加工或者辐照单独起到的影响作用可以比其中某一个影响因素更大(对比图1.3)。

EAC的危险在于它的随机性以及早期阶段不容易通过常规的无损检测技术检测到,因为裂纹的扩展过程中很少出现或者不存在宏观的塑性变形。一个受到EAC影响的合金通常不表现出不正常的力学性能(弯曲强度及拉伸强度)。这对于承压部件来说尤为重要,因为一个突发的失效可能引起灾难性的后果。人们同样不希望发生燃料包壳的破裂,因为放射性物质会释放到冷却剂中(见第4章)。但是燃料元件包壳破孔可以通过在线监测、监视取样以及测量冷却剂裂变气体活度预先监测到,如果需要移除有缺陷的燃料棒可以及时停堆。

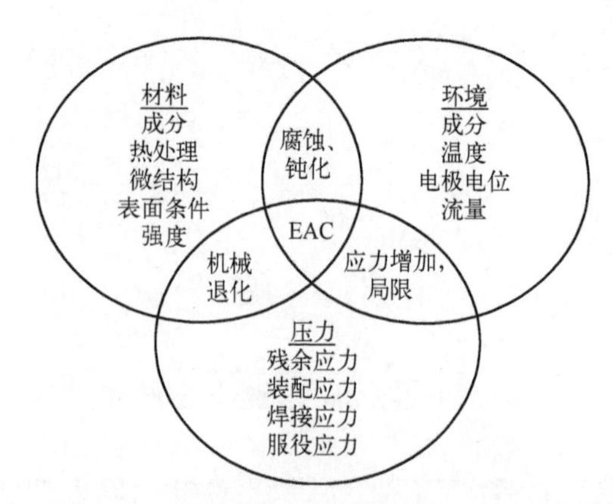


图 6.1　用于阐释拉应力条件、敏感材料条件及腐蚀性环境条件 3 种因素同时发生可引起 EAC 的维恩图

(引自文献 After Speidel, M. O., 1984. Stress corrosion and corrosion fatigue mechanics. In: Speidel, M. O., Atrens, A. (Eds.), Corrosion in Power Generating Equipment. Plenum Press, New York)

EAC 引起的裂纹通常垂直于主拉伸应力方向延伸。最终,对于同一种合金晶间(焊接金属中的枝晶间)以及穿晶裂纹的发展都可以根据环境、微结构或者应变/应力状态而被观察到,EAC 的模型可以加入其他模型或者从一种模型转化为另一种模型。对于某个特定的合金/环境系统,一种特定合金的裂纹敏感性程度可能会因为相对微小的材料、应力以及环境条件的变化而发生改变。



20 世纪 70 年代中叶,伴随着一系列晶间应力腐蚀破裂(IGSCC)事件的发生,尤其是沸水反应堆(BWR)中敏化及冷加工不锈钢的 IGSCC 事件以及压水反应堆(PWR)中蒸汽发生器传热管破裂事件的发生,EAC 现象在核能工业中开始引起关注(Combrade 等,2010;Staehle 和 Gorman,2003)。这些事件绝大多数都归咎于对结构材料选取的不合适、残余应力的存在以及不合适的水处理。特别是随着现役核电站(NPP)平均寿命的增加,尤其对于水冷反应堆(WCR),EAC 仍然是一个重要的问题。因此,EAC 在 SCWR 发展的初期就被人们意识到是一个影响 SCWR 中堆芯以及堆芯下游构件寿命的关键退化机理①。

不锈钢以及 Ni 基合金已经在 WCR 中被广泛使用,并且不锈钢燃料元件包壳也在 LWR 中被广泛使用。La Crosse BWR 采用 348 SS 燃料元件包壳已经运行了 20 年;在寿命期内,堆芯中包含超过 50000 根燃料棒(Strasser 等,1982)。在第一个工作周期(5 轮换料为一个周期)中,大约有 1.3% 的燃料棒被破坏,基本所有高于堆芯平均高度的燃料都已用完。通过改变对最大燃耗的限制以及增

① 由于各种堆内构件具有不同的使用寿命,因此环境辅助开裂原因的相对重要性在构件之间可能不相同,如燃料包壳与堆内构件。

加在反应堆操作上的限制(即通过控制棒的移动),可以减少在最后一个周期内燃料棒的破坏率。芯块和包壳相互作用产生的内应力以及冷却剂辐照分解产生的高浓度氧化物引起的外包壳腐蚀共同造成了这种破坏机制。347NG SS 已经在德国 BWR 的堆芯内部被广泛使用,并且没有发现破损,除非材料因为疏忽而敏化或者存在表面冷处理。Scott(2013)提出应变硬化不锈钢(SS)型,特别是316 SS 以及 Ni 较为稳定的 347 SS 已经在 PWR 中成功使用了几十年,以应用于需要强度适中的螺栓连接的诸多方面,并且没有出现明显的问题,除非受到了强的辐照。俄罗斯合金 1Kh18Ni10T 在 Beloyarsk 的过热器中在 8MPa、550℃条件下可以成功运行(Emel'yanov 等,1972)。不锈钢也在超临界化石燃料电厂(SCFPP)的过热器区域中被广泛使用。德国直流式锅炉使用 Nb 稳定钢(DIN 8CrNiNb163、X8CrNiMoNb1616 或 X8CrNiMoVNb1613)按照 $10^5$h 为一个周期的顺序已经运行了一段时期,并且俄罗斯 SCFPP 有着使用 Ti 稳定 12Cr18Ni12Ti 钢的重要经历。对于汽轮机(叶片、圆盘、转子),由于给水中杂质的运输以及在汽轮机上的沉积,EAC 已经成为一个重要的问题(对比图 4.15)(Zhou 和 Turnbull,2002)。EAC 的发生与蒸汽中氧气的存在有着紧密的关系。目前人们已经发现 Na、Cl 以及硫酸根离子会增加汽轮机部件发生 EAC 的风险。

Allen 等(2012)、Ru 和 Staehle(2013a ~ c)、Was 等(2007)以及 Zheng 等(2011、2014)搜集了 SCC 数据以支持 SCWR 的发展,读者可以对其进行咨询以获取更多信息。例行研讨会“核能系统中水冷反应堆材料的环境退化问题的国际研讨会”论文集中,关于当前对 WCR 的 EAC 现象研究与发展(包括 SCWR)进行了概要说明,同时总计了前期研究中得到的结论。

## 6.2 关键因素的影响

在许多合金/环境系统中,EAC 的发展进程按时间顺序由以下几步组成(Stachle,2007)。

(1) 工程裂纹起裂,为初始阶段(特定应力、冶金以及环境条件可能在金属/溶液界面发展、裂纹起裂、聚合以及较短的裂纹增长)。

(2) 单独的主裂纹的增长,裂纹增长速率(CGR)很少保持不变,并根据电厂工作条件而增加或者减小。

下面基于支持 SCWR 发展的研究结果以及其他 WCR 的操作经验讨论几个关键因素(图 6.1)对 SCW 中 EAC 的影响。重点将放在与 SCWR 发展有关的方面。

### 6.2.1 环境因素

如第 4 章所述,SCWR 堆内水化学预计与工作过程中不添加 $H_2$(常规水化学(NWC))或者添加 $H_2$(氢水化学(HWC))的 BWR① 的相同。沸水堆中,对于奥氏体不锈钢以及 Ni 基合金,控制 EAC 发展率的主要环境因素是电化学腐蚀电位(ECP),其主要受到氧化还原物质的浓度以及阴离子杂质浓度的影响,并反应在导电性上。对高温次临界水中奥氏体不锈钢以及 Ni 基合金的 EAC 现象的初始发展以及破坏增长的研究已经持续了很多年(Andresen 和 Ford,2011;Raja 和 Shoji,2011),尽管一些明显的问题仍然存在,但是基础性以及实践性的知识基础已经建立起来。这个知识基础已经让人们熟悉并且引领着与 EAC 相关的 SCWR 技术的研究与发展,这将在下面进行论述。

#### 6.2.1.1 温度

EAC 的敏感性会由于热效应而产生(Andresen 和 Ford,2011),但是在 BWR 中温度通常只是一个很小的影响因素,除了反应堆启动及停堆阶段。在电厂中,高于 250℃ 的环境中的部件大多会产生破裂(Kilian 等,2005)。然而,SCWR 堆芯中温度变化范围很大,在接近临界区域,SCW 的物性也会产生大的变化,SCWR 中温度预计会对 EAC 现象产生更大的影响。相比于常规腐蚀,材料的EAC 敏感性必须在次临界水(280℃ $<T<T_c$)、近临界水、低温 SCW(374℃ $<T<$ 450℃)以及高温 SCW($T\gg T_c$,达到包壳可承最高温度)条件下进行评估。



亚临界水中不锈钢的 EAC 敏感性/与温度的关系性很复杂,不一定遵循 Arrhenius 型关系,因为在许多情况下,在整个温度范围内存在多个速率控制过程(图 6.2)。鉴于前几章对各种化学进展速率(腐蚀产物沉积和水的辐射分解(第 4 章)和通常性腐蚀(第 5 章))与温度间复杂关系的讨论,有理由认为 SCWR 中 EAC 问题同样与温度间有复杂的关系。Allen 等(2012)对 400~650℃ 温度范围内奥氏体不锈钢的恒定拉伸速率(CERT)试验所得数据进行汇总发现,随着温度的上升,最大应力总体呈下降趋势,但是失效应变中没有显示清晰的变化趋势。

1) 近临界区-超临界水密度的影响

在近临界区,温度的影响与物质的其他理化性质快速变化混合在一起。关于 SCW 中 EAC 现象被广泛研究,这些研究对于支持超临界水氧化(SCWO)作用的发展起到重要支撑作用。但是这些研究中大多使用高浓度侵蚀性阴离

---

① 然而,在超临界水冷反应堆中,氢气不会被剥离到蒸汽相中。

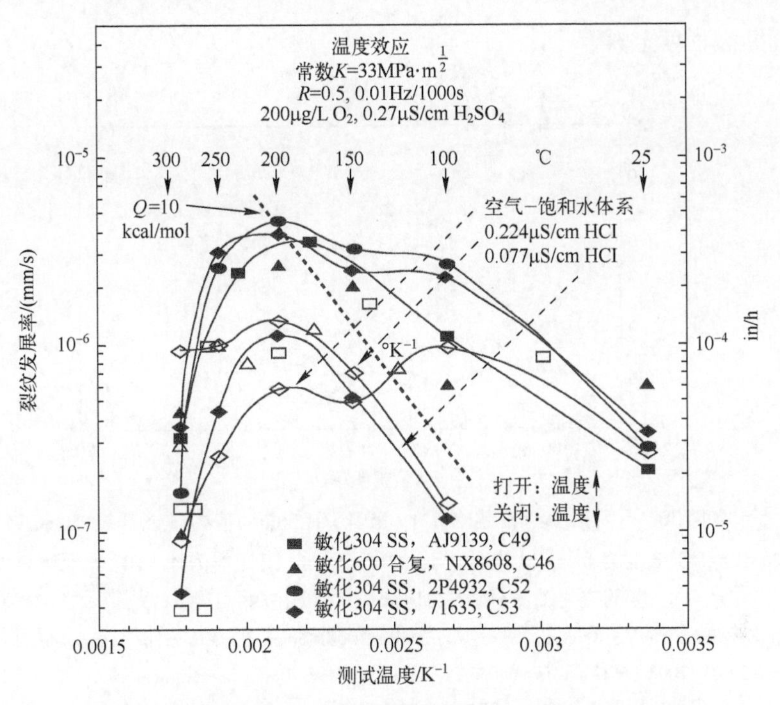


图 6.2　裂纹发展率与常应力强度因子条件下包含多种杂质的水中敏化不锈钢的测试温度的函数关系（Andresen,1992）



子（氯化物、硫酸盐），浓度远超过了 SCWR 中预期的浓度范围，与不合常规的最坏情况接近。Tsuchiya 等（2003）对敏化 304 SS 和 316 SS 拉伸试样进行了 CERT 测试，试样在 290～550℃、25MPa 条件下的拉伸速率为 $4\times10^{-7}s^{-1}$，整个试验在 DO 浓度为 8mg/L 的高纯水中进行。图 6.3 显示了晶间应力腐蚀开裂（IGSCC）率与 SCW 密度的函数关系，结果表明，SCW 密度和 SCC 的发生具有密切的联系。如果在沸水堆中，SCC 机制是由阳极溶解驱动，那么这种相关性是可以预期的。在高温，低 SCW 密度条件下，离子的离解程度是很低的（对比图 4.12），金属氧化物的溶解度也很低（图 4.10），同时溶液的侵蚀性、溶液的导电性也会降低，并且减少金属的溶解。在 400℃以上的温度条件下没有观察到 IGSCC 的现象（图 6.3），但是两种合金和 SCW 接触后，侧面都出现了小的裂纹。IGSCC 比值的变化与 SCW 密度和金属氧化物溶解度的变化趋势密切相关。316L SS 未观察到出现应力腐蚀开裂。

Watanabe 等（2003）公布了一个与 SCW 密度相似的联系，他测量了恒定温度 400℃条件下，DO 浓度为 8mg/kg 的纯水中敏化 316 SS 的破裂受 SCW 压力的影响。随着压力的上升（密度和介电常数增加），最大应力及应变破坏降低，同时晶间断裂程度上升。

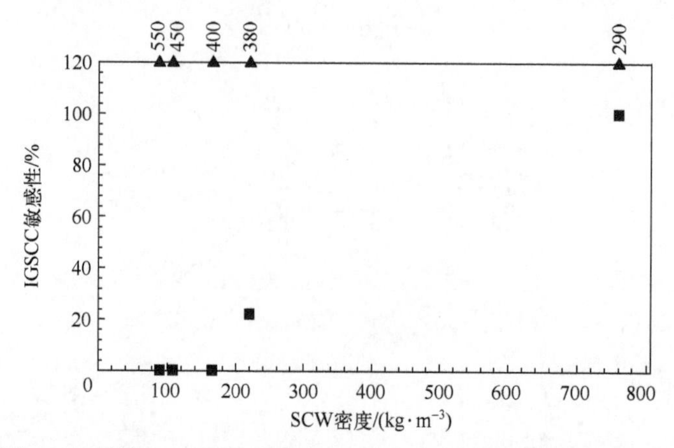


图 6.3　敏化 304 不锈钢(SS)在超临界水(SCW)中的晶间应力腐蚀开裂(IGSCC)敏感性(25MPa、DO 浓度为 8mg/kg、电导率为 0.06μS/cm、应变率为 $4\times10^{-7}s^{-1}$)随 SCW 密度的变化关系(在图的顶部显示了相应的温度(℃))

(引自文献 Tsuchiya, Y. , Kano, F. , Saito, N. , Shioiri, A. , Kasahara, S. , Moriya, K. , Takahashi, H. , September 15—19, 2003. SCC and Irradiation Properties of Metals Under Supercritical-Water Cooled Power Reactor Conditions. GENES4/ANP2003, Kyoto, Japan. Paper 1096)



图 6.4 显示了应变破坏与水的介电常数之间的关系，包括在 360℃、25MPa 条件下获得的数据，其中水的介电常数基本上与 400℃、60MPa 时相同。图 6.4 显示在这些条件下介电常数的变化(以及离子缔合、氧化物溶解度等方面)可以解释大部分观察到的与压力之间的联系。

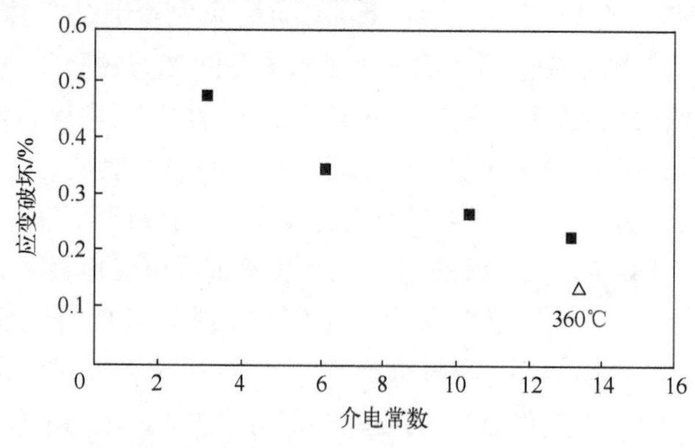


图 6.4　含 8mg/kg $O_2$ 的 400℃超临界水中应变为 $2.8\times10^{-6}s^{-1}$ 的 316 不锈钢(SS)的介电常数与应变破坏间的关系(图中空心三角形表示在 360℃、25MPa 条件下得到的试验数据)

(引自文献 Watanabe, Y. , Abe, H. , Daigo, Y. , 2003. Environmentally-assisted cracking of sensitized stainless steel in supercritical water. In: Proceedings of GENES4/NP2003, Kyoto, Japan, September 15-19, 2003, Paper 1183)

他们将其观察结果归因于离子稳定性和氧化物溶解度的提高以及随着压力变化而增加的水电导率,其增强了阳极溶解。作者提出固溶退火的 316L SS 在 400℃,即使在 60MPa 条件下也没有出现应力腐蚀开裂(SCC)。

Peng 等(2007)针对 20%冷加工的 316L SS 在 SCW(25MPa,DO=2000μg/kg)中,在应力强度因子为 27.6MPa · $m^{1/2}$ 条件下,开展了裂纹增长速率(CGR)试验。他们监测了 288~500℃条件下的 CGR(从次临界到超临界状态),以确定近临界区 SCW 中 CGR 与温度和 DO 之间的关系(图 6.5)。在 $T<T_c$ 条件下,CGR 随温度的升高而增加,其中活化能为 26kJ/mol,在 25~288℃水温下的不锈钢数据的范围内。得出的结论是,次临界水中裂纹的增长特性与现有数据库中所预期的相同。在 $T>T_c$ 条件下,CGR 随温度升高而降低,并反映水密度的降低。他们将 CGR 的降低归因于裂纹钝化,而非氧化物溶解度的变化,这是由于氧化速率随着温度升高而增加,并伴随着裂纹中离子向外输运的减少。Watanabe 等(2004a、2004b)提出在 0.01mol/kg $H_2SO_4$+800mg/kg $O_2$ 中,温度为 400℃时的开裂严重性高于 360℃时的开裂严重性。这表明溶解的机理是不成立的,在这种强氧化条件下,晶界氧化的增加非常重要,并且不同的破裂机制在纯水(阳极溶解)和强氧化条件下(氧化开裂)均有效。这个结论与 Watanabe 等(2003)在恒定温度下(消除影响氧化速率的热激活变化)试验测量的结果相符。



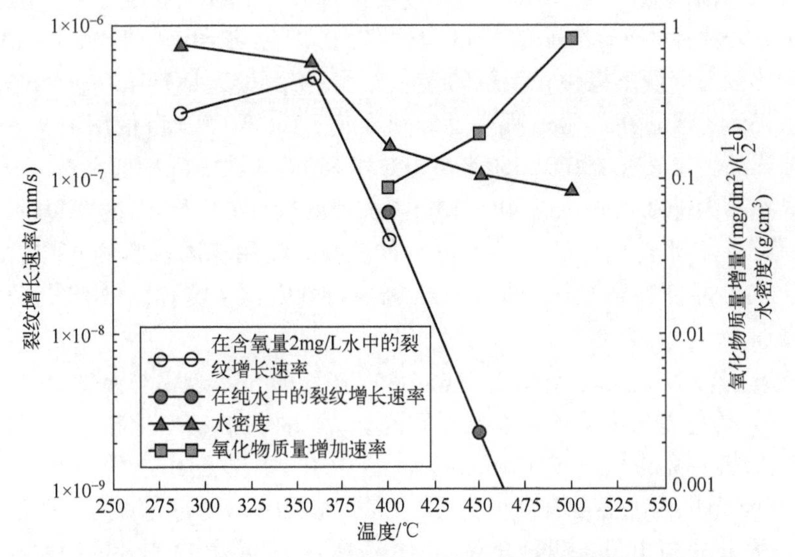


图 6.5 非敏化 316L SS 0.5T 压缩拉伸试样在纯水中从亚临界至超临界变化过程中的裂纹增长速率(CGR)与温度的关系(在超过 500h 的试验过程中,水密度和氧化物质量增加速率也显示出来)(Peng 等,2007)。

Was 等(2007)论述了温度对所选奥氏体不锈钢在 SCW 中的 SCC 敏感性的影响。在 SCW 密度较高的近临界区,介电常数越高,氧化物溶解度越大,IGSCC 敏感性就会越高。在断裂表面观察到的 SCC 百分比以及沿拉伸试样的二次断裂的密度和深度,都被认为是评价 SCC 的指标。

2) 高温特性

探究 EAC 与温度的关系以支持 SCWR 发展的测量工作大部分在 $T>T_c$ 条件下开展。Teysseyre 和 Was(2006)观察发现,对于 304 SS、316L SS、625 合金和 690 合金,二次裂纹密度随着温度的升高而降低,而在最高试验温度(550℃)下观察到的最大裂纹长度最大。在这些慢应变速率试验(SSRT)中,计算出的 CGR 随温度的升高而增加。他们建议使用一个称为裂纹严重度的参数,而不采用断裂截面上的 SCC/延性断裂比,该参数包括沿标距截面发现的裂纹的密度和长度。他们的研究报告中显示了断裂表面上以及二次断裂中出现的穿晶应力腐蚀开裂(TGSCC)迹象。然而,他们也观察到在惰性氩气体环境中 316L SS 发生应变后出现二次裂纹的 IGSCC 迹象。Janik 等(2013)以及 Penttilä 和 Toivonen(2013)评估了 08Cr18Ni10Ti、347H SS 和 316L SS 在 500℃ 和 550℃、25MPa 和 DO = 150μg/kg 条件下的抗 SCC 性。在这两种温度下,只有 316L SS 没有显示 IGSCC 的迹象,即使其暴露在 DO 浓度较高的 SCW 中。另外,在随后的试验系列(Toivonen 等,2015)中,在另一个 316L SS 的加热过程中,一个尺寸约为 8×0.5mm 的管显示出非常高的 IGSCC 敏感性。双环电化学电动势再活化(DLEPR)试验中比较与溶液不接触和接触的样品显示,在 550℃、DO = 150μg/kg 的 SCW 条件下,SSRT 样品在试验过程中迅速热敏化。Ru 等(2015)使用310S SS 的 U 形弯曲样品测定其在 290℃、380℃、550℃ 和 650℃ 的水中(pH 值在 6.5~7.5,入口电导率不大于 0.08μS/cm,DO<10μg/kg,25MPa,280℃除外)的开裂时间。接触 2000h 后,U 形弯曲样品未显示 SCC 的迹象。使用相同材料的 SSRT 试验表明,从 550℃ 的韧窝到 650℃ 的 IGSCC,断裂特性发生了变化。650℃时,样品的硬度增加。

Shen 等(2014)发现 AL6XN、HR3C 和 316Ti 的 SCC 现象与温度具有不同的关系,并认为此差异来源于破裂机理的差异。这种机制的变化与合金中的碳含量有关。Saez-Maderuelo 和 Gómez-Briceño(2016)提出了 316L 的 SCC 敏感性在 400~500℃ 范围内随温度升高而升高。Chen 等(2017)提出了在 500℃ 和 550℃ 条件下,使用直流电位差测量含有 Ar、$O_2$ 或 $H_2$ 的 SCW 中 310S SS 的 CGR。CGR 随着温度的升高而增加(图 6.6),表观活化能接近 170kJ/mol。对于 316L SS 在近临界温度范围内,其与温度间的关系与 Peng 等(2007)公开的数据结果相反。Chen 等(2017)表明,与 316L SS 相比,310S SS 较低的氧化速率将导致后一种合金的无裂纹钝化,从而增加 CGR。另一种解释是开裂机理的变化类似于一般腐蚀机理的变化。Chen 等(2017)提出了蠕变对整体裂纹扩展起主要作用的模型(见 6.3 节);他们还指出,温度升高会导致敏化作用,增加晶间腐蚀。

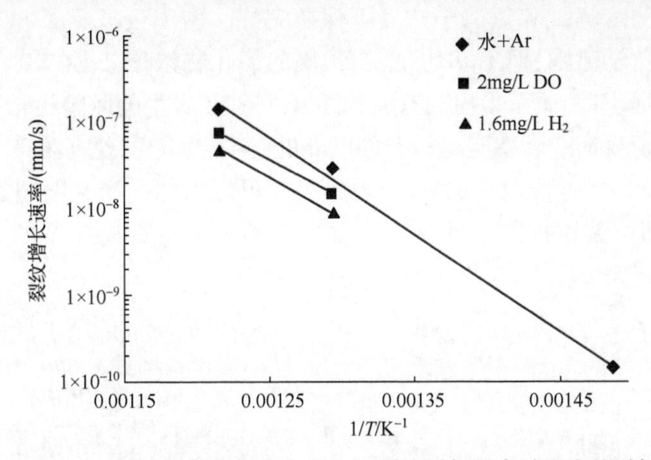


图 6.6 含 Ar、$O_2$ 或 $H_2$ 的超临界水中 310 SS 裂纹增长速率随温度的增加而递减

(引自文献 Chen, K., Du, D., Gao, W., Andresen, P. L., Zhang, L., 2017. Investigation of Temperature Effect on Stress Corrosion Crack Growth Behavior of 310S Stainless Steel in Supercritical Water Environment. ISSCWR8)



### 6.2.1.2 水化学

WCR 有严格的水化学监测和控制机制,以尽量减少杂质浓度和控制合金的 ECP(Hettiarachchi 和 Weber, 2010; Cowan 等, 2011、2012)。在沸水堆中,这一点尤其重要,因为水辐照分解不可能完全抑制氧化物质($O_2$ 和 $H_2O_2$)的净增加。压水堆中不锈钢燃料元件包壳相对于沸水堆的优越性能主要归因于压水堆中更为良性的水化学条件(Strasser 等, 1982),在这种条件下,更容易抑制氧化物中净辐射分解的形成。

虽然水电导率通常用于杂质浓度的测量,但 Andresen 和 Ford(2011)强调,裂纹尖端的阴离子活性(而非整体水电导率)至关重要,而且裂纹敏感性是阴离子特性①。强酸形成的阴离子(如 $H_2SO_4$ 和 HCl)可对 EAC 的起始和发展率产生非常重要的影响。一般来说,SCC 的发生需要未形成氧化层的合金短时间内暴露在冷却剂中(如由于拉伸应力导致的氧化膜破裂)。如果氧化物由于离子的存在而不能愈合,SCC 就可能形成和发展。因此,在不利于离子离解的 $T\gg T_c$ 条

① 尽管在 BWR 中普遍接受的 SCC 机制确实需要一种导电合理的导电解决方案(图 6.9),至少在裂纹附近的局部区域是需要的。

件下,相比于离子造成的影响,其他因素产生的影响可能更为重要。

图 6.7 显示出对于增敏或溶液退火材料,304 SS 在亚临界水中的应力腐蚀开裂与溶解氧和氯离子浓度的关系。与敏化材料相比,非敏化合金看起来能在更大范围内的溶解氧和氯离子浓度下保持完好。图中还显示了 SCWR 的预期正常工作水化学范围,但不幸的是,在预期的氯化物浓度或 SCWR 工作温度下没有相关数据,基于此图得出的有关 SCWR 中 SCC 发生的风险性结论是推测性的。然而在瞬变期或者特定运行条件以外的时期,可能出现高浓度 $Cl^-$ 的情况,由图 6.7 预测,在这段时期 EAC 出现的风险可能增加。当水温超过 93℃ 且杂质浓度较高时,裂纹扩展的重要阶段发生在沸水堆运行周期中的反应堆启动期间(Garcia 等,2012)。

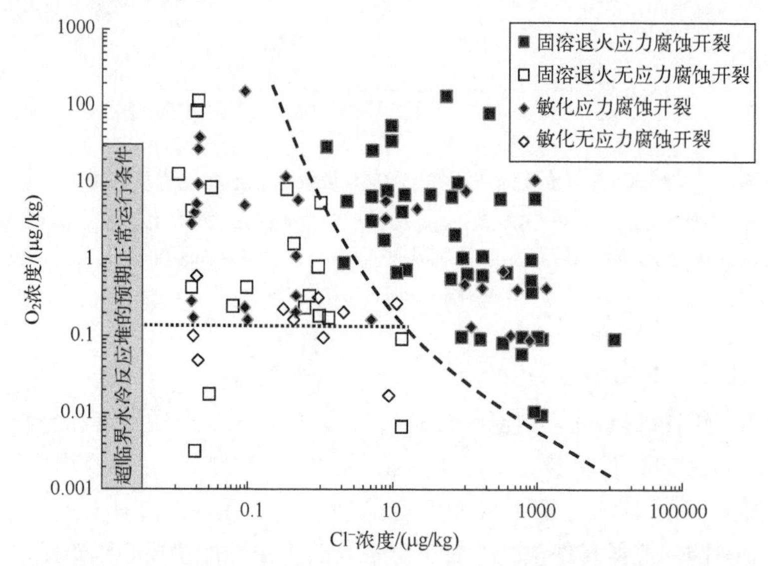


图 6.7 溶解氧和氯化物浓度对超临界水中 304 SS 的应力腐蚀开裂(SCC)的影响(图中虚线将非敏化 304 SS 的"$Cl^- - O_2$"空间划分为"无 SCC"区域(曲线右侧)和"SCC"区域(曲线左侧)。同样,水平点线将非敏化合金的"无 SCC"区域划分为"SCC"(上面)区域和"无 SCC"(下面)区域。阴影区域代表超临界水冷反应堆的预期正常运行条件(第 4 章);运行瞬变和特定工作条件以外的情况可能使水化学特性偏离该区域)

(引自文献 Gordon, G. M. , Brown, K. S. , 1989. Dependence of creviced BWR component IGSCC behavior on coolant chemistry. In: Cubicciotti, D. , Simonen, E. (Eds. ), Proc. 4th Int. Symp. on Environmental Degradation of Materials in Nuclear Power Systems—Water Reactors, Jekyll Island, August 6-10, 1989. National Association of Corrosion Engineers, pp. 14.46-14.62)

用于支持过热型核电站开发的试验(Griess 等,1962;Bevilacqua 和 Brown,1963)(第 4 章)表明,不锈钢材料会因燃料元件表面的氯化物沉积而开裂。在进行这些试验时,不可能产生氯离子浓度低于 ppb(μg/L)级别的水;由于蒸汽中存在 $O_2$,就不可避免地满足了奥氏体不锈钢材料开裂的必要条件,并且在换料、启动和停堆期间,燃料元件表面将被润湿。Beloyarsk 核电站中使用 1Kh18N10T 材料的过热蒸汽(SHS)通道元件中也得到了类似的结果。在含有氯化物的环境中进行了 144~1100h 的温度循环后,测试样品发生了开裂。有人提出,外表面的水分沉积和随后的蒸发可能导致表面高浓度氯化物的形成(Emel'yanov 等,1972)。如第 4 章所述,当 $P>P_c$ 时,NaCl 的溶解度非常高,目前 BWR 给水实践中最佳 $Cl^-$ 浓度限值(低至 0.25μg/kg)(Stellwag 等,2011)可以完全阻止堆芯内氯化物的沉积。Bull 等(2013)的报告中提出了先进气冷堆(约 500℃、15MPa)中用作过热器管的 316 SS SCC 的研究结果。当超过了所要求的过热裕度时,干涸区域会发生 SCC。研究发现,通过保持很低的杂质浓度和轻微的低碱性离子平衡(图 6.8)可以将 SCC 的风险最小化。在正常运行期间,SCWR 冷却剂为单相,但是在反应堆启动和停堆期间,冷却剂可能在两相条件下运行,阴离子可能在堆芯内沉积。在此期间,需要进行仔细的化学监测和控制,以将 EAC 风险降至最低。

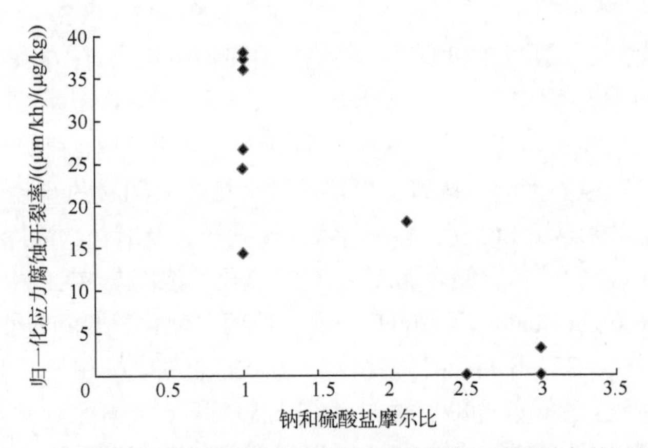


图 6.8 316 SS 的应力腐蚀开裂率与先进气冷反应堆过热区域条件(约 500℃、15MPa)下过热蒸汽的钠和硫酸盐摩尔比之间的关系

(引自文献 Bull, A., Lewis, G., Owen, J., Quirk, G., Rudge, A., Woolsey, I., 2013. Mitigating the risk of stress corrosion of austenitic stainless steels in advanced gas cooled reactor boilers. In: Proceedings of the 16th International Conference on the Properties of Water and Steam, September 1—5, 2013. University of Greenwich, London, UK)

更严重的问题可能是,由于沉积物中化学和物理参数(如温度、pH 值)的局部变化,燃料包壳氧化层中的氯化物富集,或由于化学性质不稳定(例如,冷凝器管故障)导致给水氯化物漂移。Staehle 和 Gorman(2003)强调了裂纹和压水

堆二次侧传热表面可发生 SCC 的重要性。其他杂质也值得关注,如多个发生在压水堆内的部件开裂以及重水堆蒸汽发生器传热管破裂处都发现了铅。SCW 中 $Fe(OH)_2$表面 $Cl^-$的分子动力学模拟(Kallikragas 等,2015)显示,由于表面的水密度较高,$Cl^-$优先出现在接近表面的位置(第 4 章)。对于小间隙(1~8nm)表面间 $Cl^-$和水的模拟表明,当间隙尺寸较小时,$Cl^-$扩散较快,这是由于水化膜较小;当间隙尺寸较大时,更大的水化膜阻碍了 $Cl^-$的运动。在亚临界温度下,却发生了与之相反的结果,这表明 $Cl^-$在超临界条件下可能更有害。

尽管阴离子作为 WCR 中 SCC 发生的一个诱导因素很重要,但在支持 SCWR 概念研究发展的试验中,通常不测量阴离子浓度,也没有对其影响进行系统研究的报告公开。Tan 和 Allen(2009)报告中提出,其试验循环中氯化物浓度为 300μg/kg。Gong 等(2015)提出,其回路中的 $Cl^-$浓度从 300℃时的56. 8μg/kg 下降到 450℃时的 0. 7μg/kg,作者认为这可能是溶解度下降超过临界点时氯化物沉积的结果。这些氯化物浓度明显高于工业实践最佳值(0. 25μg/kg),并且由图 6. 7 可见,其浓度之高已经足以引起人们的关注。

轻水反应堆(LWR)中 EAC 现象的第二个关键水化学因素是氧化物浓度,通常由 ECP 定量分析。现在已经确定的是在沸水反应堆(BWR)冷却剂运行条件下,奥氏体钢的 SCC 敏感性和裂纹扩展特性由腐蚀电位决定。Macdonald(2009)概述了亚临界水环境中 SCC 的电化学基础,主要现象是阳极(裂纹内)和阴极(合金外表面)的空间分离(图 6. 9)。腐蚀反应产生的 $H_2$和阴离子(如 $Cl^-$、$SO_4^{2-}$)进入裂纹,以补偿电子从裂纹尖端通过金属流向阴极的流动,可导致裂纹内的腐蚀环境的形成。ECP 由表面条件、冷却剂流量及氧化物浓度决定。



$O_2$和 $H_2O_2$造成了不同的表面氧化结构,氧化物浓度与 WCR 中 ECP 之间的关系可能较为复杂(Uchida 等,2009)。除了使用非敏化材料和最小化设计和留存应力外,通过将 ECP 降低到 230mV 以下(与标准氢电极相比)可将沸水堆中的 SCC 风险降至最低。这可以通过使用 HWC 降低水辐照分解产生的 $O_2$和 $H_2O_2$ 的浓度来实现,而贵金属的加入可以提高氢与合金表面 $O_2$和 $H_2O_2$的复合速率。Zhou 等(2004)还提出了介电(势垒)层的沉积可以抑制氧化物向金属表面的运输。

与通常腐蚀(5. 3. 3. 3 节)一样,不同作者公开的有关 DO 浓度对 SCW 开裂影响的试验数据并不总是一致的。Liu 和 Gong(2017)提出,在 620℃、25MPa 和 $7.5\times10^{-7}s^{-1}$的应变率下,310 SS(抛光至 1000 目)的最大裂纹长度随着 DO 浓度的增加而增加(图 6. 10)。Gong 等(2015)报告称,HR3C 的断裂时间随着 DO 浓度从 200μg/kg 增加到 500μg/kg 而减少,但在 DO 浓度为 1000μg/kg 时增加。在裂纹尖端观察到了 Cr 和 Mn 的损耗。他们认为,在低 DO 浓度下,氧会促进应

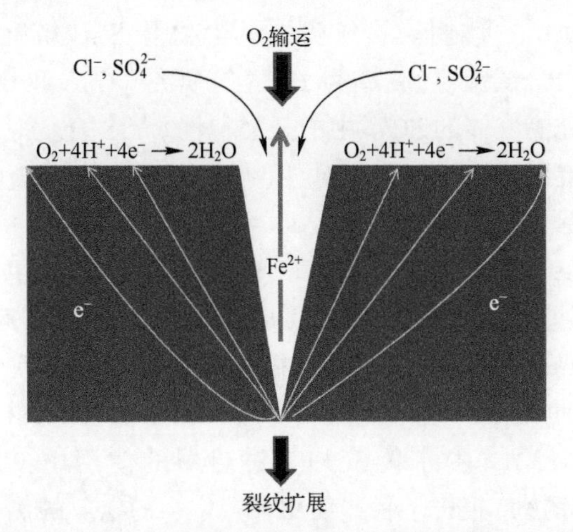


图 6.9　亚临界水溶液中应力腐蚀开裂涉及的电化学偶的示意图(说明了氧和阴离子在溶液中的作用)

(引自文献 Macdonald, D. D., 2009. The electrochemistry of stress corrosion cracking. In: Proceedings of the 14th International Conference on Environmental Degradation of Materials in Nuclear Power Systems Water Reactors. Virginia, USA, 734-744)

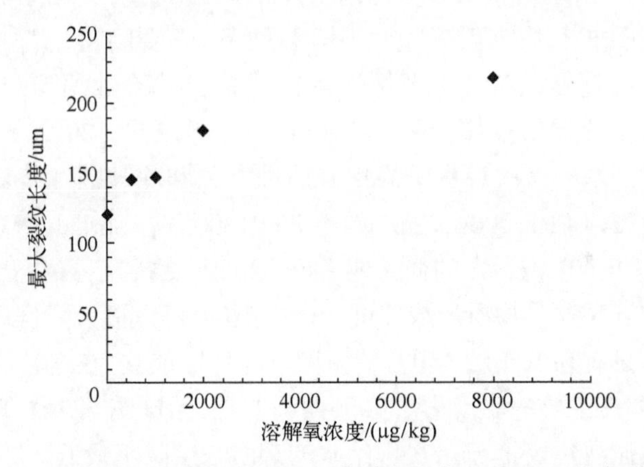


图 6.10　620℃、25MPa 和 $7.5\times10^{-7}s^{-1}$ 应变率下,310 SS 最大裂纹长度与 DO 浓度的关系(Liu 和 Gong,2017)

力腐蚀开裂,但在更高浓度下会钝化表面(裂纹钝化)。他们提出,碳氧化物的富集归因于试验结果中显示出的高浓度的有机碳,并且在试验温度中发现了合金与溶液接触后被敏感化的依据。Zhang 等(2015)提出,在 450℃时,将 DO 浓度从 0 增加到 2mg/kg 会增加 SCC 敏感性,而在 550℃时,将 DO 浓度增加到 2mg/kg 会降低 SCC 敏感性。Chen 等(2017)发现 DO 浓度对 SCW 中的 CGR 有重要影响;310S SS 在 Ar 脱气后的水中的 CGR 最高,在 DO 浓度为 2000μg/kg

的溶液中最低。DO 浓度越低，裂纹尖端氧化越慢，SCC 敏感性越高。Fournier 等(2001)通过 CERT 试验(应变率为 $10^{-6}s^{-1}$)研究了 718 和 690 合金在 400℃、25MPa 条件下的充满气体的 SCW 中的开裂特性。718 合金对 SCC 敏感，而 690 合金不敏感。作者还论述了氧脆化对 SCW 中 Ni 基合金开裂的影响，提出了温度在 $T_c$ 以上时开裂机理的变化。

Woolsey(1989)论述了德国和苏联 SCFPP 过热器使用的奥氏体不锈钢的开裂问题。这些工厂使用氧气浓度足够高的含氧给水运行，氧气可以到达锅炉。按 $10^5$h 的顺序运行周期内，德国直通式锅炉的奥氏体钢没有发生 SCC。这些工厂使用 Nb-稳定型钢(DIN8CrNiNb163、X8CrNiMoNb1616 或 X8CrNiMoVNb1 613)。在苏联 SCFPP 过热器的使用中(使用含 Ti 稳定的 12Cr18Ni12Ti 不锈钢)报告了一些使用含氧化学的电厂的例子，尽管其他电厂运行时间都较长，没有发现问题。Woolsey 认为，异常的运行条件(如无负荷运行)可能导致表面“湿润”(两相状态)，给水中杂质的存在可能是其原因(这些电厂运行中的给水的杂质浓度往往高于德国电厂)。Gruzdev 等(1970)建议将给水中氧气浓度限制在低于 150μg/kg 条件下，以防止氧气进入锅炉。Woolsey(1989)提出在启动和低负荷运行条件下使用脱氧化学剂运行。在 SCFPP 中，由于蒸汽输送的杂质沉积，涡轮叶片开裂是一个重要问题。



对于沸水反应堆(BWR)，水化学运行试验和试验数据形成了如 HWC 的 SCC 缓解策略和贵金属添加剂①的基础(Cowan 等，2011，2012)。EPRI 沸水堆水化学指南就是基于这些观察结果以及腐蚀电位和冷却剂纯度对 EAC 敏感性影响的理论预测(EPRI，2004；Fruzetti 和 Blok，2005)。Ehrlich 等(2004)分析了欧洲 SCWR-HPLWR 概念堆的候选材料可能的退化机制，并得出结论，通过适当的水化学控制对这些物质的浓度进行严格限制，可以极大降低风险(图 6.7)，尽管这一结论只有在低于约 450℃的温度下才可能成立，在这种温度下，有利于解离，阳极溶解是主要的驱动力。必须强调的是，无论图 6.7 中的限制是什么，研究人员仍然强烈建议将 SCWR 中给水的杂质浓度降至最低。

Ru 和 Staehle(2013c)将 Was 和 Teysseyre(2005)报告的工作与早期对轻水反应堆(LWR)环境中相同合金的研究(Coriou 等，1969)进行了比较，发现一些结果难以根据轻水堆相关经验进行解释。例如，304L SS 和 625 合金在脱气超临界水中显示出裂纹深度超过 300μm 的 IGSCC，并且随着温度从 400℃升高到 550℃，其对 SCC 的敏感性增加，这与 350℃下纯净亚临界水的结果相矛盾。在

---

① 当在沸水反应堆中使用氢水化学(HWC)时，堆芯几何形状的复杂性导致堆芯中 $H_2$ 的不均匀分布，并且某些区域中 $H_2$ 浓度不足，需要引入诸如添加贵金属之类的方法来提高效率。

不会引起 SCC 的化学条件下(DO 和阴离子浓度低)对超临界水中未敏化奥氏体不锈钢进行试验,仍然发现了开裂现象。然而,如第 5 章所述,涉及空间上分离的阳极和阴极位置的电化学过程不会发生在 $T \gg T_c$ 的条件下,发生的主要是局部的化学反应。这表明 $T \gg T_c$ 的条件下,影响 SCW 中 EAC 环境因素发生了变化,特别是氧和阴离子的作用。Scenini 等(2008)的研究中提出电抛光和机械抛光对于 400℃、20.7MPa 条件下氢化蒸汽中反 U 形样品,以及 600℃的 C 形合金环的 SCC 特性的影响具有相关性。加氢蒸汽可以加速材料的开裂,以研究原生水的 SCC 过程。他们报告提出,与机械抛光样品相比,电抛光样品显示出更大的 SCC 敏感性,这是由于短路扩散路径的存在,机械抛光样品可以在所有 $H_2$ 分压下形成外部氧化物(第 5 章)。在低于 Ni/NiO 平衡势($H_2$ 分压)的无应变表面上,没有形成外部氧合物。作者提出,保护性外部氧化物的形成减少了 $O_2$ 和 $H_2$ 向表面的迁移。观察结果发现,表面 CW 有利于降低氢化 SCW 中 SCC,这与 LWR 条件下获得的数据显示的结果相反,但与第 5 章中讨论的腐蚀机理的变化一致。下一节将对这个问题再次讨论。

### 6.2.2 材料因素

与高温水中 EAC 相关的材料特性包括以下内容(Andresen 和 Ford,2011):

(1) 合金成分;

(2) 合金微观结构,特别是晶界敏化程度;

(3) 屈服强度;

(4) 表面损伤/冷加工程度。

从历史上看,由于 $M_{23}C_6$ 碳化物的生长,晶界处 Cr 的损耗程度与沸水反应堆(BWR)中奥氏体不锈钢部件的 IGSCC 情况最为相关(Bruemmer 等,1993)。通常,碳含量超过 0.03%的奥氏体不锈钢如果加热或缓慢冷却到 520~820℃,则被视为“敏化”;如第 3 章所述,这会导致 Cr 的碳化物在晶界形成。结果,与析出的 Cr 碳化物相邻区域的 Cr 含量将减少到 12%以下(图 6.11),从而导致这些区域的耐腐蚀性丧失。沸水堆中最常见的敏化原因是焊接,其中敏化包含在焊接热影响区内。如前面关于超临界水堆的叙述,这一范围的下限是在超临界水反应堆堆芯的预期温度范围内,这表明在运行期间发生敏化是可能的。Cao 等(2011)证实了含碳量高的 304 SS,比含碳量低的同一合金更易发生 SCC。通过在敏化温度范围内快速冷却,控制碳含量在 0.03%以下,可以防止晶界 Cr 的耗尽,通过固溶退火敏化材料及快速冷却,或者在钢上加入强碳化物如 Ti 或 Nb。在后一种方法中,钢中的碳优先与 Ti 或 Nb 反应形成 Ti 或 Nb 碳化物,而不是与 Cr 反应。添加 Ti 或 Nb 的不锈钢称为“稳定型”。但是,如果碳含量过高,稳定不锈钢中仍然会发生 IGSCC。

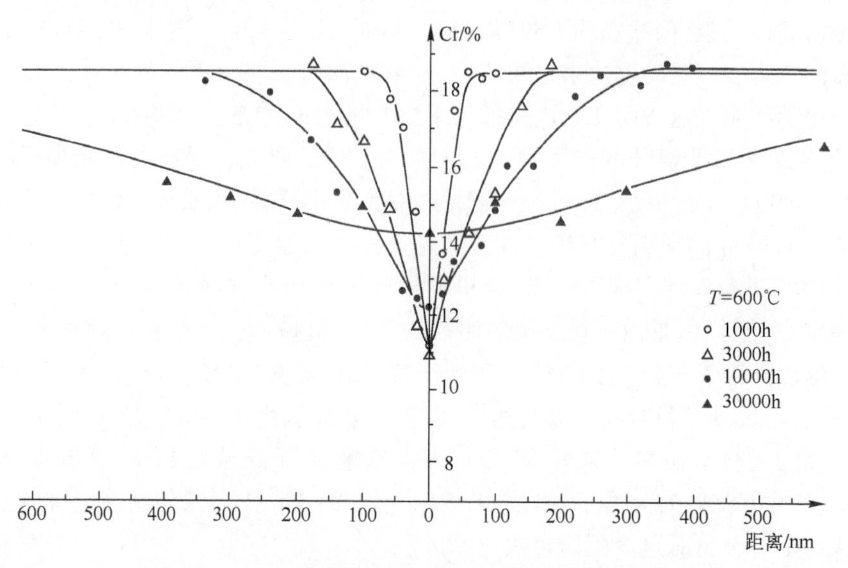


图 6.11　在 550℃条件下材料老化过程中晶界附近 Cr 的形成及增长过程(Sahlaoui 等,2004)



晶界敏化可导致沿晶间裂纹的形成和扩展,裂纹扩展速率取决于其他材料属性(屈服强度等)以及水化学和载荷特性。裂纹对少量晶界敏化敏感,特别是对 Cr 含量敏感(图 6.12)。这一认识使核工业能够在最大限度地降低晶界敏化程度的基础上,确定行之有效的缓解措施(Andresen 和 Ford,2011)。

为支持 SCWR 的发展中对 Ni 基合金中 EAC 的研究较少,并且研究结果在某种程度上不太一致。几个小组研究了在侵蚀性远远高于 SCWR 所预期的化学条件下,SCWO 工艺中使用的 625 合金的 SCC 特性(Kim 等,2011;Fujisawa 等,2005;Bosch 和 Delafosse,2005)。Bosch 和 Delafosse(2005)在 400℃和 500℃的温度下,使用慢应变率和恒定加载方法在 SCW 中对 625 合金进行了 SCC 试验。他们发现,在含有 10%(质量分数)的 $H_2O_2$ 的 SCW 里,SSRT 中的合金对 SCC 敏感,而恒定加载试验没有显示出任何明显的裂纹扩展迹象。Teysseyre 和 Was(2006)在 400~500℃温度范围内,开展了在排气后 SCW 中 625 合金的慢速应变率试验。在所有温度条件下都观察到了晶间裂纹,平均裂纹长度随温度的升高而增大,与类似条件下的奥氏体合金中发现的裂纹相比,其裂纹扩散范围更大。Was 等(2007)根据裂纹深度和总试验时间的测量,提出该温度范围内 SCC 增长率在 $3\times10^{-9}$ mm/s 和 $3\times10^{-8}$ mm/s 之间。文章中提到,其裂解活化能与 316L SS 的相似。

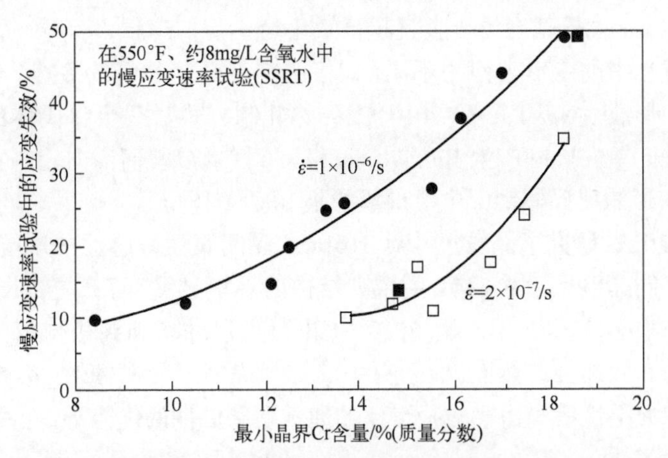


图6.12　在288℃水中以两种应变速率对敏化不锈钢进行缓慢应变速率试验时晶界Cr含量对应变失效百分率的影响值得注意的是,松散材料中在晶界附近的Cr含量从18%(质量分数)下降了不到1%便可以对晶间应力腐蚀开裂敏感性产生显著影响(Bruemmer等,1993)

如第3章所述,超临界水冷反应堆面临的主要困难之一是由于热效应下的材料老化(扩散和次级相沉淀,特别涉及晶界时)引起的材料结构的缓慢变化,导致SCC敏感性的长期增加(Zheng等,2011;Jiao等,2013)。这两种材料的使用寿命为数十年,材料的老化将是它们共同面临的问题。燃料包壳的使用寿命相对较短,但其峰值温度很高(对于加拿大SCWR的概念,最高可达800℃)。Ru和Staehle(2013a~c)预测,动态热范围下会引起的材料显著老化现象(3.5节),其改变了材料的晶界化学和结构,从而改变其对SCC的敏感性。直到最近,试验中才开始使用装配材料,而金属间相(χ、η和σ相)的生成需要更长的老化时间,其影响很少被人们所关注。这些相对于含有Nb和Ti以主动沉淀碳化物的不锈钢来说特别重要(Jiao等,2013)。Udy和Boulger(1953)在732℃条件下使用静压胶囊开展了SCC试验,除了他们开展的研究工作外,高温下合金微观结构演变的长期过程中产生的对SCC的协同效应目前了解甚少。

研究人员已经提出了几种减缓材料老化作用,从而降低IGSCC敏感性程度的方法。Parvathavarthini等(2009)提出通过使用晶间工程产生较低的有效晶界能(EGBE)①,从而减缓敏感化的趋势。Parvathavarthini和Dayal(2002)提出,对于不容易被敏感化的钢材料,其含碳量的限值与稳定钢中其他合金成分的存在密切相关,比如Cr、Mo、Ni、N、Mn、B、Si以及Ti和Nb。特别是Cr对不锈钢的钝化特性有显著的影响。随着Cr含量的增加,晶界达到Cr耗减阻碍极限的时间

① 有效晶界能是晶界性质几个方面的综合因素。

变长。因此,Cr 含量越高的合金就越抗敏化。

生产过程中的变形类型的不同(如板的交叉轧制与管的冷拔)会导致不同的微观结构形成。Nezakat 等(2014)报告称轧制方式对残余奥氏体的形变纹理有直接影响。单向轧制产生 Brass、Goss 和 γ-纤维纹理,而横向轧制在厚度上减少 90%后变形的奥氏体钢中主要形成的是 Brass 织构。

Li 等(2013)研究了晶粒尺寸对 316L SS 晶间腐蚀敏感性的影响。DL-EPR 试验以及不同晶粒尺寸的 316L 的微观结构的观察显示,随着晶粒尺寸的增加,晶间腐蚀的敏感性下降。Li 等(2013)提出晶粒尺寸增加到最佳值是提高抗晶间腐蚀性的有效途径。然而,由于更小的粒径提高了耐腐蚀性(第 5 章),在不损失良好的力学性能和由小粒径形成的通常耐腐蚀性的优势下,提高 EAC 抗性可能是一个挑战。

Terada 等(2008)公开了 316L(N)SS 蠕变试验在 600℃下进行 7500h 和 85000h 的 DL-EPR 试验结果。DL-EPR 试验表明,对晶间腐蚀有明显的敏化作用。然而,他们没有在晶界观察到 $M_{23}C_6$ 碳化物,并将晶间腐蚀的敏化作用归因到金属间相(主要是 σ 相)。

已知在轻水堆环境(如日本沸水堆堆芯罩)中,表面粗糙度和冷加工程度会影响奥氏体不锈钢的腐蚀敏感性。最近的研究表明,在模拟的 BWR 环境中,冷或热加工会增加低碳不锈钢的应力腐蚀开裂增长率(Shoji 等,2008)。图 6.13 显示出不同形式冷加工相关参数之间的相互作用以及破裂增长率(Andresen 和 Ford,2011)。正如 Garud 和 Ilevbare(2009)所指出的,事实上,CW 影响到多个 SCC 的诱发因素(应力和材料),这使得关于 CW 影响的结论明显自相矛盾。



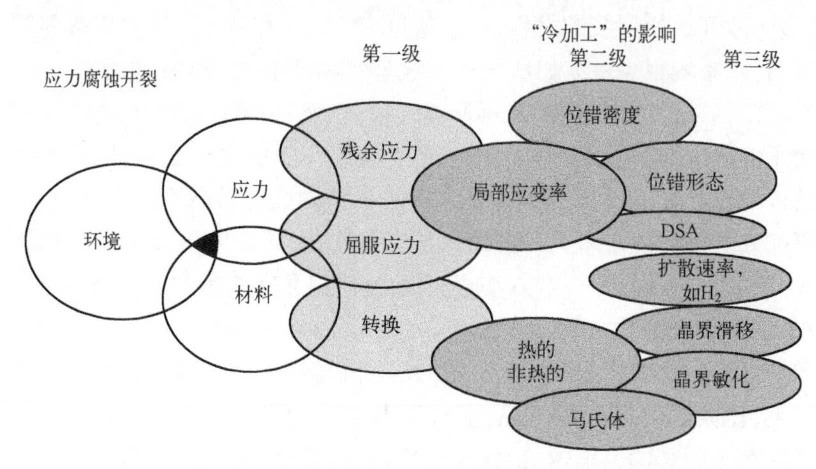


图 6.13 "冷加工"相关参数之间的相互作用及其对应力腐蚀裂纹扩展的"材料"和"应力"共同作用条件的影响

冷加工在压水堆和沸水堆环境中奥氏体不锈钢 SCC 中起到的作用一直是许多研究的主题(Andresen,2002;Garud 和 Ilevbare,2009;Hou 等,2011;Devrient 等,2011)。在典型压水堆工况下,当合金发生一定量的永久塑性变形时,非敏化奥氏体钢会发生 IGSCC 和/或 TGSCC,以提高微观结构上的硬度和屈服强度。由于冷加工造成的沿晶界的应变局部化是这些 SCC 免疫微观结构中裂纹萌生的主要原因。CW 的作用与中子辐照的作用类似,后者在形变集中时促进形变过程以变形带的形式向周围扩展,在没有障碍物的平面上进行错位运动(Lee 等,2001)。Lozano Perez 等(2009)观察到,在冷加工合金 304 的应力样品中,变形带的氧化程度比在与无应力表面相交的带(即无应力样品)中的氧化程度更深,这表明形变的微观结构会由于应力的存在而促进氧化过程。裂纹尖端高应力区域的氧化增强现象被认为是由于试验合金先前冷加工形成的高度错向和复杂塑性单元(亚边界)中存在快速扩散路径。

CW 对亚临界水中奥氏体钢和 Ni 基合金的应力腐蚀开裂的影响相同,但在 SCW 条件下可能不再相同,因为 400℃或者更高温度条件下,热激活扩散将有助于微观结构的恢复过程。在 $1/3T_m \sim 1/2T_m$ 的温度下,再结晶开始发生,导致具有低位错密度的小等轴"软"晶粒形成。事实上,在 650℃的超临界水下,具有重型机械加工相关变形层的表面已显示出显著的再结晶过程,这对耐腐蚀性提升会产生有利影响(第 5 章)。必须考虑奥氏体钢中"形变诱发的易发生 SCC 的微观结构"形成的可能性;即使在 SCW 的温度条件下,SSRT 试样在动态应变(即拉伸冷加工)过程中也可能产生这种易发型微观结构。在这种情况下,通过过度的塑性应变,通常 SCC 免疫微观结构可以转化为"SCC 易发"微观结构;在以往的 SSRT 情况下,试样通常被拉断,然后在试样的断裂表面或测量长度中检查是否存在 SCC。如果试验过程中不使合金受到这种极端塑性的影响,则这种试验与此就没有相关性。不幸的是,大多数公开的 SSRT 或 CERT 试验结果都不是基于重加工性、试验诱发性及塑性的条件下得到的。

关于这些冶金因素对 SCW 中 SCC 的影响,相关的研究较少。DL-EPR 试验结果(Oh 和 Hong,2000)表明,在材料的整个老化过程中,20%CW 会加剧材料的敏感化,而 40%CW 会抑制材料的敏感化。同样,在 CW 低于 20%时,在基本相同的碳含量条件下,氮含量的增加加速了材料的敏感化。

Novotny 等(2011)公开了 316L SS 在 550℃、25MPa 条件下,氧含量可控的纯净超临界水中的 SSRT 试验结果。SSRT 试样在制造过程中被无意地冷加工。应变率(延伸率)和氧含量不同。结果表明,以 IGSCC 裂纹扩展为特征的 SCC 敏感性没有显著增加。对样品断面的分析证实,断裂是 TGSCC 和韧性断裂共同作用的结果。尽管 CW 在一定程度上增强了 SCC 敏感性,后续试验中,SCW 中过快的氧化会导致裂纹钝化。Sáez-Maderuelo 和 Gómez-Briceño(2016)测量了退火和冷加工 316L SS 在 400℃和 500℃下的 SCC 敏感性(25MPa、溶解氧浓度小于 10μg/kg),他们发现塑性形变的存在增加了这种合金的 SCC 敏感性。Scenini 等(2008)提出了一个某种程度上矛盾的结果,其发现在氢化 SHS 中测试的 600 合金的电抛光样品比机械抛光样品对 SCC 的敏感性大得多,这归因于机械抛光样品由于存在短路扩散路径而形成外部氧化物的能力(第5 章)。电位在 Ni/NiO 平衡点以下($H_2$分压)的无应变表面上,没有外部氧化物形成。对于 SCW 中的奥氏体不锈钢,CW 的积极作用(增强耐腐蚀性(5. 3. 2 节)提高屈服强度)可能会被 SCC 敏感性的增加(当然还有延展性的降低)而抵消。

### 6. 2. 3 机械因素

应力和应变、残余应力和残余应变以及应变率等机械因素是影响 EAC 的关键参数,它们与特定的合金/环境组合有着较强的相互作用。迄今为止,为支持 SCWR 的发展,EAC 的材料研究主要集中在利用 SSRT 和 CERT 测试评估各种材料对 SCC 的敏感性上,即材料的裂纹产生倾向及其扩展此类裂纹的能力。对 SCW 中各种因素对 SCC 的影响还不存在更深入的认识,也没有对这些因素的定量分析。例如,研究人员广泛、定量地研究了应力强度因子 $K_i$对 WCR 奥氏体不锈钢堆芯内构件的影响,以理解 CGR 配置方程中的速率控制过程。一般来说,对于 WCR,CGR 与 $K$ 的关系可以用以下经验幂律方程表示,即

$$\mathrm{CGR}(K_i) = C_0 \cdot K_i^n \tag{6.1}$$

图 6. 14(Shoji,2011)显示了模拟沸水反应堆含氧水环境中非敏感性奥氏体不锈钢的 CGR 与 $K_i$间关系的结果。

目前,在 SCW 中进行此类研究还很少。Janik(2015)利用可用数据构建了一个现象学图用于指示 SCW 中 SCC 敏感性与 DO 浓度和 316L SS 的应变率之间的关系(图 6. 15)。他使用了断裂后拉伸试样断裂表面测量的 IGSCC/TGSCC 百分比值,以及被忽视的试样量规中最后的二次断裂结果。尽管结果有些不确定,数据没有恰巧落在“SCC”和“无 SCC”两个区域,正如一些作者所指出的,当 DO 浓度高于 2000μg/kg 时,SCC 很少出现或者不会出现(6. 2. 1. 2 节)。

然而,在 SSRT 中,裂纹尖端的应变率是已知的,对于实际的应变率,当裂纹扩展率较低时,失效的局部化可能与 SCC 裂纹无关。因此,在断裂表面观察到的 SCC 数量可能不能代表试样对 SCC 的敏感性。这种方法的关注点侧重于裂纹的扩展方面。其他人则更关注对裂纹产生的敏感性,即规范面上裂纹的范

围,并且使用裂纹密度、平均裂纹长度/深度以及每单位面积的裂纹长度作为参数。

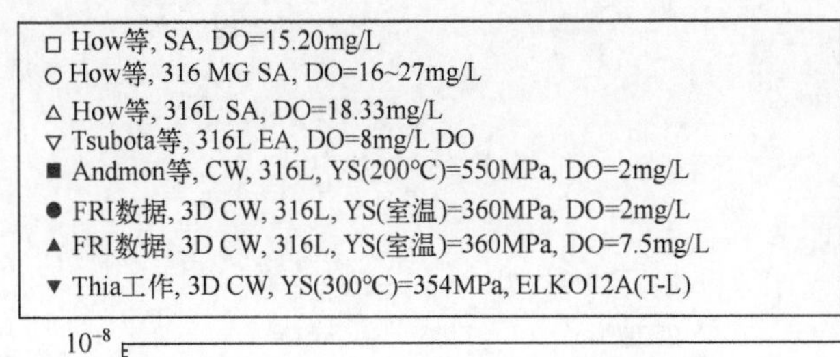


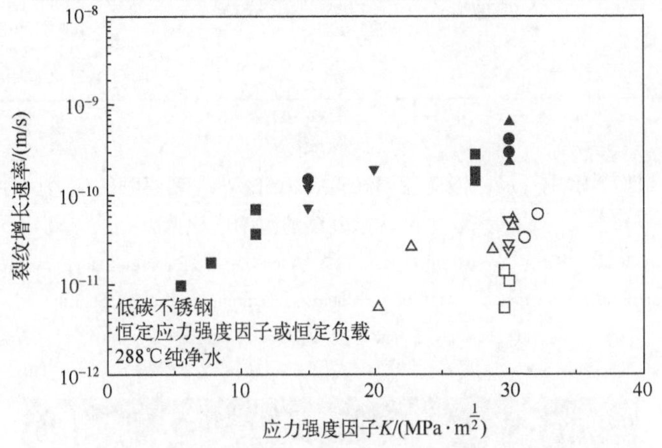


图 6.14 288℃的含氧的纯净水中非敏化不锈钢的裂纹增长速率与应力强度因子的关系(Shoji,2011)

由候选合金制成并含有不同初始 DO 浓度的 SCW 静态加载胶囊试样,在 500℃、25MPa 下进行长达 19000h 的不同的环向应力实验①(Behnamian 等,2014;Zheng 等,2014)。图 6.16 显示了与溶液接触 19000h 后 310S 胶囊样品内表面 O、Fe、Cr 和 Ni 的 TEM-EDX 元素图分析。选择性内氧化与晶间穿透有关,特别是在晶界上。一些试验在更高的环向应力下进行,发现环向应力与屈服应力之比越高,最大和平均裂纹长度值越高(图 6.17)。将胶囊试样的环向应力从约 62MPa 增加到接近屈服强度时,在 500℃时 316L 的最大受击深度增加了 39%。在相同条件下,310S 的最大受击深度增加了 35%。625 合金的最大贯穿深度增加了 30%。

① 在这些加载胶囊中,加载是双向的:同时存在环向应力和轴向应力,轴向应力是环向应力的 50%。

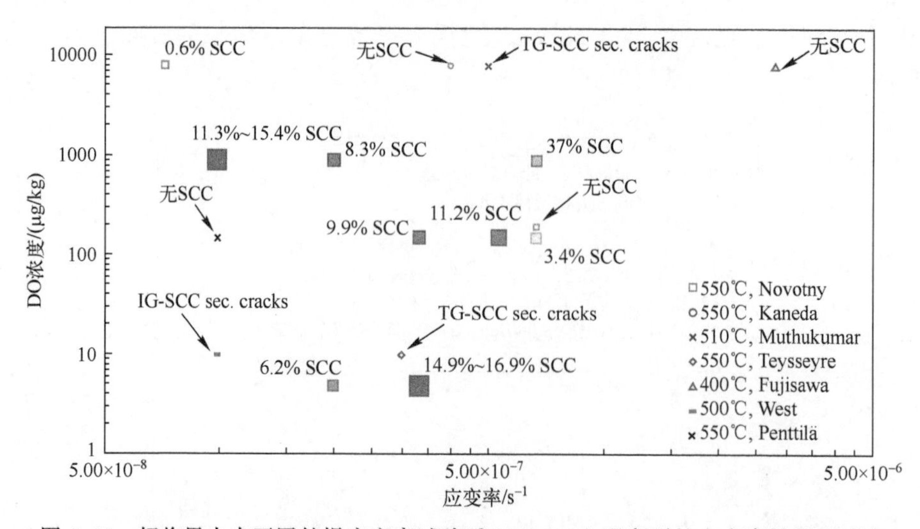


图 6.15　超临界水中开展的慢应变率试验后 316L SS 上观察到的应力腐蚀开裂现象（符号的大小与 SCC 的范围成比例）

（引自文献 Janik，P.，2015. Acoustic Emission — A Method for Monitoring of Environmentally Assisted Cracking Processes（Doctor thesis）. Institute of Chemical Technology Prague，Faculty of Environmental Technology，Department of Power Engineering）



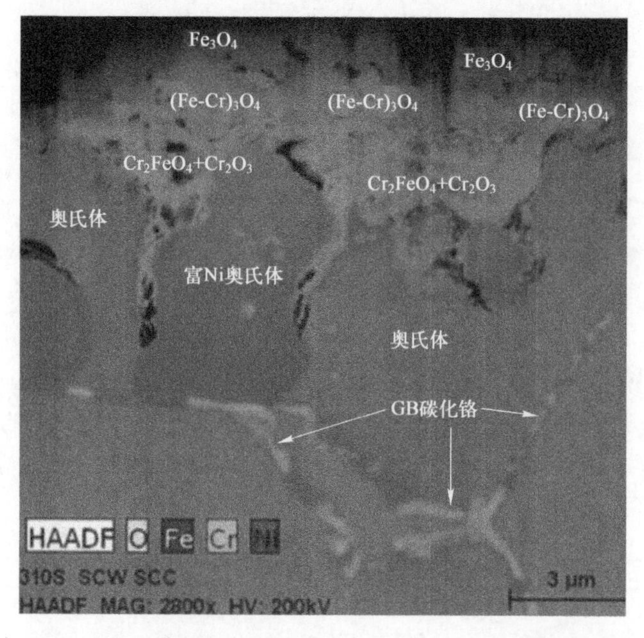


图 6.16　310S 胶囊样品与超临界水在 500℃、25MPa 下接触 19000h 后的内截面 TEM-EDX 元素图分析（Zheng 等，2014）（见彩插）

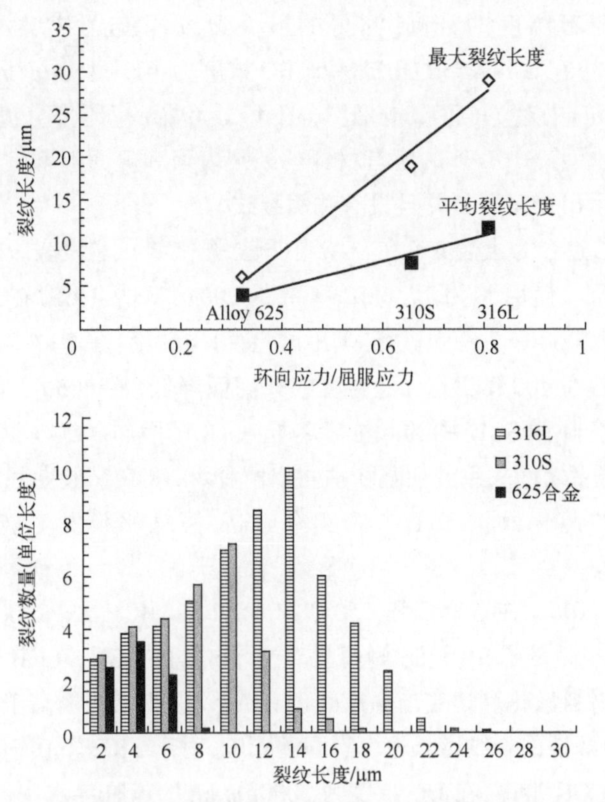




图 6.17　选择性内氧比与晶间穿透有关(原始的、5000h、25MPa、500℃、去离子水，胶囊试验开始时 DO 浓度为 8mg/L,试验结束时浓度小于 5μg/L)

(引自文献 Behnamian, Y., Dong, Z., Kohandehghan, A., Zahiri, R., Mitlin, D., Chen, W., Luo, J. L., Zheng, W., Guzonas, D., 2014. Oxidation and stress corrosion cracking of austenitic alloys in neutral pH extreme hydrothermal environments(supercritical water). In: NACE Northern Area Western Conference, Edmonton, Alberta, January 27-30, 2014)

长期(19000h)试验中观察到一个有趣的观察,沿晶界的晶间破坏在最前沿区域非常钝,这表明在试验终止时存在某种休眠性。很明显,富 Cr 沉淀的形成明显早于腐蚀破坏的前端。由于在 500℃下,缺乏动态应变和微观结构的热恢复性,最初以沿晶方式开始的腐蚀侵蚀不能继续增长,19000h 后,所有沿晶穿透深度均小于 15μm。IG 沉淀物还可以提高对晶界滑移和滑动的抗力,与敏化合金 800H 的特性相同(Janssen 等,2009)。

Teysseyre 和 Was 等(2006)利用测量表面收集的裂纹密度、平均裂纹长度和横截面最大裂纹深度的结果,评估了纯净除氧的 SCW 中退火 304 SS 和 316L SS 的 SCC 敏感性。他们注意到裂纹密度随着温度的升高而降低,而平均裂纹长度则增加。随着温度的升高,测得的最大裂纹深度显著增加。Penttilä 和 Toivonen(2013)在 SCW(550℃、25MPa、DO 浓度为 100~150μg/kg、入口电导率小于 0.1μS/cm)中对 316NG、316Ti、347H 和 1.4970(标距部分抛光至 600 目)进行了 SSRT;在除 316L 外的所有材料的断裂表面和侧面上都发现了穿晶 SCC 迹象,它只显示出韧窝开裂并且没有侧面裂纹。

如前所述,迄今为止很少有在 SCW 中开展的 CGR 试验发表。Peng 等(2007)在 288~500℃下,使用 $K_i$ 为 27.6MPa · m 的 20%CW 的 316L,在 SCW(25MPa、溶解氧浓度为 2000μg/kg)中进行 CGR 试验,得出结论:在 $T<T_c$ 时,CGR 随温度的升高而增加,在 400℃以上时随温度的升高而降低(图 6.5)。在 400℃时,与除氧情况相比,降低 SCW 中的氧浓度将导致 CGR 增加。CGR 试验与 SSRT 测得的 SCC 与温度之间关系的明显矛盾可以解释为,CGR 试验期间快速的氧化作用导致裂纹的钝化,而 SSRT 中应变率更高,使氧化膜发生破裂(Peng 等,2007)。

Chen 等(2017)在 500℃和 550℃的 SCW(25MPa、Ar、2000μg/kg DO 或 1.6mg/L $H_2$)中测量了 310S SS 侧槽紧凑拉伸试样的 CGR。CGR 随温度升高而增加,且由于材料敏化和蠕变与温度间也有关系,使得活化能高于亚临界水。氧气对应力腐蚀开裂的裂纹增长速率有显著影响,对于 310S 不锈钢,在用 Ar 除氧后的水中,其 CGR 最高,在 DO 浓度为 2000μg/kg 的超临界水中,其 CGR 最低。在此基础上,作者提出了一种改进的滑移性氧化还原机理以解释 SCW 中 310S 不锈钢的 SCC 特性。



Arioka 等(2011)研究了在模拟压水堆环境中以及高于 450℃的空气中,冷加工的 690、304SS 和 316L SS 合金的 SCC 和蠕变特性。作者基于对所有试验材料的 IG 裂纹尖端和裂纹壁前方的晶界空穴的观察,提出了蠕变氧化机理。冷加工后的扩散作用,即裂纹尖端位置处应力梯度的存在诱导出空穴,促使空穴的形成。

#### 6.2.3.1　辐照因素

除了微观结构缺陷(第 3 章中讨论的点和线,即错位和空洞)外,辐照还引起微观水平上的材料化学变化,如辐照诱导的偏析(RIS)和沉淀,这两个都对 SCC 敏感性有影响。损伤程度取决于辐照剂量、通量和辐照光谱,由于所涉及的扩散过程的热性质,其与金属受到辐照时的温度也有关。奥氏体合金中辐照促使的应力腐蚀开裂的形成已经成为所有型号的轻水反应堆(LWR)面临的共同问题。在 20 世纪 70 年代中期,由越来越多的观察发现,当超过“阈值”影响水平(5×

$10^{20}$个/$cm^2$)时,在快中子(大于 1MeV)辐照下,高应力堆芯部件中出现了辐照促进应力腐蚀开裂(IASCC),这个问题开始引起人们的关注(图 6.18)(Gordon 和 Brown,1989)。

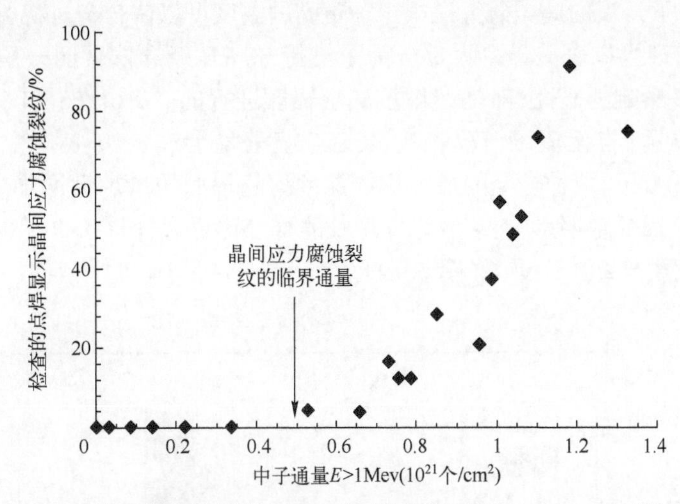


图 6.18 高电导率沸水反应堆(BWR)的开槽控制叶片中 IASCC 受到快中子通量的影响(此处由于点焊和膨胀的 B4C 吸收管引起的动态效应,拉伸应力很高)(引自文献 Gordon, G. M., Brown, K. S., 1989. Dependence of creviced BWR component IGSCC behavior on coolant chemistry. In: Cubicciotti, D., Simonen, E. (Eds.), Proc. 4th Int. Symp. on Environmental Degradation of Materials in Nuclear Power Systems-Water Reactors, Jekyll Island, August 6-10, 1989. National Association of Corrosion Engineers, pp. 14.46-14.62)



在影响低于 $5\times10^{-20}$个/$cm^2$时,在靠近大型不锈钢堆芯结构焊接处也观察到 IASCC,裂纹程度取决于某几种因素的共同作用,包括制造工序中的预先热敏化、快中子通量及影响、焊接残余应力和冷却剂纯度(Andresen 等,1989; Nelson 和 Andresen,1991; Andresen 和 Ford,1995; Bruemer 等,1996; Was 等,2006)。

辐照造成的影响包括腐蚀电位(由于水辐照,第 4 章),施加的拉伸应力,晶界成分和屈服强度的变化。中子通量(即腐蚀电位)引起的这些变化对开裂敏感性有直接影响,而由通量引起的变化(RIS、辐照硬化、应力松弛)是在较长时间内产生积累性影响。辐照对控制不锈钢裂纹扩展参数的具体影响如下(Andresen 等,1989; Nelson 和 Andresen,1991; Andresen 和 Ford,1995; Bruemmer 等,1996; Was 等,2006):

(1) 腐蚀电位及其随辐照通量的变化;

(2) 晶界成分由于受到辐照产生的变化;

(3) 辐照引起的硬化;

(4) 对于替换加载结构,辐照引起的蠕变应力松弛。

相比于在当前的 WCR 运行条件下,关于亚临界水中 IASCC 的大量研究,SCWR 条件下开展的此类研究还很少。更高的温度可以使合金对机械载荷的微观结构响应产生显著差异。大部分为了支持 SCWR 概念发展而开展的有关辐射对 IGSCC 影响的研究,都是利用重离子辐射进行的。316L SS 和 690 合金的质子辐照显示,与无辐照情况相比,IGSCC 显著增加(Teysseyre 等,2007)。在 400℃和 500℃下,两种合金的最大裂纹深度均随辐照的增加而增加。在 400℃下,690 合金因受辐照影响,裂纹深度增长的百分比更大,而在 500℃时,316L SS 的裂纹深度增长百分比更大(图 6.19)。



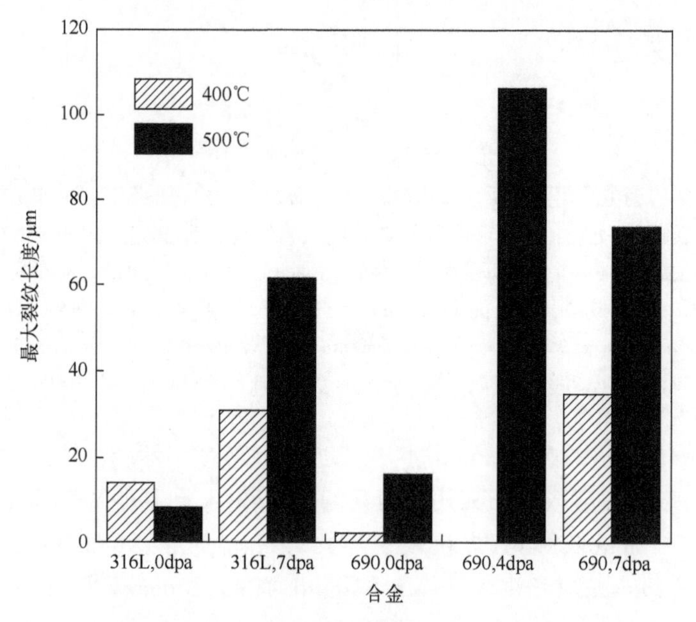


图 6.19 在 400℃和 500℃条件下辐照对超临界水中 316L SS 和 690 合金裂纹扩展深度的影响(Teysseyre 等,2007)

West 和 Was(2013)研究了 IASCC 中应变不相容性的作用。拉伸试样在 400℃下受质子辐照到 7dpa,然后在 SCW 中 400℃下被应变 5%。晶间裂纹主要发生在晶界最易受到滑移不连续性影响的位置,这表明应变不相容性促进了裂纹的形成。

Zhou 等(2009)观察到,质子辐照试样会导致 SCW 中的奥氏体合金发生 SCC,裂纹的严重程度随着剂量和温度的增加而增加。对于 316L SS,裂纹随硬度的增加而增加，与辐照和试验温度无关。316L 不锈钢在不同剂量辐照后观察到的二次裂纹的典型样貌如图 6.20 所示。

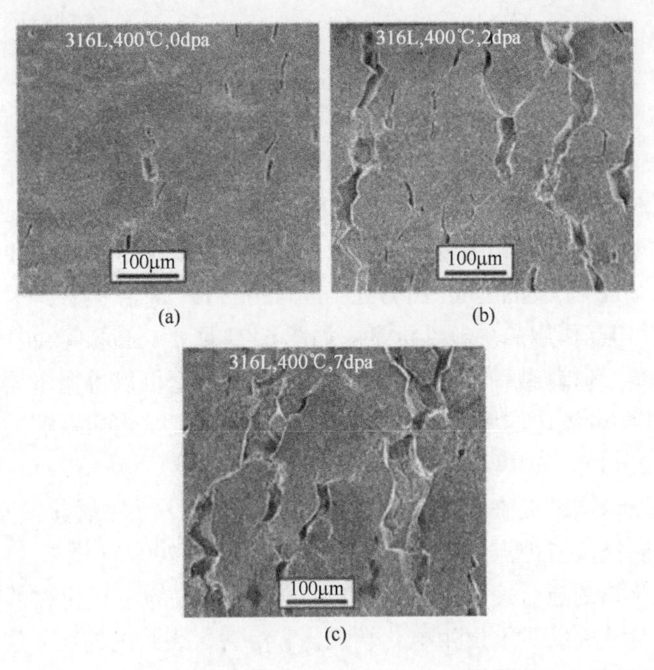


(a) (b) (c)

图 6.20　未辐照、2dpa 和 7dpa 条件下在 400℃超临界水中应变至破坏的 316L 试样的测量表面的微观图像（Zhou 等，2009）
（a）未辐照；（b）辐照到 2dpa；（c）辐照到 7dpa。



Teysseyre 等（2007）展示了在 400℃和 500℃下，用高达 7dpa 的质子预辐照的 316L SS 和 690 合金的裂纹密度和裂纹深度，然后在 SCW（25MPa、DO< μg/kg、电导率<0.1μS/cm）中进行测试（CERT），测试温度与未辐照材料的辐照增加温度相同。在 500℃下材料的开裂程度高于在 400℃下的开裂程度。辐射使得这两种合金晶界处 Cr 和 Fe 损耗，但提高了 Ni 含量；500℃下辐照后，损耗区尺寸为±3.4nm，最低 Cr 浓度为 13.8%（质量分数）（对于未辐照材料为 19%（质量分数））。在相同的辐照和试验条件下，发现 F/M 合金具有抗裂性。

在美国核能过热蒸汽项目期间共有 496 个燃料元件在过热蒸汽环境中辐照超过 2 年时间，其中只有 4 个燃料元件发生破裂。试验表明，燃料包壳厚度为 0.406mm 的燃料组件在温度高达 738℃的过热蒸汽中暴露 10292h，没有出现 SCC（Comprelli 等，1969）。一个燃料棒在 493℃通电 6188h 后发生破裂，但是辐照后试验显示其原因可能是低周疲劳(Rabin 等,1967)。Beloyarsk 核电站使用 Kh18Ni10T① 和 EI-847② 不锈钢(Emel'yanov 等,1972),其在 550℃、8MPa 压力下已经运行了许多年。

## 6.3 机理和模型

如图 6.1 所示,SCC 是材料、环境和力学的综合和协同效应的结果。人们已经得到不同 SCC 机理和控制参数,如滑移氧化(或内氧化)、应力-应变强化型固态氧化、与氢相关机理以及去合金化(Newman,2002;Andresen 和 Ford,1988;Shoji,2003;Scott 和 Combrade,2003;Das 等,2008)。高温亚临界水中发生 SCC 的氧化/力学相互作用的一般性示意图如图 6.21 所示(Shoji 等,2010)。

根据材料/环境/载荷不同因素的组合,瞬态氧化可以遵循不同的动力学速率规律,裂纹尖端氧化的增强可以通过物理降解、物理/化学降解或两者兼有来实现(Shoji,2011)。

人们普遍认同的亚临界水中的 SCC 增长机理是滑移-氧化模型,它基于分离的阳极和阴极反应位点的电化学模型(Macdonald,2009)(图 6.9)。如果氧化速度快于氧化物的愈合速度,则由于应力或应变导致的钝氧化物的破裂引起的裂纹扩展会导致裂纹尖端发生电化学氧化和溶解。裂纹内的离子腐蚀产物作为阳极和阴极过程的电荷载体。在更高温度以及低 SCW 密度下,离子离解程度较低(图 4.12),金属氧化物的溶解度较低(图 4.10),降低了溶液的侵蚀性,降低了溶液电导率,减少了金属溶解。如第 5 章所述,当 $T \gg T_c$ 时,涉及空间上分离的阳极和阴极位点的电化学过程不再发生,并且局部化学反应占主导地位。图 6.21 所示的所有过程都受到水物性变化的影响。这表明,当 $T > T_c$ 时,影响 SCW 中 EAC 的环境因素发生变化,特别是氧气和阴离子的作用。当 $T \gg T_c$ 时,固相内扩散传质、蠕变和化学氧化将会是影响裂纹扩展的更加重要的因素。



Arioka 等(2011)公开了其对 SCW 中奥氏体不锈钢 SCC 机理进行的系统研究,关于 SCC 和蠕变产生及发展过程的机理。研究发现,在温度升高到 500℃时,SCC 和蠕变具有极大的相似性。该研究主要关注由局部应力引起的空位累积产生空穴,从而导致 SCC 的产生和扩展。Ru 和 Staehle(2013a~c)指出,由于

---

① 成分(质量分数):C 最大 0.1%;Si 最大 0.8%;Mn1%~2%;Ni10%~11%;S 最大 0.02%;P 最大 0.035%;Cr17%~19%;5(C0.02%)<Ti<0.6%。

② 成分(质量分数):C0.04%~0.06%;Mn0.4%~0.8%;Si≤0.4%;S≤0.010%;P≤0.15%;Cr15.0%~16.0%;Ni15.0%~16.0%;Mo2.7%~3.2%;Nb≤0.9%;N≤0.025%;B≤0.001%;Co≤0.02%;Cu≤0.05%;Bi≤0.01%;Pb≤0.001%;Ti≤0.05%。

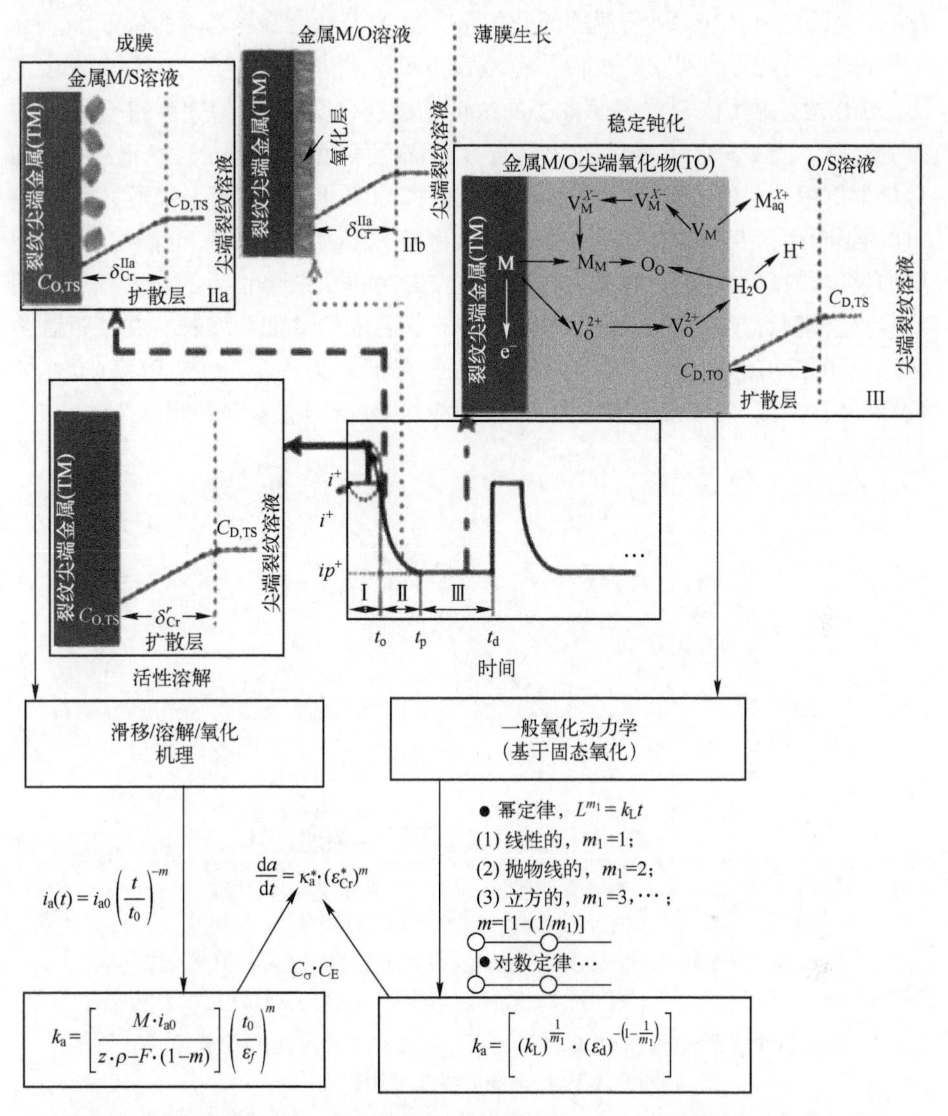


图 6.21　高温亚临界水中奥氏体合金应力腐蚀开裂子过程示意图(Shoji 等,2010)

SCW 环境可能具有很强的氧化性,如果开裂速度很高,就与 Peng 等(2007)的结果以及 6.2.1.2 节中讨论的与 DO 关系的数据一致,这将令人感到吃惊。Chen 等(2017)提出了一个修正后的、包含蠕变的滑移-氧化模型,在该模型中,在裂纹尖端形成蠕变裂纹之前,外加应力引起沿晶界的蠕变空洞形成核。主应力腐蚀开裂(SCC)裂纹与蠕变裂纹相交,导致裂纹扩展速率增加。

第 5 章中描述的现象学腐蚀模型可以进行修改,以解释以上几点(图 6.22)。在 $T_c$ 以下时,图 6.9 所示的基于阳极溶解的 SCC 模型(此处表示滑移-阳极氧化)是主要的,SCC 速率随着温度的升高而增加,直至接近临界区域的位置,此时水物性的快速变化导致裂纹扩展显著减少(图 6.5)。当 $T$ 低于 $T_c$ 时,SCC 的发生取决于 DO 浓度和 $Cl^-$、$SO_4^{2-}$ 等阴离子的浓度。蠕变也会发生,但速度很慢(Arioka 等,2011)。当 $T \gg T_c$ 时,蠕变速率更高,当仍然发生滑移-氧化时,氧化是直接通过化学作用发生的,水、DO 和 $H_2O_2$ 都可以作为氧化剂(如第 5 章所述),这使 DO 的影响因素变得复杂。提供短路扩散途径的因素如冷加工和晶粒尺寸变得更加重要。在此温度范围,诸如 Chen 等(2017)提出的修正后的包含蠕变的滑动-氧化模型看起来更加合理,并且裂纹随着温度的增加而增长。在接近临界区,试验测量出的开裂特性应该与环境因素有密切关系,如 SCW 密度(介电常数)和氧化剂浓度;以及材料因素,如表面冷加工、Cr 含量和晶粒尺寸。

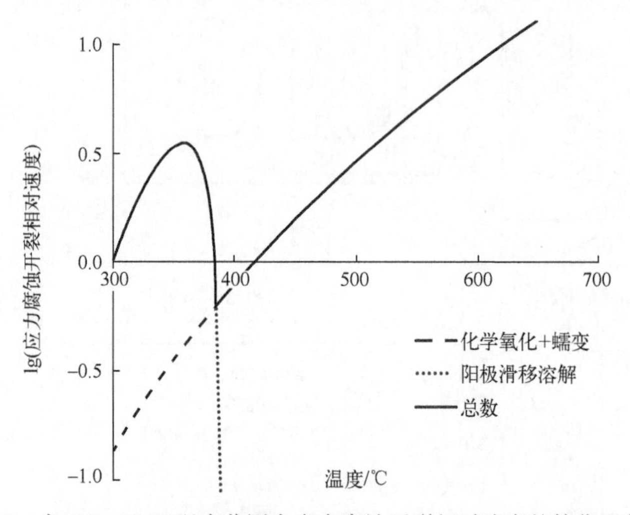


图 6.22　在 300~650℃温度范围内应力腐蚀开裂相对速度的简化现象学模型(认为温度低于 $T_c$ 条件下,阳极滑移-氧化是其主要机理)以及在 $T \gg T_c$ 条件下包含蠕变修正后的滑移-氧化模型(在近临界区域,这两个模型具有相同的重要性,两条曲线在 $T_c$ 附近交汇)

对奥氏体不锈钢和 Ni 基合金在 SCW 中的 EAC 特性这样的复杂体系中基本力学和协同效应的评估需要开展系统的工作,特别是基于断裂力学的 CGR 试验需要进行系统的研究。而在其他方面,基于机理的多尺度模拟研究也仍然需要开展。

## 参考文献

Allen, T. R., Chen, Y., Ren, X., Sridharan, K., Tan, L., Was, G. S., West, E., Guzonas, D. A., 2012. Material performance in supercritical water. In:Konings,R. J. M. (Ed.),Compre-hensive Nuclear Materials,vol. 5. Elsevier,Amsterdam,pp. 279-326.

Andresen,P. L.,April 1992. Effect of Temperature on Crack Growth Rate in Sensitized Type 304 Stainless Steel and Alloy 600. Paper 89,NACE Corrosion/92,Nashville.

Andresen,P.,2002. Stress Corrosion Crack Growth Behavior of Cold Worked 304l and 316L Stainless Steels. Electric Power Research Institute Report 1007379.

Andresen,P. L.,Ford,F. P.,1988. Life prediction by mechanistic modeling and system moni-toring of environmental cracking of iron and nickel-alloys in aqueous systems. Mater. Sci. Eng. A 103,167-184.

Andresen,P. L.,Ford,F. P.,1995. Modeling and prediction of irradiation assisted cracking. In:Gold,R.,McIlree,A. (Eds.),Proc. 7th Int. Symp. on Environmental Degradation of Ma-terials in Nuclear Power Systems—Water Reactors,Breckenridge,August 7-10,1995. National Association of Corrosion Engineers,pp. 893-908.

Andresen,P. L.,Ford,F. P.,September 2011. Prediction of stress corrosion cracking(SCC) in nuclear power systems. In:Raja,V. S.,Shoji,T. (Eds.),Stress Corrosion Cracking Theory and Practice. Woodhead Publishing,ISBN 978-1-84569-673-3.

Andresen,P. L.,Ford,F. P.,Murphy,S. M.,Perks,J. M.,1989. State of knowledge of irradiation effects on EAC in LWR core materials. In:Cubicciotti,D.,Simonen,E. (Eds.),Proc. 4th Int. Symp. on Environmental Degradation of Materials in Nuclear Power Systems—Water Reactors,Jekyll Island,August 6-10,1989. National Association of Corrosion Engineers,pp. 1.83-1.121.

Arioka,K.,Yamada,T.,Miyamoto,T.,Terachi,T.,2011. Dependence of stress corrosion cracking of alloy 600 on temperature,cold work,and carbide precipitation,the role of diffusion of vacancies at crack tips. Corrosion 67(3),035006-11-035006-18.

Behnamian,Y.,Dong,Z.,Kohandehghan,A.,Zahiri,R.,Mitlin,D.,Chen,W.,Luo,J. L.,Zheng,W.,Guzonas,D.,2014. Oxidation and stress corrosion cracking of austenitic alloys in neutral pH extreme hydrothermal environments(supercritical water). In:NACE Northern Area Western Conference,Edmonton,Alberta,January 27-30,2014.

Bevilacqua,F.,Brown,G. M.,1963. Chloride Deposition from Steam onto Superheater Fuel Clad Materials. General Nuclear Engineering Corporation Report GNEC 295.

Bosch,C.,Delafosse,D.,2005. Stress corrosion cracking of Ni-based and Fe-based alloys in supercritical water. In:*Corrosion 2005*. NACE International,Houston,TX. Paper 05396.

Bruemmer,S. M.,Arey,B. W.,Charlot,L. A.,August 1-5,1993. Grain boundary chromium concentration effects on the IGSCC and IASCC of austenitic stainless steels. In:Simonen,E.,Gold,R. (Eds.),Proc. 6th Int. Symp. on Environmental Degradation of Materials in Nuclear Power Systems—Water Reactors. National Association of Corrosion Engineers,San Diego,CA,pp. 227-287.

Bruemmer,S. M.,Cole,J. I.,Garner,F. A.,Greewood,L. R.,Hamilton,M. L.,Reid,B. D.,Simonen,E. P.,Lucas,G. E.,Was,G. S.,Andresen,P. L.,Pettersson,K.,1996. Critical Issues Reviews for the Understanding and Evaluation of Irradiation Assisted Stress Corrosion Cracking. Electric Power Research Institute Report TR-107159.

Bull,A.,Lewis,G.,Owen,J.,Quirk,G.,Rudge,A.,Woolsey,I.,2013. Mitigating the risk of stress corrosion of austenitic stainless steels in advanced gas cooled reactor boilers. In:Proceedings of the 16th International Conference on the Properties of Water and Steam, September 1 - 5, 2013. University of Greenwich, London, UK.

Cao, G., Firouzdor, V., Allen, T., 2011. Stress corrosion cracking of austenitic alloys in supercritical water. In: Busby, J. T., Ilevbre, G., Andresen, P. L. (Eds.), Proceedings of the 15th International Conference on Environmental Degradation of Materials in Nuclear Power Systems Water Reactors. TMS (The Minerals, Metals & Materials Society), pp. 1923-1935.

Chen, K., Du, D., Gao, W., Andresen, P. L., Zhang, L., 2017. Investigation of temperature effect on stress corrosion crack growth behavior of 310S stainless steel in supercritical water environment. In: 8th Int. Symp. on Super-Critical Water-Cooled Reactors, March 13-15, 2017, Chengdu, China.

Combrade, P., Ford, P., Nordmann, F., 2010. Key Results from Recent Conferences on Structural Materials Degradation in Water Cooled Reactors. Nuclear Technology International. LCC6 Special Report Advanced.

Comprelli, F. A., Busboom, H. J., Spalaris, C. N., 1969. Comparison of Radiation Damage Studies and Fuel Cladding Performance for Incoloy - 800, Irradiation Effects in Structural Alloys for Thermal and Fast Reactors. ASTM STP 457. American Society for Testing and Materials, pp. 400-413.

Coriou, H., Grall, L., Olivier, P., Willermoz, H., 1969. Influence of carbon and nickel content on stress corrosion cracking of austenitic stainless alloys in pure or chlorinated water at 350℃ In: Staehle, R. W., Forty, A. J., van Rooyen, D. (Eds.), Proceedings of 1967 Conference: Fundamental Aspects of Stress Corrosion Cracking. NACE, Houston, TX, pp. 352-359.

Cowan, R., Ruhle, W., Hettiarachchi, S., 2011. Introduction to Boiling Water Reactor Chemistry, vol. 1. Advanced Nuclear Technology International, Sweden.

Cowan, R., Ruhle, W., Hettiarachchi, S., 2012. Introduction to Boiling Water Reactor Chemistry, vol. 2. Advanced Nuclear Technology International, Sweden.



Das, N. K., Suzuki, K., Takeda, Y., Ogawa, K., Shoji, T., 2008. Quantum chemical molecular dynamics study of stress corrosion cracking behavior for fcc Fe and Fe-Cr surfaces. Corros. Sci. 50, 1701-1706.

Devrient, B., Kilian, R., Kuster, K., Widera, M., 2011. Influence of bulk and surface cold work on crack initiation and crack growth of austenitic stainless steels under simulated BWR environmemt. In: Busby, J. T., Ilevbare, G., Andresen, P. L. (Eds.), Proc. 15th Int. Conf. on Environmental Degradation of Materials in Nuclear Power Systems Water Reactors. TMS.

Ehrlich, K., Konys, J., Heikinheimo, L., 2004. Materials for high performance light water reactors. J. Nucl. Mater. 327(2-3), 140-147.

Emel'yanov, I., Shatskaya, O. A., Rivkin, E. Yu., Nikolenko, N. Y., 1972. Strength of construction elements in the fuel channels of the Beloyarsk power station reactors. At. Energiya 33(3), 729-733 (in Russian) (Translated in Soviet 1972. Atomic Energy 33(3) 842-847.

EPRI, October 2004. BWR Water Chemistry Guidelinese—2004 Revision. EPRI Technical Report TR-1008192, VIP-130.

Fournier, L., Delafosse, D., Bosch, C., Magnin, T., 2001. Stress corrosion cracking of nickel base superalloys in aerated supercritical water. In: Proceedings of the 2001 NACE Corrosion Conference, March 2001, Houston, TX, United States. NACE, Houston. Paper 01361.

Fruzzetti, K. P., Blok, J., 2005. A review of the EPRI chemistry guidelines. In: Proc. Int. Conf. on Water Chemistry of Nuclear Reactor Systems, San Francisco.

Fujisawa, R., Nishimura, K., Nishida, T., Sakaihara, M., Kurata, Y., Watanabe, Y., 2005. Cracking susceptibility of Ni base alloys and 316 stainless steel in less oxidizing or reducing SCW. In: Corrosion 2005. NACE International, Houston, TX. Paper 05395.

Garcia, S. E., Giannelli, J. F., Jarvis, M. L., 2012. Advances in BWR water chemistry. In: Proc. NPC2012 Nuclear Plant Chemistry Conference, Paris. Paper 80 001.

Garud, Y. S., Ilevbare, G. O., 2009. A review and assessment of cold-work influence on SCC of austenitic stainless steels in light water reactor environment. In: Proceedings of the 14th International Conference on Environmental Degradation of Materials in Nuclear Power Systems Water Reactors. Virginia, USA, pp. 780-790.

Gong, B., Jiang, E., Huang, Y., Zhao, Y., Liu, W., Zhou, Z., 2015. Effect of oxidation chemistry of SCW on stress corrosion cracking of austenitic steels. In: 7th International Symposium on Supercritical Water-cooled Reactors (ISSCWR-7), March 15-18, 2015, Helsinki, Finland.

Gordon, G. M., Brown, K. S., 1989. Dependence of creviced BWR component IGSCC behavior on coolant chemistry. In: Cubicciotti, D., Simonen, E. (Eds.), Proc. 4th Int. Symp. On Environmental Degradation of Materials in Nuclear Power Systems—Water Reactors, Jekyll Island, August 6-10, 1989. National Association of Corrosion Engineers, pp. 14.46-14.62.

Griess, J. C., Martin, J. M., Mravca, A. E., Impara, R. J., Weems, S., September 7, 1962. Task Force Review on Feasibility of BONUS Reactor Stainless Steel Clad Superheater Elements. memo to David F. Cope, Director. Reactor Division, ORNL.

Gruzdev, N. I., Shchapov, G. A., Tipikin, S. A., Boguslavskii, V. B., 1970. Investigating the water conditions in the second unit at Beloyarsk. Therm. Eng. 17(12), 22-25 (Translated from Russian).

Hettiarachchi, S., Weber, C., 2010. Water chemistry improvements in an operating boiling water reactor (BWR) and associated benefits. In: Proceedings of the Nuclear Plant Chemistry Conference 2010 (NPC 2010), October 3-8, 2010, Quebec City, Canada.

Hou, J., Peng, Q. J., Shoji, T., Wang, J. Q., Han, E.-H., Ke, W., 2011. Effects of cold working path on strain concentration, grain boundary microstructure and stress corrosion cracking in alloy 600. Corros. Sci. 53, 1137-1142.

Janik, P., 2015. Acoustic Emissione—A Method for Monitoring of Environmentally Assisted Cracking Processes (Doctor thesis). Institute of Chemical Technology Prague, Faculty of Environmental Technology, Department of Power Engineering.

Janik, P., Novotny, R., Nilsson, K. F., Siegl, J., Hausild, P., March 3-7, 2013. Pre-qualification of cladding materials for SCWR fuel qualification testing facility - stress corrosion cracking testing. In: 6th International Symposium on Supercritical Water-Cooled Reactors (ISSCWR-6), Shenzhen, Guangdong, China. Paper ISSCWR6-13034.

Janssen, J. S., Morra, M. M., Lewis, D. J., 2009. Sensitization and stress corrosion cracking of alloy 800H in high-temperature water. Corrosion 65, 67-78.

Jiao, Y., Kish, J., Cook, W., Zheng, W., Guzonas, D. A., 2013. Influence of thermal aging on the corrosion resistance of austenitic Fe-Cr-Ni alloys in SCW. In: 6th International Symposium on Supercritical Water-cooled Reactors (Shenzhen, Guangdong, China), March 03-07, 2013.

Kallikragas, D. T., Plugatyr, A. Yu., Guzonas, D. A., Svishchev, I. M., 2015. Effect of confinement on the hydration and diffusion of chloride at high temperatures. J. Supercrit. Fluids 97, 22-30.

Kilian, R., Hoffmann, H., Ilg, U., Küster, K., Nowak, E., Wesseling, U., Widera, M., 2005. German experience with intergranular cracking in austenitic piping in BWRs and assessment of parameters affecting the in-service IGSCC behaviour using an artificial neural network. In: Nelson, L., King, P. (Eds.), Proc. 12th Int. Conf. on Environmental Degradation of Materials in Nuclear Power Systems—Water Reactors, Skamania Lodge, August 5-9, 2005. The Metallurgical Society, pp. 803-812.

Kim, H., Mitton, D. B., Latanision, R. M., 2011. Stress Corrosion Cracking of Alloy 625 in PH 2 Aqueous Solution at High Temperature and Pressure. Corrosion 67, Paper No. 035002-2.

Lee, E. H., Yoo, M. H., Byun, T. S., Hunn, J. D., Farrell, K., Mansur, L. K., 2001. On the origin of deformation microstructures in austenitic stainless steel: Part I—Microstructures. Acta Mater. 49, 3269-3276.

Li, S. -X., He, Y. -N., Yu, S. -R., Zhang, P. -Y., 2013. Evaluation of the effect of grain size on chromium carbide precipitation and intergranular corrosion of 316L stainless steel. Corros. Sci. 66, 211-216.

Liu, J., Gong, B., 2017. The effect of dissolved oxygen on stress corrosion cracking of 310 SS in SCW. In: 8th Int. Symp. on Super-critical Water-cooled Reactors, March 13-15, 2017, Chengdu, China.

Lozano-Perez, S., Yamada, T., Terachi, T., Schröder, M., English, C. A., Smith, G. D. W., Grovenor, C. R. M., Eyre, B. L., 2009. Multi-scale characterization of stress corrosion cracking of cold-worked stainless steels and the influence of Cr content. Acta Mater. 57, 5361-5381.

Macdonald, D. D., 2009. The electrochemistry of stress corrosion cracking. In: Proceedings of the 14th International Conference on Environmental Degradation of Materials in Nuclear Power Systems Water Reactors. Virginia, USA, pp. 734-744.

Nelson, J. L., Andresen, P. L., 1991. Review of current research and understanding of irradiation assisted stress corrosion cracking. In: Cubicciotti, D., Simonen, E. (Eds.), Proc. 5th Int. Symp. on Environmental Degradation of Materials in Nuclear Power Systems—Water Reactors, Monterey, August 25-29, 1991. American Nuclear Society, pp. 10-26.

Newman, R. C., 2002. Stress Corrosion Cracking Mechanisms. Marcel Dekker, New York.

Nezakat, M., Akhiani, H., Hoseini, M., Szpunar, J., 2014. Effect of thermo-mechanical pro-cessing on texture evolution in austenitic stainless steel 316L. Mater. Charact. 98, 10-17.

Novotny, R., Hähner, P., Siegl, J., Hausild, P., Ripplinger, S., Penttilä, S., Toivonen, A., 2011. Stress corrosion cracking susceptibility of austenitic stainless steels in supercritical water conditions. J. Nucl. Mater. 409(2), 117-123.

Oh, Y. J., Hong, J. H., 2000. Nitrogen effect on precipitation and sensitization in cold-worked type 316L(N) stainless steels. J. Nucl. Mater. 278, 242-250.

Parvathavarthini, N., Dayal, R. K., 2002. Influence of chemical composition, prior deformation and prolonged thermal aging on the sensitization characteristics of austenitic stainless steel. J. Nucl. Mater. 305, 209-219.

Parvathavarthini, N., Mulki, S., Dayal, R. K., Samajdar, I., Mani, K. V., Raj, B., 2009. Sensitization control in AISI 316L(N) austenitic stainless steel: defining the role of the nature of grain boundary. Corros. Sci. 51, 2144-2150.

Peng, Q. J., Teysseyre, S., Andersen, P. L., Was, G. S., 2007. Stress corrosion crack growth in type 316 stainless steel in supercritical water. Corrosion 63(11), 1033-1041.

Penttilä, S., Toivonen, A., 2013. Oxidation and SCC behaviour of austenitic and ODS steels in supercritical water. In:6th International Symposium on Supercritical Water-Cooled Re-actors(ISSCWR-6),Shenzhen,Guandong,China,March 3-7,2013. Paper ISSCWR6-13029.

Rabin,S. A.,Atraz,B. G.,Bader,M. B.,Busboom,H. J.,Hazel,V. E.,1967. Examination and Evaluation of Rupture in EVESR Superheat Fuel Rod with 0.012-inch-thick Incoloy-800 Cladding. AEC Research and Development Report GEAP-5416.

Raja,V. S.,Shoji,T.(Eds.),September 2011. Corrosion Cracking Theory and Practice. Woodhead Publishing,ISBN 978-1-84569-673-3.

Ru,X.,Staehle,R. W.,2013a. Historical experience providing bases for predicting corrosion and stress corrosion in emerging supercritical water nuclear technology:Part 1 — review. Corrosion 69(3),211-229.

Ru,X.,Staehle,R. W.,2013b. Historical experience providing bases for predicting corrosion and stress corrosion in emerging supercritical water nuclear technology:Part 2 — review. Corrosion 69(4),319-334.

Ru,X.,Staehle,R. W.,2013c. Historical experience providing bases for predicting corrosion and stress corrosion in emerging supercritical water nuclear technology:Part 3 — review. Corrosion 69(5),423-447.

Ru,X.,Zhang,Q.,Zhao,Y.-X.,Tang,R.,Xiao,Z.,2015. The stress corrosion cracking(SCC)behavior of Type 310S stainless steel in supercritical water(SCW). In:Proceedings 7th International Symposium on Supercritical Water-Cooled Reactors(ISSCWR-7),March 15-18,2015,Helsinki,Finland. Paper No. ISSCWR7-2093.

Sáez-Maderuelo,A.,Gómez-Briceño,D.,2016. Stress corrosion cracking behavior of annealed and cold worked 316L stainless steel in supercritical water. Nucl. Eng. Des. 307,30-38.

Sahlaoui,H.,Makhlouf,K.,Philibert,J.,2004. Effects of ageing conditions on the precipitates evolution, chromium depletion and intergranular corrosion susceptibility of AISI 316L:experimental and modeling results. Mater. Sci. Eng. A 372,98-108,15.

Scenini,F.,Newman,R. C.,Cottis,R. A.,Jacko,R. J.,2008. Effect of surface preparation on intergranular stress corrosion cracking of alloy 600 in hydrogenated steam. Corrosion 64,824-835.

Scott,P.,2013. SCC of stainless steels in PWRs. In:Scott,P. M.(Ed.),INL Seminar on SCC in LWRs Idaho Falls,Idaho,USA,2013 March 19-20. INL.

Scott,P. M.,Combrade,P.,2003. On the mechanism of stress corrosion crack initiation and growth in alloy 600 exposed to PWR primary water. In:Proc. of the 11th Int. Symp. Environmental Degradation Materials Nuclear Power Systems—Water Reactors. ANS,pp. 29-38.

Shen,Z.,Zhang,L.,Tang,R.,Zhang,Q.,2014. The effect of temperature on the SSRT behavior of austenitic stainless steels in SCW. J. Nucl. Mater. 454,274-282.

Shoji,T.,August 10-14,2003. Progress in the mechanistic understanding of BWR SCC and its implication to the prediction of SCC growth behavior in plants. In:Proc. 11th Int. Conf. on Environmental Degradation of Materials in Nuclear Power Systems—Water Reactors,Skamania,Stevenson,WA. ANS,pp. 588-598.

Shoji,T.,September 2011. Factors affecting stress corrosion cracking(SCC)and fundamental mechanistic understanding of stainless steels. In:Raja,V. S.,Shoji,T.(Eds.),Stress Corrosion Cracking Theory and Practice. Woodhead Publishing,ISBN978-1-84569-673-3.

Shoji,T.,Lu,Z. P.,Takeda,Y.,Murakami,H.,Fu,C. Y.,2008. Deterministic prediction of stress corrosion crack growth rates in high temperature water by combination of interface oxidation kinetics and crack tip asymptotic field. In:2008 ASME Pressure Vessels and Piping Division Conference(ASME PVP 2008). ASME, pp. PVP2008-61417.

Shoji, T., Lu, Z. P., Murakami, H., 2010. Formulating stress corrosion cracking growth rates by combination of crack tip mechanics and crack tip oxidation kinetics. Corros. Sci. 52, 769-779.

Speidel, M. O., 1984. Stress corrosion and corrosion fatigue mechanics. In: Speidel, M. O., Atrens, A. (Eds.), Corrosion in Power Generating Equipment. Plenum Press, New York.

Staehle, R. W., February 2007. Predicting Failures Which Have Not yet Been Observed. Expert Panel Report on Proactive Materials Degradation Assessment, USNRC NUREG Report, CR-6923.

Staehle, R. W., Gorman, J. A., 2003. Quantitative assessment of submodes of stress corrosion cracking on the secondary side of steam generator tubing in pressurized water reactors: Part 1. Corrosion 59, 931-993.

Stellwag, B., Landner, A., Weiss, S., Huttner, F., 2011. Water chemistry control practices and data of the European BWR fleet. Powerpl. Chem. 13(3), 167-173.

Strasser, A., Santucci, J., Lindquist, K., Yario, W., Stern, G., Goldstein, L., Joseph, L., 1982. An Evaluation of Stainless Steel Cladding for Use in Current Design LWRs. Electric Power Research Institute Report NP-2642.

Tan, L., Allen, T. R., 2009. Localized corrosion of magnetite on ferritic-martensitic steels exposed to supercritical water. Corros. Sci. 51, 2503-2507.

Terada, M., Escriba, D. M., Costa, I., Materna-Morris, E., Padilha, A. F., 2008. Investigation on the intergranular corrosion resistance of the AISI 316L(N) stainless steel after long time creep testing at 600℃. Mater. Charact. 59, 663-668.

Teysseyre, S., Was, G. S., 2006. Stress corrosion cracking of austenitic alloys in supercritical water. Corrosion 62(12), 1100-1116.

Teysseyre, S., Jiao, Z., West, E., Was, G. S., 2007. Effect of irradiation on stress corrosion cracking in supercritical water. J. Nucl. Mater. 371(1-3), 98-106.

Toivonen, A., Penttilä, S., Novotny, R., 2015. SCC tests in SCWat 550℃ on two heats of 316L. In: 7th International Symposium on Supercritical Water-cooled Reactors (ISSCWR-7), March 15-18, 2015, Helsinki, Finland.

Tsuchiya, Y., Kano, F., Saito, N., Shioiri, A., Kasahara, S., Moriya, K., Takahashi, H., September 15-19, 2003. SCC and Irradiation Properties of Metals Under Supercritical-Water Cooled Power Reactor Conditions. GENES4/ANP2003, Kyoto, Japan. Paper 1096.

Uchida, S., Satoh, T., Tsukada, T., Miyazawa, T., Satoh, Y., Ishaa, K., 2009. Evaluation of the effects of oxide film on electrochemical corrosion potential of stainless steel in high temperature water. In: Proceedings of the 14th International Conference on Environmental Degradation of Materials in Nuclear Power Systems Water Reactors. Virginia, USA, pp. 1100-1109.

Udy, M. C., Boulger, F. W., 1953. Survey of materials for supercritical-water reactor. Battelle Memorial Inst, Columbus, Ohio. OSTI ID: 4230973, Report Number(s), BMI-890.

Was, G. S., Busby, J., Andresen, P. L., 2006. Effect of irradiation on stress corrosion cracking and corrosion in light water reactors. In: ASM Handbook on Corrosion; Environments and Industries, 13C, pp. 386-414.

Was, G. S., Ampornrat, R., Gupta, G., Teysseyre, S., West, E. A., Allen, T. R., Sridharan, K., Tan, L., Chen, Y., Ren, X., Pister, C., 2007. Corrosion and stress corrosion cracking in supercritical water. J. Nucl. Mater. 371, 176-201.

Was, G. S., Teysseyre, S., 2005. Challenges and recent progress in corrosion and stress corrosion cracking of alloys for supercritical water reactor core components. In: Allen, T. R., King, P. J., Nelson, L. (Eds.), Proceedings of the 12th International Conference on Environmental Degradation of Materials in Nuclear Power Systems Water Reactors. TMS (The Minerals, Metals&Materials Society), pp. 1343-1357.

Watanabe, Y., Abe, H., Daigo, Y., 2003. Environmentally-assisted cracking of sensitized stainless steel in supercritical water. In: Proceedings of GENES4/NP2003, Kyoto, Japan, September 15-19, 2003. Paper 1183.

Watanabe, Y., Abe, H., Daigo, Y., Nishida, T., 2004a. Corrosion Resistance and Cracking Susceptibility of 316L Stainless Steel in Sulfuric Acid-Containing Supercritical Water. Corrosion 2004, Paper 04493.

Watanabe, Y., Abe, H., Daigo, Y., Fujisawa, R., Sakaihara, M., 2004b. Effect of physical property and chemistry of water on cracking of stainless steels in sub-critical and supercritical water. Key Eng. Mater. 261-263, 1031-1036.

West, E., Was, G. S., 2013. Strain incompatibilities and their role in intergranular cracking of irradiated 316L stainless steel. J. Nucl. Mater. 441, 623-632.

Woolsey, I. S., 1989. Review of Water Chemistry and Corrosion Issues in the Steam-Water Circuits of Fossil-Fired Supercritical Plants. Central Electricity Generating Board Report RD/L/3413/R88, PubID 280118.

Zhang, Q., Tang, R., Xiong, R., Zhang, L., 2015. Effect of temperature and dissolved oxygen on the SCC behavior of 316Ti in supercritical water. In: 7th International Symposium on Supercritical Water-Cooled Reactors (ISSCWR-7), March 15-18, 2015, Helsinki, Finland.

Zheng, W., Luo, J., Li, M., Guzonas, D. A., Cook, W., 2011. Stress corrosion cracking of SCWR candidate alloys: a review of published results. In: 5th International Symposium on SCWR (ISSCWR-5), Vancouver, BC, Canada. Paper P095, March 13-16, 2011.

Zheng, W., Zeng, Y., Luo, J., Novotny, R., Li, J., Shalchi Amirkhiz, B., Guzonas, D., Matchim, M., Collier, J., Yang, L., 2014. Stress-corrosion cracking properties of candidate fuel cladding alloys for the Canadian SCWR: a summary of literature data and recent test results. In: 19th Pacific Basin Nuclear Conference (PBNC 2014) Vancouver, British Columbia, Canada, August 24-28, 2014.

Zhou, S., Turnbull, A., 2002. Steam Turbine Operating Conditions, Chemistry of Condensates, and Environment Assisted Cracking—A Critical Review. National Physical Laboratory Report MATC(A)95.

Zhou, Z.-F., Lvov, S. N., Thakur, S., Zhou, X., Chou, P., Pathania, R., 2004. Hydrothermal deposition of zirconia coatings on BWR materials for IGSCC protection. In: Proc. Int. Conf. Water Chemistry of Nuclear Reactor Systems, October 11-14, 2004, SanFrancisco, CA, USA. Paper 3-4.

Zhou, R., West, E. A., Jiao, Z., Was, G. S., 2009. Irradiation-assisted stress corrosion cracking of austenitic alloys in supercritical water. J. Nucl. Mater. 395, 11-22.

# 内容简介

压水反应堆和超临界火电机组在世界范围内已经积累了丰富的设计和运行经验,然而当超临界条件与强辐照环境共同存在时将会产生新的挑战,尤其是对反应堆内的材料特性提出了更严苛的要求。本书专门针对超临界水冷反应堆内的材料辐照损伤机理展开了较为系统的阐述,从内容上看,首先综述了超临界水冷反应堆的发展历史和研究现状,概括了现有研究中的不足,这些不足主要体现在材料辐照损伤、常规化学腐蚀和水化学等几方面。基于研究现状,原著者分别从3个方面开展了深入论述:首先,以试验研究为手段,探究了带电粒子、$\gamma$射线及中子对材料的辐照损伤机理,在试验研究的基础上借助蒙特卡罗方法建立了定量描述材料辐照损伤过程的动力学模型,为进一步分析计算和设计提供了便利条件;其次,针对超临界水的化学特性、辐照条件下的感生放射性以及反应堆运行过程中的水质控制方法进行了深入、细致的研究,从机理上诠释了超临界水的化学动力行为;最后,本书还对超临界水冷反应堆中材料的常规化学腐蚀问题进行了较为全面的阐述。本书的特点在于大部分研究结果均是以试验为基础获得的,特别是机理性的研究成果具有较好的开拓性和创新性,无论是科学价值还是工程应用价值都较高,而这些宝贵的研究成果为堆内材料的设计改进、运行策略制定以及反应堆工程化应用提供了重要支撑。

本书内容涉及材料科学、热力学、腐蚀科学、化学动力学及核能等交叉学科知识,读者对象以从事超临界水冷反应堆研究设计工作的科研人员、工程技术人员和高校教师、研究生为主,部分内容也适用于有专业基础的本科生。

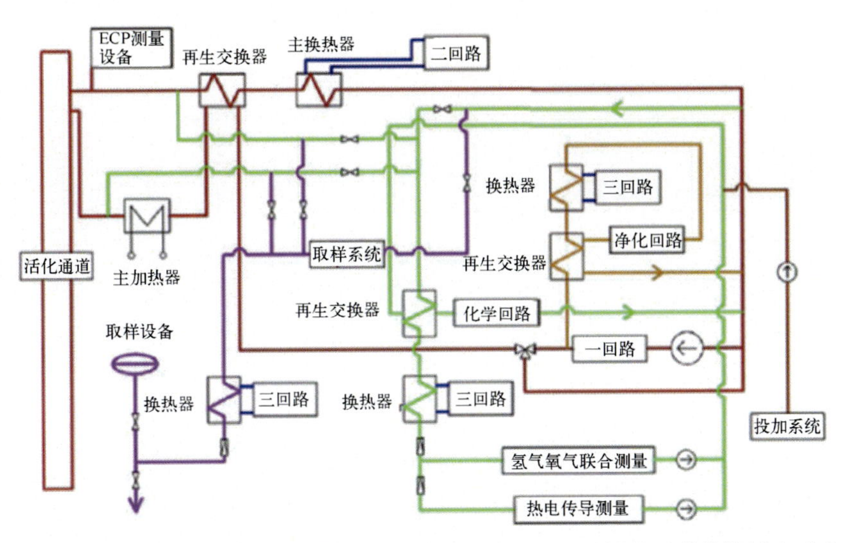


图 2.8　CVR 中 SCW 回路的流程图（Ruzickova 等，2011）。活性通道将被插入反应堆核芯，并包含试验部分。辅助回路包括主回路（红色）、溶剂添加系统（暗橙色）、测量系统（绿色）、净化系统（橙色）和冷却回路。溶剂添加系统可以添加化学物质（气体和溶液）。主回路中的化学监测（蓝色）由侧流回路提供。回路中的循环介质通过机械过滤器和离子交换器净化，以保持主回路中所需的化学条件

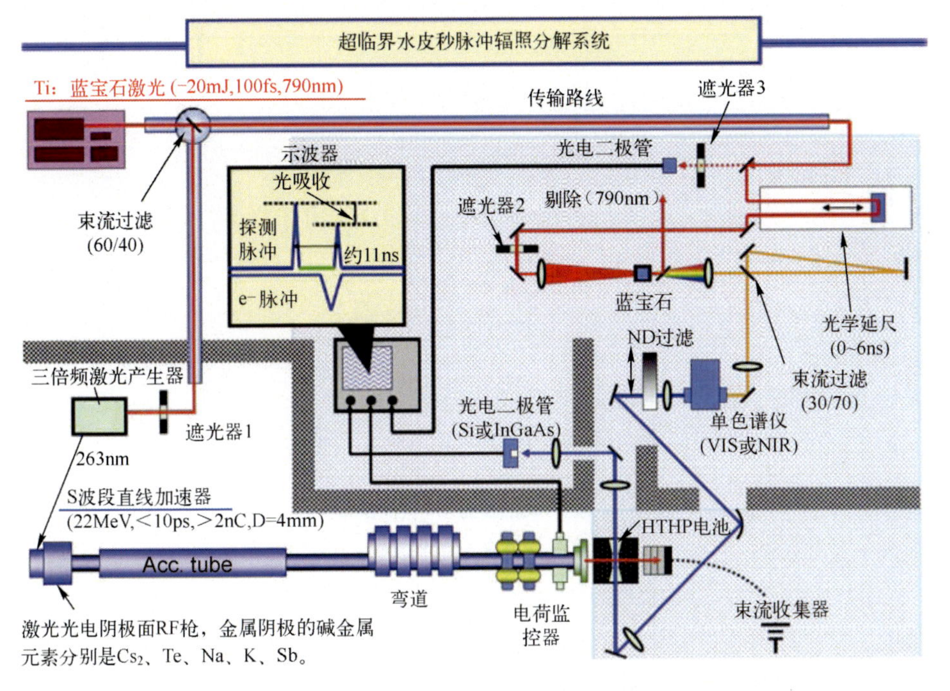


图 2.10　东京大学核学院（Katsumura 等，2010）皮秒时间分辨率的脉冲辐照分解系统原理

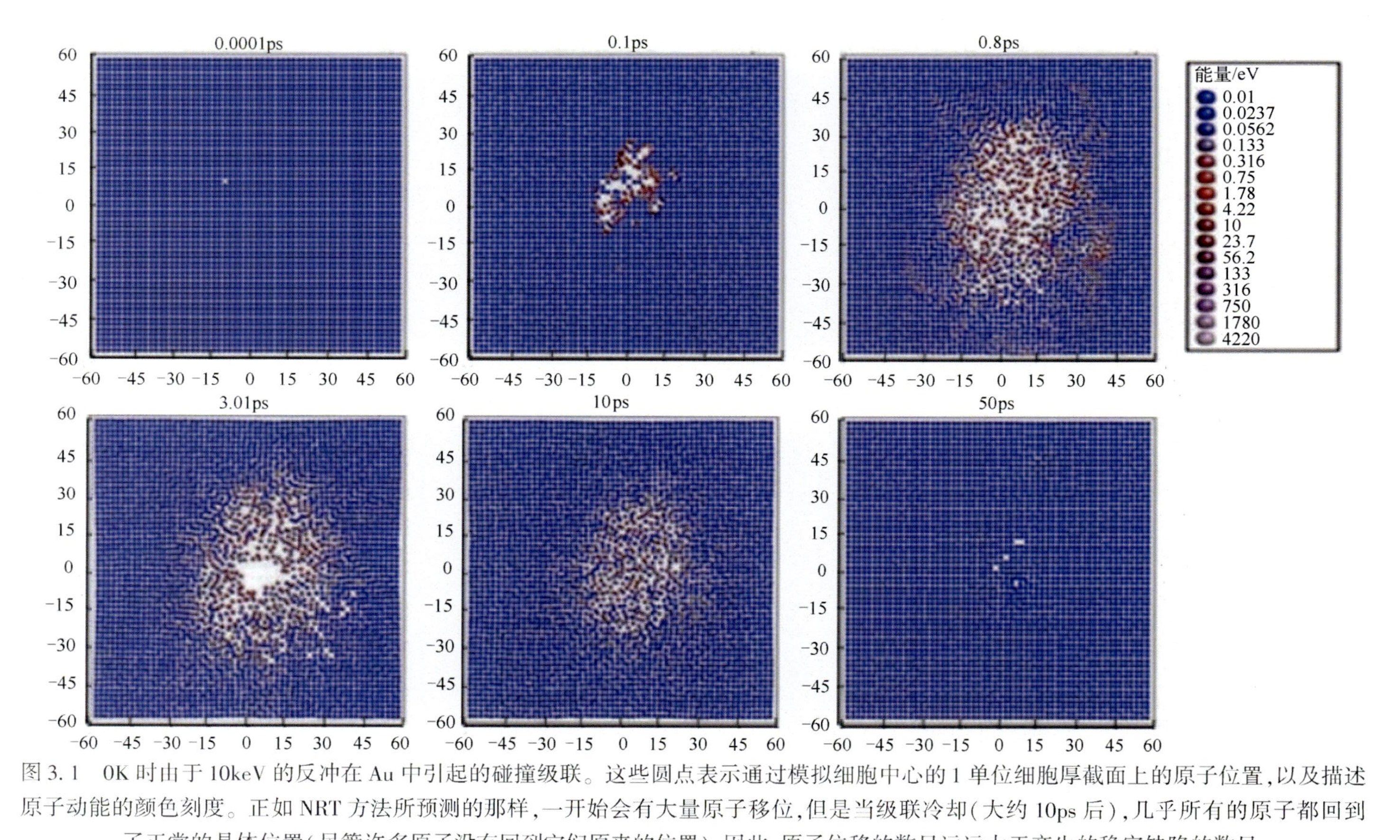


图 3.1　0K 时由于 10keV 的反冲在 Au 中引起的碰撞级联。这些圆点表示通过模拟细胞中心的 1 单位细胞厚截面上的原子位置，以及描述原子动能的颜色刻度。正如 NRT 方法所预测的那样，一开始会有大量原子移位，但是当级联冷却（大约 10ps 后），几乎所有的原子都回到了正常的晶体位置（尽管许多原子没有回到它们原来的位置），因此，原子位移的数目远远大于产生的稳定缺陷的数目

（引自文献 OECD, 2015. Primary Radiation Damage in Materials, NEA/NSC/DOC (2015) 9. https://www.oecd-nea.org/science/docs/2015/nsc-doc2015-9.pdf.）

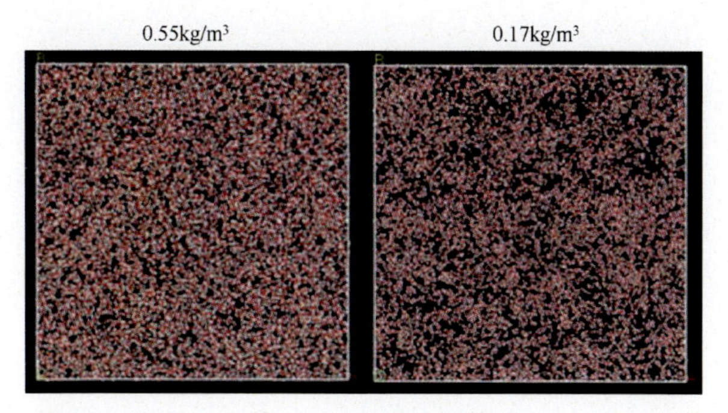


图 4.2 密度为 $55kg/m^3$ 和 $17kg/m^3$ 超临界水的分子动力学模拟(显示氧(红色)和氢(白色)原子的三维分布,$T=400℃$)

(引自文献 Metatla N, Jay-Gerin J-P, Soldera A, 2011. Molecular dynamics simulation of sub-critical and supercritical water at different densities. In: 5th International Symposium on Supercritical Water-cooled Reactors (ISSCWR-5), Vancouver, British Columbia, Canada, March 13-16, 2011)



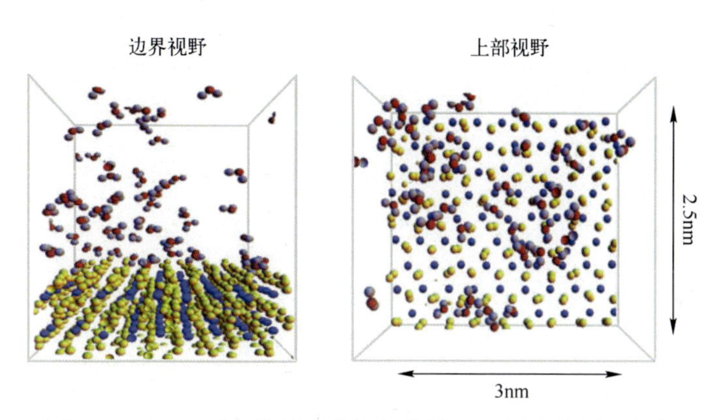


图 4.25 715K 时超临界水在氢氧化铁(Ⅱ)界面的原子密度分布(表面铁原子用深蓝色表示,表面氧原子用黄色表示,水氧原子用红色表示,水氢原子用紫色表示)

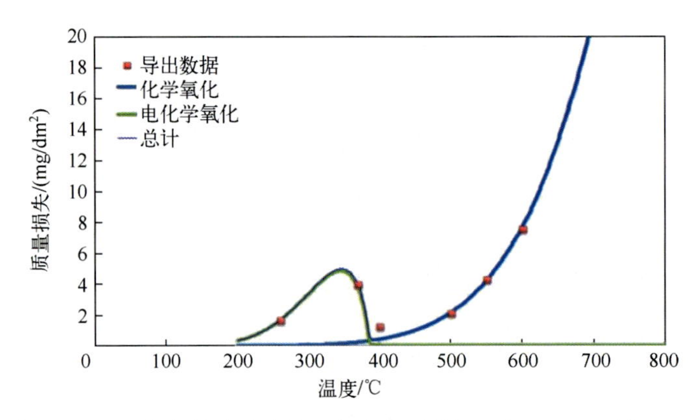

图 5.48　625 合金的试验腐蚀速率和模型腐蚀速率(Steeves 等,2015)

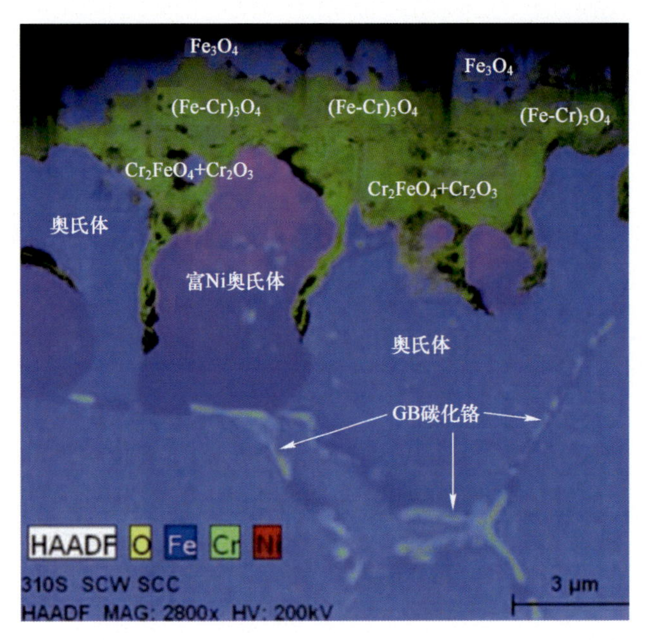

图 6.16　310S 胶囊样品与超临界水在 500℃、25MPa 下接触 19000h 后的内截面 TEM-EDX 元素图分析(Zheng 等,2014)